# PROGRESS IN SWEETENERS

# ELSEVIER APPLIED FOOD SCIENCE SERIES

Biotechnology Applications in Beverage Production
C. CANTARELLI and G. LANZARINI

*Forthcoming titles in this series:*

Microstructural Principles of Food Processing and Engineering
J. M. AGUILERA and D. W. STANLEY

Food Refrigeration Processes: Analysis, Design and Simulation
A. C. CLELAND

Food Gels
P. HARRIS (Editor)

Food Antioxidants
B. J. F. HUDSON (Editor)

Development and Application of Immunoassay for Food Analysis
J. H. R. RITTENBURG (Editor)

Food Irradiation
S. THORNE

# PROGRESS IN SWEETENERS

*Edited by*

## T. H. GRENBY

B.Sc., Ph.D., C.Chem., F.R.S.C.

*Department of Oral Medicine and Pathology,
United Medical and Dental Schools,
Guy's Hospital, London, UK*

ELSEVIER APPLIED SCIENCE
LONDON and NEW YORK

ELSEVIER SCIENCE PUBLISHERS LTD
Crown House, Linton Road, Barking, Essex IG11 8JU, England

*Sole Distributor in the USA and Canada*
ELSEVIER SCIENCE PUBLISHING CO., INC.
655 Avenue of the Americas, New York, NY 10010, USA

WITH 58 TABLES AND 93 ILLUSTRATIONS

© 1989 ELSEVIER SCIENCE PUBLISHERS LTD

© 1989 FÉDÉRATION DENTAIRE INTERNATIONALE—Chapter 14

**British Library Cataloguing in Publication Data**

Progress in sweeteners.
1. Artificial sweeteners
I. Grenby, T. H.
664'.5

ISBN 1-85166-364-9

**Library of Congress Cataloguing in Publication Data**

Progress in sweeteners/edited by T. H. Grenby.
    p. cm.—(Elsevier applied food science series)
Bibliography: p.
Includes index.
ISBN 1-85166-364-9
1. Sweeteners. I. Grenby, T. H. (Trevor H.) II. Series.
TP421.P76 1989
664'.5—dc20               89-12054
                                 CIP

Printed in Great Britain by Galliard (Printers) Ltd, Great Yarmouth

# PREFACE

This book has evolved from the same publisher's *Developments in Sweeteners* series. The last of the *Developments* volumes appeared in 1987. Shortly after that Elsevier Science Publishers Ltd adopted the policy of replacing them by a new set of titles, but by then it had become plain that the interest generated by some of the topics dealt with in *Developments in Sweeteners—3* would lead to further articles from different viewpoints and in related areas. In addition, the subject of sweeteners and their use has been expanding so fast that a demand exists both from those developing new products, to communicate the advances they have made, and from those who want to know more about these advances, in order to apply them and keep in touch with the latest state of progress.

This new volume should provide something useful for both parties. In size and scope it exceeds every one of the earlier *Developments* series. Some of the material is of unique topical interest, bringing us right up to the state-of-the-art frontiers in scientific research. The opening chapter, for example, deals with the major advances that have been made over the last few years in peptide sweeteners. Following the success of aspartame, there was every incentive for the development of 'second generation' peptide sweeteners, with the prospect of very great rewards for the recognition of product(s) showing improvements over aspartame. The extent to which this has been achieved can be judged from Chapter 1.

The next chapter tells the fascinating story of how the sophisticated techniques of genetic engineering have been applied to produce and derive specific, complex sweeteners from living organisms. The scope and intricacy of these techniques will be new to many readers and the authors have done their best to make them intelligible to those of us unfamiliar with the advanced methodology.

v

Certain names are bandied about with great regularity as potentially useful new sweeteners without very much serviceable information coming to light on them. A major object in this book has been to dispel some of the obscurity surrounding them and to collect what is known about them into a helpful review form. The first of these reviews is on glycyrrhizin, summarising basic data and its medical and dental attributes. Another is on sucralose, unveiling its properties and applications, which should be of special interest to many who have been seeking published information about it, which has hitherto been scarce. Sucralose is also mentioned in the chapter displaying the painstaking craft of the organic chemist in the enhancement of the sweetness of sucrose by conversion to chloro-deoxy-derivatives. Both this chapter and the one on peptide sweeteners shed light on the molecular structures that can be perceived as sweet by the taste receptors on the tongue and illustrate the considerable advances that have been made in this field in recent years.

The last contribution in this descriptive section on various novel sweeteners deals with palatinose. This differs from all the other materials considered so far in that it is a bulk, not an intense, sweetener. Details of its properties and the research done on it will be new to many readers.

The remaining chapters are on selected applications of sweeteners and on their health and nutritional characteristics. One of the main uses of intense sweeteners is of course in soft drinks, but manufacturers have been reluctant to release any technical information on this. We are fortunate in this volume, however, to have a comprehensive account of this important application, relating primarily to experience in the UK, but plainly also of interest to readers in many other countries.

Turning to the controversial subject of health benefits, the next chapter explores the need of diabetics for non-glucose-containing sweeteners. Although a range of foods formulated with sorbitol, etc., instead of sucrose are marketed for diabetics, opinions vary on whether they serve any useful purpose. It has not been easy to secure an authoritative view on this, but this chapter, from one of the country's major diabetology units, should provide a helpful source of information. Another controversial area is the energy supplied to the body by polyols, as bulk sweeteners replacing sugars in the diet. There has been conflict in the past on the value of such sweeteners for reduced-calorie and other medical purposes. The careful assessment of the methodology given here should provide a valuable insight into the subject.

Another nutritional advantage put forward for low-calorie sweeteners is to aid weight control. Opinions are again divided on the benefits that can be derived by these means, with a diversity of beliefs voiced by groups of psychologists applying various techniques to the problem. The point of view of the group at Leeds University is expressed here. This is followed by an article on a topic that arouses great interest within the food industry, and has particular importance in the formulation of foods and drinks containing non-sugar sweeteners, providing an up-to-date insight on how the perceptual characteristics of different sweeteners are established.

The last three chapters tread on firmer ground than the role of sweeteners in diabetic and slimming products. Their benefit in assisting dental health is well established by now, but there is competition between the promoters of the various sugar substitutes to demonstrate their efficacy in curbing dental caries and to call attention to their associated dental properties. The chapter on the role of Lycasin® reviews the evidence for this unusual material in a succinct and clear way, the next contribution summarises some of the more recent studies on xylitol, on which it has hitherto been difficult to find information, going on to review theories on the mechanism of action of xylitol in great detail, while the final chapter summarises the dental research that has been done on lactitol.

It is to be hoped that among this wealth of topics and expert data, readers wanting to keep up to date on the subject of sweeteners will find something to interest and inform them. Without the committed cooperation of the authors and their patience in coping with the vagaries of the postal services to and from their various corners of the globe (contributions from twelve separate institutions in five different countries), assembling the material for this book would have become an overwhelming task. With thanks to all of them for their untiring efforts,

TREVOR GRENBY

# CONTENTS

# LIST OF CONTRIBUTORS

GILLIAN ANANTHARAMAN

*Nestlé Research Centre, Nestec Limited, Vers-chez-les-Blanc, 1000 Lausanne 26, Switzerland*

LINDLEY C. BLAIR

*INGENE, 1545-17th Street, Santa Monica, California 90404, USA*

JOHN E. BLUNDELL

*BioPsychology Group, Psychology Department, University of Leeds, Leeds LS2 9JT, UK*

N. FINER

*Division of Medicine, United Medical and Dental Schools, Guy's Hospital, London SE1 9RT, UK*

PRADIP GHOSH-DASTIDAR

*INGENE, 1545-17th Street, Santa Monica, California 90404, USA*

T. H. GRENBY

*Department of Oral Medicine and Pathology, United Medical and Dental Schools, Guy's Hospital, London SE1 9RT, UK*

LESLIE HOUGH

*Department of Chemistry, Kensington Campus, King's College London, Campden Hill Road, London W8 7AH, UK*

J. M. JANUSZ

*The Procter & Gamble Company, Miami Valley Laboratories, PO Box 398707, Cincinnati, Ohio 45239-8707, USA*

M. R. JENNER

*Tate & Lyle Speciality Sweeteners, PO Box 68, Whiteknights, Reading, Berkshire RG6 2BX, UK*

RIAZ KHAN

*Tate & Lyle Research & Technology, Philip Lyle Memorial Research Laboratory, PO Box 68, Whiteknights, Reading, Berkshire RG6 2BX, UK*

RAJU K. KODURI

*INGENE, 1545-17th Street, Santa Monica, California 90404, USA*

JAR-HOW LEE

*INGENE, 1545-17th Street, Santa Monica, California 90404, USA*

KAUKO K. MÄKINEN

*Department of Biochemistry, University of Michigan School of Dentistry, Ann Arbor, Michigan 48109-1087, USA*

PETER J. ROGERS

*BioPsychology Group, Psychology Department, University of Leeds, Leeds LS2 9JT, UK*

ANDREW J. RUGG-GUNN

*Department of Oral Biology and Child Dental Health, University of Newcastle upon Tyne Dental School, Framlington Place, Newcastle upon Tyne NE2 4BW, UK*

M. N. SELA

*Department of Oral Biology, Faculty of Dental Medicine, Hadassah Medical School, Jerusalem, Israel*

D. STEINBERG

*Department of Oral Biology, Faculty of Dental Medicine, Hadassah Medical School, Jerusalem, Israel*

ICHIRO TAKAZOE

*Department of Microbiology, Tokyo Dental College, 1-2-2 Masago, Chiba City, 260 Japan*

ANNE TUNALEY

*College of Agricultural and Environmental Sciences, Department of Food Science and Technology, 1480 Chemistry Annex, University of California, Davis, California 95616, USA*
*Present address: KFC Corporation, PO Box 3455, Louisville, Kentucky 40232-4550, USA*

JOACHIM L. WEICKMANN

*INGENE, 1545-17th Street, Santa Monica, California 90404, USA*

A. G. WELLS

*Beecham Products Research Department, The Royal Forest Factory, Coleford, Gloucestershire GL16 8JB, UK*

PIERRE WÜRSCH

*Nestlé Research Centre, Nestec Limited, Vers-chez-les-Blanc, 1000 Lausanne 26, Switzerland*

Chapter 1

# PEPTIDE SWEETENERS BEYOND ASPARTAME

J. M. JANUSZ

The Procter & Gamble Company, Miami Valley Laboratories,
Cincinnati, Ohio, USA

## SUMMARY

The accidental discovery of the sweet dipeptide L-aspartyl-L-phenylalanine
methyl ester (aspartame) at G. D. Searle and Co. spawned considerable
efforts to relate the structure of dipeptides and dipeptide analogs to sweetness.
This review attempts to summarize much of this effort. The groundwork laid
by Searle workers led to the first structural model for sweet dipeptide analogs.
This required L-aspartic acid coupled to an amine substituted at the α carbon
atom with one small group ($R_1$) and one large group ($R_2$) in the proper
orientation. Refinement of this simple model by optimization of the R groups
led to sweeteners up to 50 000 times sweeter than sucrose. Recent work has
determined that appropriate substitution on the aspartyl amino group
provides novel conjugates with high sweetness intensity. Conformational
analysis by NMR spectroscopy, calculation and synthesis of rigid analogs
has begun to define the shape of the sweet receptor site.

## 1 INTRODUCTION

In 1965 James Schlatter at G. D. Searle accidentally discovered the sweet
dipeptide derivative L-aspartyl-L-phenylalanine methyl ester (1), now
known as aspartame. While many structurally diverse compounds were
known to be sweet,[1] aspartame was the first sweet dipeptide. The sweetness
intensity was about 180 times that of sucrose. Since then more than 1000
related dipeptides and dipeptide analogs have been prepared[2,3] with
aspartame as the prototype. This review makes no attempt to cover each of
these sweeteners. Instead, progress in the field will be described in three

1

parts: the groundwork at Searle and elsewhere leading to a simple structural model for dipeptide sweeteners, work on novel analogs which adds new structure–activity information, and recent modeling work to refine the structural requirements for dipeptide sweeteners.

**1**

## 2 THE GROUNDWORK

### 2.1 Discovery and Early Analogs

As for every new class of non-nutritive sweetener, aspartame was discovered by accident.[3] James Schlatter was recrystallizing L-aspartyl-L-phenylalanine methyl ester, an intermediate in the synthesis of the C-terminal tetrapeptide of gastrin, when the mixture bumped, splashing some of the dipeptide on his fingers. He discovered the sweetness shortly afterwards on licking his fingers to pick up a piece of weighing paper. From this serendipitous beginning, aspartame began a long and tortuous road to market. In the USA it was approved for dry uses in 1974, approval was suspended in 1975, restored in 1981 and extended to beverages in 1983. Additional uses continue to be approved, and under the brand name Nutrasweet it is found today in a myriad of products.

Searle's selection of aspartame came after evaluating many dipeptide analogs. Attempts to replace L-aspartic acid with other naturally occurring amino acids led to bitter dipeptides even when the replacement was glutamic acid, the higher homolog of aspartic acid.[4] In addition, it was found that the aspartyl $\alpha$-amino and $\beta$-carboxyl groups must remain unsubstituted since the corresponding $\alpha$-dimethylamino and $\beta$-methyl ester analogs are tasteless. The observation that the N-terminal amino acid must be zwitterionic with the charged groups a fixed distance apart suggested that these dipeptides fit the Shallenberger and Acree sweetener model.[5] This model suggests that all sweet substances possess a hydrogen-bond donor (AH) and acceptor (B) $2 \cdot 5$–$4 \cdot 0$ Å apart. This AH–B pair hydrogen-bonds with a complementary pair on the sweet receptor. For aspartyl dipeptides the ammonium group is the donor (AH) and the carboxylate group the acceptor (B). Stereochemical constraints also apply to both amino acids. Of the 4 possible diastereomers of aspartame, only the L,L is sweet.

In contrast to the strict structural requirements for the N-terminal part of the sweetener, the C-terminal part can be varied considerably. The phenylalanine can be substituted on the ring, although a loss of sweetness intensity was observed except for the ortho OMe derivative.[6] L-Methionine and L-tyrosine can replace phenylalanine. Higher homologs of the ester group were sweet although sweetness intensity decreased with increasing size. The early discoveries are summarized in Table 1.

The Searle group made the first major structural departure from these dipeptides when they found that the C-terminal amino acid could be

TABLE 1
EARLY STRUCTURAL MODIFICATIONS OF ASPARTAME

| Asp replacements | Stereochemistry | | Taste[a] | Phe-OMe replacements | Taste[a] |
|---|---|---|---|---|---|
| Ala, Gly, His, iso-Leu, Leu, | Asp-Phe-OMe | | | L-Asp-L-Met-OMe | 180 × |
| Lys, nor-Leu, nor-Val, Phe, | L | L | 180 × | L-Asp-L-Tyr-OMe | 180 × |
| Pro, Sar,[b] Ser, Thr, | L | D | Bitter | | |
| Try, Trp, Val, Glu | D | L | Bitter | | |
| Ox | D | D | Bitter | | |

[a] Values given refer to the number of times sweeter than sucrose, i.e. sweetness potency.
[b] N-Me-Gly.

replaced by a variety of simple amines[7] (Table 2). Note that for clarity throughout this review structures will be drawn with $R_1$ and $R_2$ in the plane of the paper as shown in the table. $R_1$ will commonly be referred to as the 'upper' ground and $R_2$ as the 'lower' group. While the analog 2 derived from benzylamine (lacking the α-substituent $R_1$) was not sweet, the amide 3 derived from L-amphetamine was sweet, demonstrating that the ester group is not essential. The amide 4 from D-amphetamine was not sweet, reflecting the same stereochemical bias observed with L- and D-phenylalanine in aspartame. Of the $R_1$ groups

## TABLE 2

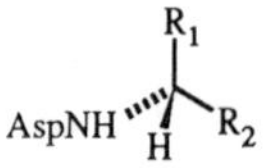

$$\text{AspNH}^{\text{\tiny ...}}\underset{\text{H}}{\overset{R_1}{\diagup}}R_2$$

| Cmpd | $R_1$ | $R_2$ | Intensity |
|------|-------|-------|-----------|
| 2 | H | $CH_2Ph$ | 0 |
| 3 | Me | $CH_2Ph$ | 50 |
| 4 | Me (D) | $CH_2Ph$ | 0 |
| 5 | Et | $CH_2Ph$ | 5 |
| 6 | $CH_2OH$ | $CH_2Ph$ | 1 |
| 7 | (Me)[a] Me | $CH_2Ph$ | 20 |
| 8 | (NMe)[b] Me | $CH_2Ph$ | 0 |
| 9 | H | $CH(CH_3)Ph$ (D,L) | 0 |
| 10 | Me | $CH_2$—furyl 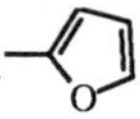 | 10 |
| 11 | Me | $CH_2$—cyclohexyl 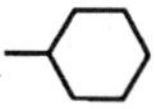 | 50 |
| 12 | Me (D,L) | n-Bu | 30 |
| 13 | Me | n-Am | 50 |
| 14 | Me (D,L) | n-Hex | 10 |

[a] H=Me

[b] AspNH=AspNMe

used (H, $CH_3$, $C_2H_5$, $CH_2OH$), the methyl group led to the highest sweetness intensity, again suggesting that a small $R_1$ group is optimal. Interestingly, the analog **7** with two methyl groups at the former chiral center retained significant sweetness, suggesting that the critical interaction of the methyl group in the 'pro-L' configuration is not eliminated by the second methyl group. Methyl substitution on nitrogen (**8**) or a shift of the methyl group to the $\beta$-carbon (**9**) resulted in a loss of sweetness. The $R_2$ group can vary considerably, with furyl (**10**) and cyclohexyl (**11**) acceptable replacements for the phenyl group. Simple acyclic alkyl groups can be used (**12–14**), with an optimal length of 5 carbons. Having established that $R_1$ can be an ester and $R_2$ an alkyl group (as in **I**), Searle workers examined analogs with the $R_1$ and $R_2$ groups reversed (as in **II**),[8] i.e. L-aspartyl-D-amino acid esters (Table 3). This arrangement requires a D-amino acid ester with a small side-chain, as this group occupies the $R_1$ position. For $R_1 =$ methyl (D-alanine esters) sweetness varied with the size of the ester group (**15–20**), with the propyl compound (**17**) being optimal. Increasing the size of $R_1$ from methyl to butyl (**18, 21–23**) decreased sweetness intensity to zero.

$R_1$ (small)

CO$_2$R

AspNH ⁗/ alkyl

L

H

$R_2$ (large)

**I**

$R_1$ (small)

alkyl

AspNH ⁗/ CO$_2$R

D

H

$R_2$ (large)

**II**

The fact that ester and alkyl groups are interchangeable at $R_1$ and $R_2$ suggests that receptor interactions at these sites are determined by size and shape rather than by electrostatic properties. When $R_1$ and $R_2$ are hydrophobic and of dissimilar size, with $R_1$ small relative to $R_2$, the resulting $\alpha$-aspartyl amides (**III**) are likely to be sweet.

| | | $R_1$ (small) | $R_2$ (large) |
|---|---|---|---|
| | L | CO$_2$Me | CH$_2$Ph, alkyl |
| | D | CH$_3$ | CO$_2$alkyl |

$R_1$

AspNH ⁗/ $R_2$

H

**III**

## TABLE 3

$$\text{AspNH} \overset{R_1}{\underset{H}{\langle}} CO_2R$$

| Cmpd | $R_1$ | R | Intensity |
|------|-------|------|-----------|
| 15 | Me | Me | 25 |
| 16 | Me | Et | 80 |
| 17 | Me | n-Pr | 170 |
| 18 | Me | i-Pr | 125 |
| 19 | Me | n-Bu | 10 |
| 20 | Me | n-Am | 6 |
| 21 | Et | i-Pr | 170 |
| 22 | Pr (D,L) | i-Pr | 17 |
| 23 | Bu | i-Pr | 0 |

These observations were explicitly formalized by Ariyoshi[9] using Fischer projection formulas (Fig. 1). Ariyoshi expanded on Kaneko's[10] stereochemical model for amino acids (**IV**). In this model the $NH_3^+$ and $CO_2^-$ groups bind to the receptor. The side-chain R is essential for modulating sweetness intensity. For example, glycine (R = H) and D-alanine **24** (R = $CH_3$) have only a low level of sweetness while D-tryptophan **25** (X = H) and 6-chloro-D-tryptophan **25** (X = Cl) are 35 and 1300 times as sweet as sucrose respectively. Kier has suggested that this increase in sweetness is due to hydrophobic or dispersion bonding between the R groups and the receptor.[11] For sweet dipeptides (**V**) the second amino acid (the bottom half of **V**) occupies the same position as the amino acid side-

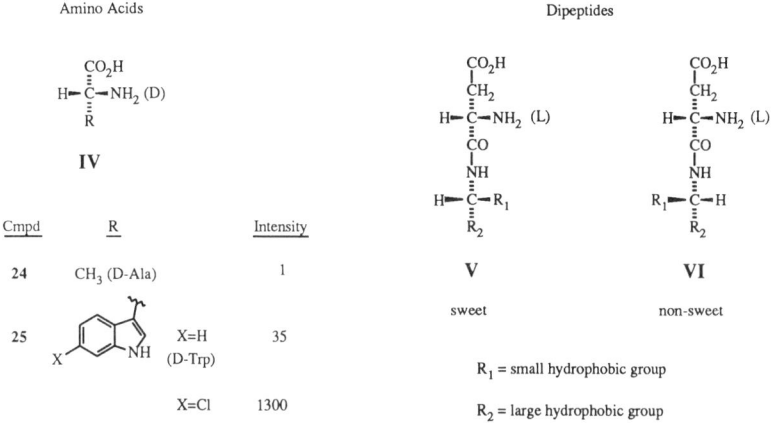

FIG. 1.    Ariyoshi model for sweet dipeptides.

chain R in **IV**. The AH–B groups are again the $NH_3^+$ and $CO_2^-$ groups which, while further apart, are still within the $2·5–4·0\,\text{Å}$ range defined by Shallenberger and Acree.[5] Note that the orientation of the groups about the amino acid chiral center in **IV** and the aspartic acid chiral center for the dipeptides **V** and **VI** is the same although the former is D and the latter L. The fact that the carboxylate in **V** and **VI** is actually part of the side-chain (the R group in **IV**) accounts for the change in stereochemical assignment.

The second critical feature for sweetness of dipeptide analogs is the arrangement of groups about the lower chiral center. Sweet compounds bear a small substituent at $R_1$ and a large substituent at $R_2$. In some cases the configuration will be designated as D, as for the L-aspartyl-D-alanine esters, and in others it will be L, as for L-aspartyl-L-phenylalanine methyl ester (aspartame). In both cases the sweeteners conform to Ariyoshi's model **V**. If the orientations of $R_1$ and $R_2$ are reversed as for **VI**, the resulting compounds are not sweet. For example, L-aspartyl-D-phenylalanine methyl ester, a diastereomer of aspartame, is not sweet. Even though the aspartyl portion has the correct stereochemical configuration, the incorrect orientation of the $R_1$ and $R_2$ groups eliminates sweetness.

The Fischer projection formulas show clearly the stereochemical requirements for sweetness. However, the equivalent but more compact structure **III** will be used throughout. A similar representation was proposed by Goodman and Gilon[12] for L-aspartyl-L-phenylalanine (and D-alanine) esters.

The dipeptide analogs prepared in these early publications suggested

J. M. JANUSZ

TABLE 4

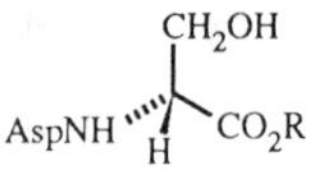

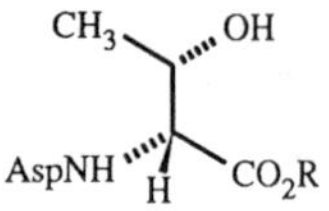

|  | CH$_2$OH<br>AspNH''''—CO$_2$R<br>H |  |  | CH$_3$''''—OH<br>AspNH''''—CO$_2$R<br>H |  |
|---|---|---|---|---|---|
| Cmpd | Intensity | R | Cmpd | Intensity |
|---|---|---|---|---|
| 26 | 45 | Me | 31 | 25 |
| 27 | 115 | Et | 32 | 110 |
| 28 | 320 | Pr | 33 | 150 |
| 29 | 120 | i-Pr | 34 | 105 |
| 30 | 60 | 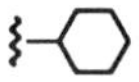 ⬡— | 35 | 30 |

strict structural requirements for interaction with the sweet receptor. The requirements were refined in an empirical way in the studies outlined below.

## 2.2 Substantiation of the Ariyoshi Model

The requirement for a small R$_1$ group somewhat limits the variety of substituents at this position. As seen above, the smallest member of a given class is generally best (e.g. methyl among alkyl groups, carbomethoxy among ester groups). Hydroxylated R$_1$ groups, although larger than their alkyl counterparts, can provide intensely sweet compounds. Ariyoshi *et al.*[13] prepared a series of L-aspartyl-D-serine and -D-threonine esters (Table 4). The D-serine esters (**26–30**) were more intensely sweet than their non-hydroxylated counterparts (e.g. L-aspartyl-D-alanine propyl ester **17** is 170 times as sweet as sucrose). The D-threonine esters (**31–35**) were about as sweet as the D-alanine esters but the allo-threonine esters (not shown) were considerably less sweet. This stereochemical dependence suggests that not only the size but also the shape of the upper group is important.

A variety of amino alcohol derived sweeteners have been prepared (Table 5). These are hydroxylated in either the upper or lower group. They lack the ester moiety of the compounds in Table 4 which results in improved

TABLE 5

$$\text{AspNH} \overset{\overset{\displaystyle R_1}{|}}{\underset{H}{\mid}} R_2$$

| Cmpd | $R_1$ | $R_2$ | Intensity |
|------|-------|-------|-----------|
| 36 | $CH_2OH$ | Iso-amyl | 25 |
| 37 | $CH(OH)CH_3$ | Iso-amyl | 60 |
| 38 | $CH(OH)CH_3$ | | 500 |
| 39 | $CH_3$ | $CH(OH)CH_2$— | 100 |
| 40 | $CH_3$ | $CH(OH)CH_2$ | 360 |
| 41 | $CH_3$ | $CH(OH)CH_2$ | 600 |

stability. Compounds 36 and 37 with acyclic lower groups show modest intensity.[14] The *S,S* isomer† of 37 was the sweetest of the 4 possible diastereomers. Compounds 38–41,[15,16] which bear cyclic or substituted bicyclic alkyl groups, are more intensely sweet.

L-Serine was the basis for two new classes of dipeptides (Table 6). Ariyoshi *et al.*[13] prepared the L-serine esters 42–45 where the hydroxy group was esterified to give the requisite large lower group. These compounds were considerably less sweet than the related D-serine esters

† The D/L nomenclature commonly used to designate the absolute configuration of amino acids and carbohydrates has been replaced by the *R/S* nomenclature for other classes of compounds. The *R* and *S* designations are assigned using a set of sequence rules to rank the four groups on the asymmetric carbon (Cahn–Ingold–Prelog system).

TABLE 6

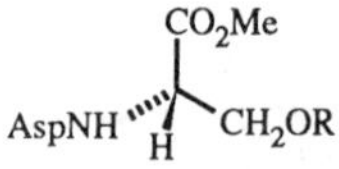

CO$_2$Me

AspNH ''''/CH$_2$OR
          H

| Cmpd | R | Intensity |
|------|---|-----------|
| 42 | COMe | 10 |
| 43 | COEt | 40 |
| 44 | COPr | 30 |
| 45 | CO-i-Bu | 50 |
| 46 |  | 800 |
| 47 | 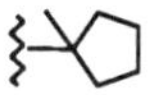 | 308 |
| 48 | CH(t-Bu)$_2$ | 400 |
| 49 | 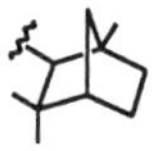 | 1000[a] |

[a] Derived from 2-methylserine

**26–30.** However, the L-serine ethers[17] **46–48** were quite sweet. The cyclic and highly methylated character of these R groups is important for intensity. The sweetness of the fenchyl ether **49** derived from 2-methylserine was also quite intense.[18]

The requirement for a relatively large lower group in the model structure **III** provides for more structural variety. Miyoshi *et al.*[19] examined homologs of the Searle D-alanine esters (Table 7). Insertion of a methylene group as shown avoids ready formation of diketopiperazines (DKP) by reaction of the aspartyl amino group with the ester carbonyl. Cyclization to DKP is one of the major decomposition pathways for aspartame, especially at neutral and basic pH. However, none of the resulting compounds **50–56** was intensely sweet.

## TABLE 7

AspNHCH$_2$–C(R$_1$)(CO$_2$R) with CH$_2$ / structure

| Cmpd | R$_1$ | R | Intensity |
|------|-------|---|-----------|
| 50 | CH$_3$ | i-Pr | 2 |
| 51 | CH$_3$ | cyclohexyl | 10 |
| 52 | CO$_2$Me | CH$_2$Ph | 0 |
| 53 | CO$_2$Me | cyclohexyl | 0 |

AspNH–CH(CH$_3$)–CH$_2$CO$_2$R

| Cmpd | R | Intensity |
|------|---|-----------|
| 54 | Et | 20 |
| 55 | i-Pr | 9 |
| 56 | cyclopentyl | 5 |

TABLE 8

$$AspNH \overset{R_1}{\underset{H}{\wedge}} CH_2O_2CR$$

| Cmpd | $R_1$ | R | Intensity |
|------|-------|---|-----------|
| 57 | $CH_3$ | Me | 60 |
| 58 | $CH_3$ | Et | 100 |
| 59 | $CH_3$ | i-Pr | 120 |
| 60 | $CH_3$ | $\S$—◇ | 220 |
| 61 | $CH_3$ | $\S$—◇✕ | 0 |
| 62 | $CH_2OH$ | Me | 25 |
| 63 | $CH_2OH$ | Et | 135 |
| 64 | $CH_2OH$ | $\S$—◁ | 120 |
| 65 | $CH_2OH$ | $\S$—◇ | 65 |
| 66 | Et (D,L) | Me | 10 |
| 67 | Pr (D,L) | Me | 3 |

Higher intensity was observed for the D-alaninol and D-serinol esters **57–65** (Table 8).[19] As seen before, intensity passes through a maximum and then decreases as the size of R increases. At the upper group position, intensity decreases with increasing size of $R_1$ (**57, 66, 67**).

Table 9 shows a comparison of the sweetness intensity of the Searle D-alanine propyl ester versus two of the related Miyoshi compounds. The lower group of each is of similar size. Insertion of a methylene group in the D-alanine ester lowers the intensity about 10-fold while insertion plus

TABLE 9

| Cmpd | Structure | Intensity |
|------|-----------|-----------|
| 17 | AspNH—C(CH$_3$)—CO$_2$Pr | 170 |
| 54 | AspNH—C(CH$_3$)—CH$_2$CO$_2$Et | 20 |
| 58 | AspNH—C(CH$_3$)—CH$_2$O$_2$CEt | 100 |

reversal of the ester group lowers the intensity only by half. Conformational changes are probably responsible for the differences in intensity.

## 3 NOVEL DIPEPTIDE SWEETENER ANALOGS

### 3.1 High Intensity Dipeptide Analogs

In the first few years after Searle's original publication,[4] all the dipeptide sweeteners disclosed had intensities less than twice that of aspartame. The first breakthrough to high sweetness intensity resulted from the work of Fujino et al.[20] with L-aspartyl-D,L-aminomalonic acid diesters. Table 10 gives examples which clearly demonstrate important structural features in the lower ester. Rigid groups which are appropriately substituted gave the highest intensity. For example, methylation of the cyclohexyl group at C-2 gave a large increase in intensity (70 versus 71). Substitution at the 3 or 4 position did not (72, 73), presumably because the receptor cannot sterically accommodate substitutions as well at these positions. The rigid bicyclic fenchyl ester 74 with the carbon atoms $\beta$ to the ester oxygen fully substituted was very intensely sweet. A systematic comparison of sweeteners derived from each of the 4 possible fenchol isomers 74–77 showed an intensity range of 1000–50 000 × .[21]

Insertion of a methylene group,[22] as shown for 78, results in a complete

## TABLE 10

$$AspNH-\begin{array}{c} CO_2Me \\ CO_2R \end{array}$$

| Cmpd | R | Intensity |
|------|---|-----------|
| 68 | $CH(Et)_2$ | 335 |
| 69 | $CH(i\text{-}Pr)_2$ | 575 |
| 70 | cyclohexyl | 620 |
| 71 | cyclohexyl (substituted) | 6375 |
| 72 | methylcyclohexyl | 530 |
| 73 | 4-methylcyclohexyl | 40 |
| 74 | bicyclic, (-)-α | 27600 |
| 75 | bicyclic, (+)-α | 1000 |
| 76 | bicyclic, (-)-β | 5000 |
| 77 | bicyclic, (+)-β | 50000 |
| 78 | $CH_2CO_2$-bicyclic | 0 |
| 79 | bicyclic | 645 |
| 80 | cubyl/cage | 10000 |

## TABLE 11

AspNH—CH(CH$_3$)(H)—CONHR

| Cmpd | R | Intensity |
|------|---|-----------|
| 81 | Pr | 25 |
| 82 | Bu | 125 |
| 83 | | 100 |
| 84 | | 250 |
| 85 | | 600 |
| 86 | | 800 |
| 87 | | 500 |
| 88 | | 2000 |
| 89 | | 1000 |
| 90 | | 1200 |
| 91 | | 320 |

loss of sweetness. The isobornyl compound **79** differs from the fenchyl compound **77** only in the position of the geminal methyl groups but has about 1/80 of the sweetness, presumably owing to interference in receptor site binding by the methyls on the one-carbon bridge. Recently a few more examples (e.g. **80**) have been added to this class.[23]

With the publication of the Fujino sweeteners, the era of high intensity, which we will define rather arbitrarily as starting at 1000 times sweeter than sucrose, was opened. Fujino's observation of the importance of rigid,

TABLE 12

$$\text{AspNH}^{\backslash\backslash\backslash}\underset{\text{H}}{\overset{\text{CH}_3}{|}}\text{NHCOR}$$

| Cmpd | R | Intensity |
|------|---|-----------|
| **92** | (cyclopentyl) | 85 |
| **93** | (methylcyclopentyl) | 350 |
| **94** | (dimethylcyclopentyl) | 900 |
| **95** | (S)[a] (dimethylcyclopentyl) | 700 |
| **96** | (dimethyl, S-substituted cyclobutyl) | 175 |
| **97** | [AspNH—] (dimethylcyclopentyl) | 450 |
| **98** | [AspNH, CH₂OH] (cyclopentyl) | 75 |

[a] S stereochemistry at chiral center shown. All others are R.

appropriately branched 'lower groups' was a refinement of the Ariyoshi model and the basis for other high-intensity dipeptides.

Sukehiro et al.[24] investigated a series of L-aspartyl-D-alanine amides (Table 11) which are more stable variants of Searle's L-aspartyl-D-alanine esters. A limited number of acyclic amides and the cyclohexyl amide were prepared (81–83), all of similar intensity ($\sim 100 \times$). The lack of any difference in intensity led Sukehiro to conclude that only the size and not the shape of the amide group was important. However, workers at Pfizer showed that this conclusion was premature.[25] By incorporating rigid, branched lower groups the intensity was increased significantly (84–91). Simple methyl substitution on the 2 and 6 positions of the cyclohexyl ring gave a 6-fold increase in intensity over the parent. Optimization of the R group resulted in the selection of the 2,2,4,4-tetramethylthietane amide 88 for further development. This sweetener has high solubility and is more stable than aspartame. A food additive petition for this sweetener, known as Alitame, has been submitted. The corresponding D-serine and O-methyl-D-serine amides[26] were also prepared but were less intensely sweet. It is interesting that the fenchyl amide 91 in this series has only modest intensity, as opposed to the Fujino methyl fenchyl malonic acid diesters which were the most intensely sweet.

Goodman and co-workers[27] have applied a 'retro-inverso' peptide modification to the Pfizer amides. This nomenclature refers to a reversal of the usual carbonyl and nitrogen groups of an amide structure. For the present case, a normal amide bond sequence is replaced by an acylated 1,1-diaminoalkane coupled to a malonic acid derivative. This retro-inverso modification of the D-alanine amide bond led to compounds with sweetness intensities up to $900 \times$ (Table 12).

normal peptide                                retro-inverso peptide

In contrast to the Pfizer sweeteners, the 2,2,5,5-tetramethylcyclopentyl compound 94 was the most intensely sweet, while the 2,2,4,4-tetramethylthietane 96 had lower intensity. The R or S stereoisomers of the 1,1-diaminoalkane moiety gave about equal intensity (94 versus 95) and the dimethylated sweetener **97** retained half the intensity of **94**. For small groups such as methyl, either stereochemical form can be tolerated, as can disubstitution. This could have practical significance in that less costly racemic 1,1-diaminoalkanes could be used to prepare these sweeteners. Evidently the sweet receptor has some modest degree of flexibility in accommodating the upper group. The Goodman compounds are claimed to be highly soluble and considerably more stable than aspartame.

The amide group in the D-alanine amides and retro amides, which is the larger $R_2$ group in our generalized sweetener structure **III**, can also occupy the small $R_1$ group position. Compounds **99–102** (Table 13) are D,L-aminomalonic acid amide esters.

High intensity is again perceived with branched, cyclic ester groups **(99–101)**.[28] A related monomethyl amide **102** was prepared at Procter & Gamble but was much less sweet.[29] The lower intensity may be due to the isomer of fenchol used $((+)$-$\alpha)$ or to a detrimental effect of the amide hydrogen. By way of comparison, replacement of the methyl ester of aspartame by an amide, monomethyl or dimethyl amide leads to non-sweet compounds.[4]

TABLE 13

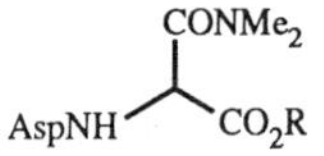

CONMe$_2$

AspNH — CO$_2$R

| Cmpd | R | Intensity |
|------|---|-----------|
| 99 | 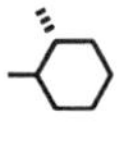 | 500 |
| 100 | 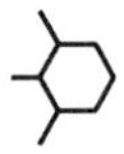 | 4000 |
| 101 |  | 5000 |
| 102 | 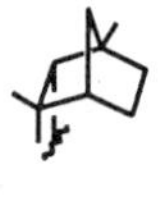 | 20[a] |

[a] With CONHMe upper group

Sweetener **102** was prepared as a stabilized version of the Fujino aminomalonic acid diesters where the labile methyl ester group was replaced by the isosteric $N$-methyl-amide. The loss of intensity prompted consideration of other simple groups which might mimic the critical structural features of the ester group. Planar groups connected via an $sp^2$ center (i.e. a carbon atom with 3 substituents all in the same plane) were selected. The D,L-furylglycine ($+$)-$\alpha$-fenchyl ester **103** was prepared, and the sweetness intensity of this mixture of diastereomers increased considerably over that of the corresponding $N$-methyl-amide **102** (Table 14).[30] Subsequent extension of this work to phenylglycine and other heterocyclic glycine esters **104–106** resulted in very high intensity sweeteners, especially as the ($-$)-$\alpha$- or ($+$)-$\beta$-fenchyl esters.[31]

Phenyl and heteroaromatic groups are larger than common $R_1$ upper groups, and the planarity of the aromatic groups is apparently crucial for

TABLE 14

AspNH$\overset{R_1}{\underset{H}{\mid}}$CO$_2$R

| Cmpd | $R_1$ | R | Intensity |
|------|-------|---|-----------|
| 103 | | | 600 |
| 104 | | | 16450 |
| 105 | | | 1750 |
| 106 | | | 2400 |
| 107 | | | 0 |

TABLE 15

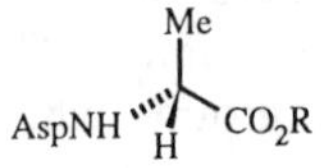

$$\text{AspNH} \overset{\text{Me}}{\underset{\text{H}}{\diagdown}} \text{CO}_2\text{R}$$

| Cmpd | R | Intensity |
|------|---|-----------|
| 108 |  | 1200 |
| 109 | 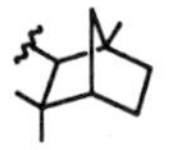 | 6000 (1200)[a] |

[a] ( ) value is vs 9% sucrose.[32a] Others are threshold values.

their sweetness. For example, reducing the furylglycine sweetener **103** to the tetrahydrofuryl derivative **107** destroys the sweetness.

In the past few years, several groups have optimized existing sweeteners using Fujino's discovery of the fenchyl group as an extremely potent lower group. For example, Searle's L-aspartyl-D-alanine esters were re-examined. Fenchyl alcohols again gave high intensity, with the $(+)$-$\beta$ isomer the most intense at $6000 \times$ (Table 15).[32]

A large number of patents have recently been issued to General Foods Corporation in addition to those already cited for amino-alcohols,[15] L-serine ethers[17] and D-alanine esters.[32a] For example, the novel olefinic compounds **110**[33] (and the corresponding alkanes[33f,34]), where the *trans* double bond might be considered an amide isostere (i.e. having similar size and shape), were claimed. The amides[35] and retro-inverso amides[36] **111**, both known classes of sweeteners discussed above (Tables 11 and 12), were claimed as patentable for specific combinations of $R_1$, $R_2$ and $R_3$. No intensity data were provided.

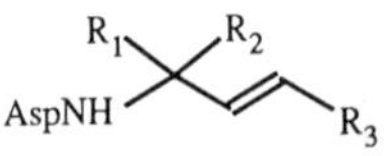

$$\text{AspNH} \overset{R_1 \quad R_2}{\diagup} R_3$$

$$\text{AspNH} \overset{R_1 \quad R_2}{\diagup} \underset{(\text{NHCO})}{\text{CONHR}_3}$$

110 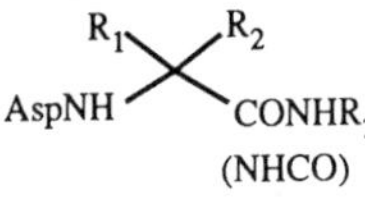        111

## 3.2 Aspartic Acid Modifications

The original Searle work[4] showed that no naturally occurring amino acid could replace aspartic acid and that the α-amino and β-carboxyl groups should remain unsubstituted. Replacement of the β-carboxyl group with phosphonate[37] (112), sulfonate[38] (113) or cyano[39] (114) resulted in a complete loss of sweetness. Cyclic analogs 115–120 of the aspartyl portion of aspartame were likewise not sweet.[40] The absence of a charged carboxyl oxygen to function as a H-bond acceptor probably prevents effective interaction with the receptor.

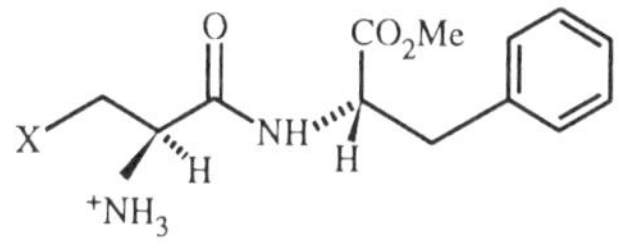

| Cmpd | X |
|------|------|
| 112 | $PO_3^-$ [a] |
| 113 | $SO_3^-$ |
| 114 | CN |

[a] Asp isostere is D,L

(non-sweet)

The peptide bond also is critical for sweetness.[4,41,45] The amide grouping cannot be methylated (121), reversed (122) or replaced by an ester (123) or hydrazide (124).[42] Goodman and co-workers[41,45] interpret these observations as evidence that the amide bond interacts with the sweet receptor. The observation that the thioamide 125 has reduced sweetness[43] versus aspartame is relevant to this point.

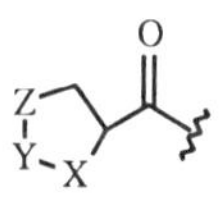

| Cmpd | X-Y-Z | Cmpd | X-Y-Z |
|------|---------|------|-----------|
| 115 | NHNHCO | 118 | $NHCOCH_2$ |
| 116 | $NHNHCH_2$ | 119 | $NHCH_2CH_2$ |
| 117 | NHCOCO | 120 | $CH_2CH_2CH_2$ |

(non-sweet)

The use of D,L-aminomalonic acid (Ama) in place of L-aspartic acid was one of the first modifications of the N-terminal amino acid which retained sweetness (Table 16). The relationship of the amino and carboxyl groups for Ama is the same as in ordinary amino acids, so Ama conforms to the Shallenberger and Acree AH–B hypothesis. Briggs and Murley[44] prepared the D,L-Ama analog of aspartame 126 and claimed it to be 300–400 times as sweet as sucrose. Fujino *et al.*[20] and Chorev *et al.*[45] subsequently reported the same compound as 200 × and 'intensely sweet' respectively. Predictably, only one of the diastereomers is sweet, so this compound is at least as sweet as aspartame. Ariyoshi[9] has pointed out that D-Ama-L-Phe-OMe is probably the sweet diastereomer rather than the L,L diastereomer. The absolute configuration of D-Ama conforms to that of L-Asp because of a change in the sequence priority of the substituents at the chiral center. Seltzman[46] extended this lead by preparing D,L-Ama-D-alanine esters. The isopropyl ester **127** had a sweetness of 58 × but curiously the *trans*-2-methylcyclohexyl and other esters were not sweet. MacDonald *et al.*[41] reported that the adamantyl ester **128** of D,L-Ama-D,L-Ama-OMe was intensely sweet.

| Cmpd | X | Intensity |
|------|------|-----------|
| 121 | CONMe | 0 |
| 122 | NHCO[a] | 0 |
| 123 | $CO_2$ | 0 |
| 124 | CONHN[b] | 0 |
| 125 | CSNH | "sweet" |

[a] From $^-O_2CCH(NH_3)^+$ —X—}

[b] From }—X— $(CO_2Me)Ph$

One notable exception to the observation that only unmodified L-aspartyl or aminomalonic acids are acceptable N-terminal amino acids was published in 1973. After accidentally discovering the sweetness of L-3-trifluoroacetamidosuccinanilides **129** and **130**, Lapidus and Sweeney[47]

TABLE 16

| Cmpd | $R_1$ | $R_2$ | Intensity |
|------|-------|-------|-----------|
| 126 | $CO_2Me$ | $CH_2Ph$ | 350  (200) |
| 127 | $CH_3$ | $CO_2$-i-Pr | 58 |
| 128 | $CO_2Me$ | $CO_2$⟨adamantyl⟩ | intensely sweet |

prepared a variety of N-acyl-L-aspartyl-α-anilides and amides. Among these was the N-trifluoroacetyl aspartame **131**, which has an intensity of sweetness close to aspartame itself. Italian workers[48] soon showed that trifluoroacetylation can in some cases eliminate sweetness, as with the aspartyl amphetamine derivative **132**. Trifluoroacetylation of the glutamyl derivative **134** did not result in sweetness, and sweet taste was lost on trifluoroacetylation of the malonyl derivative **133**.[49,50] Kawai et al.[50] offered a conformational explanation for these differences. This one successful modification of the α-amino group remained a curiosity for a decade before a second example was discovered. N-(N'-Formylcarbamoyl)-aspartame **135** was about as sweet as aspartame.[51]

| Cmpd | X | Intensity |
|------|---|-----------|
| 129 | H | 12 |
| 130 | CN | 3000 |

| Cmpd | $R_1$ | Intensity |
|------|-------|-----------|
| 131 | $CO_2Me$ | 120 |
| 132 | $CH_3$ | 0 |

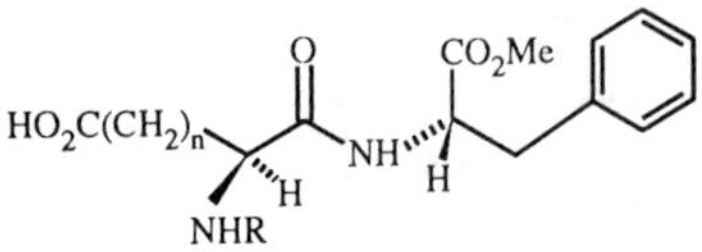

| R \ n | 0 | 1 | 2 |
|---|---|---|---|
| H | ++ | ++ | 0 |
| $CF_3CO$ | 0 | ++ | 0 |
| | 133 | 131 | 134 |

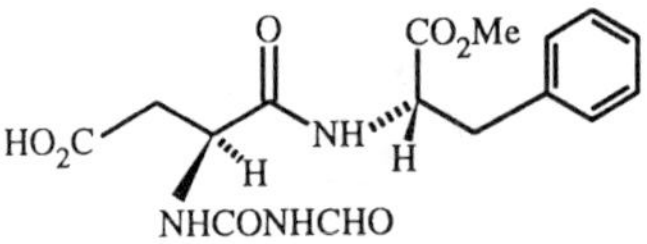

$HO_2C(CH_2)_n$ — structure with $CO_2Me$, phenyl, NHR

$HO_2C$ — structure with $CO_2Me$, phenyl, NHCONHCHO

**135**

Both structures **131** and **135** offered a potential stability advantage by preventing cyclization to a diketopiperazine, but neither was any more intensely sweet than aspartame itself. It was therefore quite surprising when French workers disclosed a group of $N$-(4-substituted phenylcarbamoyl/ thiocarbamoyl)-L-aspartyl dipeptides[52] and the corresponding cyano-imino compounds[53] with intensities up to 100 times that of aspartame when compared to 10% sucrose. Table 17 gives representative examples.

These aspartyl modified dipeptides can be considered as a combination of two sweeteners—aspartame (or related dipeptides) and suosan[54] **144.**

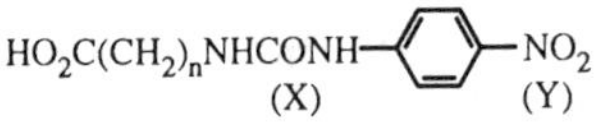

$HO_2C(CH_2)_n NHCONH$—⟨ring⟩—$NO_2$
(X)                      (Y)

**144** (n=2)      Suosan, 700x

Analogs:

n = 1, 2

X = O, S, NCN, C(CN)$_2$, NH, NSO$_2$Ph, NHR

Y = NO$_2$, CN

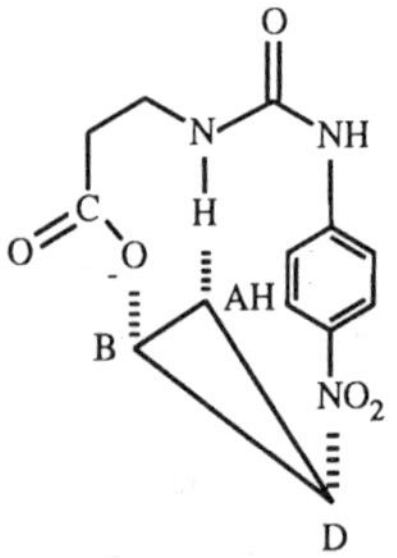

TABLE 17

| Cmpd | X | Y | $R_1$ | $R_2$ | Intensity | |
|------|---|---|-------|-------|-----------|---|
| 136 | O | H | $CO_2Me$ | $CH_2Ph$ | 130 | |
| 137 | O | CN | $CO_2Me$ | $CH_2Ph$ | 10000 | |
| 138 | O | $NO_2$ | $CO_2Me$ | $CH_2Ph$ | 14000 | $(4500)^a$ |
| 139 | S | CN | $CO_2Me$ | $CH_2Ph$ | 50000 | $(17000)^a$ |
| 140 | S | $NO_2$ | $CO_2Me$ | $CH_2Ph$ | 55000 | |
| 141 | O | $NO_2$ | $CO_2Me$ | n-Bu | 14000 | |
| 142 | O | $NO_2$ | $CH_3$ | CONHPr | 300 | |
| 143 | NCN | CN | $CO_2Me$ | $CH_2Ph$ | 40000 | $(6400)^a$ |

[a] ( ) values are vs 10% sucrose by wgt. Others are
vs 2% sucrose on a molar basis.

The French workers had studied suosan in detail and proposed a model for the interaction of suosan with the sweet receptor.[55] Their model included a binding site (D) for the *para* substituent in addition to the AH–B sites. The combination of suosan and aspartame into a single compound (**138**) may result in four binding sites—the Shallenberger–Acree–Kier AH–B-$\delta$ and the Tinti *et al.* D site.

Recent analog work has uncovered some extremely potent compounds.[56] While they are not dipeptides, compounds **145–147** are the most intensely sweet compounds known. Another successful aspartyl modification was recently reported by workers at Ajinomoto.[57] Sweet tripeptides were prepared with a third amino acid added to the N-terminal aspartyl residue (versus the C-terminal residue as in the tri-pentapeptides discussed in Section 3.3). Of the 24 tripeptides prepared, the most intense was the aspartame derivative **148** which was about as sweet as aspartame itself. Curiously, the tripeptide **149** derived from the aminomalonic acid analog of aspartame was not sweet. Tetrapeptides such as **150** were bitter, suggesting there may not be enough room for 2 additional amino acids at the sweet receptor site. Clearly, there is room for 1 additional amino acid bound to L-aspartic acid but this space is sterically restricted. Thus far, tripeptides are no sweeter than the parent dipeptides. The tripeptide **151**, extended via the aspartyl β-carboxyl group, was not sweet.

### 3.3 Tri-pentapeptides

The variety of structures that can be accommodated as lower groups prompted the investigation of the tripeptides[13,24,58-61] (Table 18) made by addition of an amino acid to the C-terminal end of sweet dipeptides. While **152** was not sweet, substitution of D-Ala for Gly as the second amino acid (**153**) or L-Ala for Gly as the third amino acid (**154**) gave some sweetness. The preferred stereochemistry for the second and third amino acids was D and L respectively. Optimization gave tripeptides **155** and **156** of modest intensity. The tripeptides were not nearly as sweet as the best L-aspartyl-D-alanine amides in Table 11.

Goodman and co-workers[59] examined the effects of size and hydrophobicity in greater detail by using conformationally constrained

| Cmpd | R | Intensity | |
| --- | --- | --- | --- |
| | | vs 2% sucrose | vs 10% sucrose |
| 145 | $c\text{-}C_8H_{15}$ | 170000 | 100000 |
| 146 | $c\text{-}C_9H_{17}$ | 200000 | -- |
| 147 | | 120000 | 50000 |

$HO_2CCH_2NH-\overset{\underset{|}{NHR}}{C}=N$ —⟨⟩— CN

$HO_2CCH_2NH-\overset{\underset{|}{NHCH(CH_3)Ph}}{C}=N$ —⟨Cl, Cl⟩

| Cmpd | Structure | Intensity |
|------|-----------|-----------|
| **148** | D-Ala-L-Asp-L-Phe-OMe | 180 |
| **149** | D-Ala-D,L-Ama-L-Phe-OMe | bitter |
| **150** | L-Ala-D-Ala-L-Asp-L-Phe-OMe | bitter |
| **151** | L-Asp-L-Phe-OMe <br> \| <br> D-Ala | bitter |

C-terminal amino acids. Disubstitution is tolerated on the α carbon atom. Methyl (**157**), ethyl (**158**) and cycloalkyl groups up to cyclohexyl (**159**) all gave compounds about as sweet as those in Table 18 when the differences in tasting procedures are taken into account. No large difference in intensity was observed as the size and hydrophobicity of the C-terminal amino acid increased. The sudden change from sweet to bitter on expanding the ring size from 6 to 7 suggests that the sweet and bitter receptors are closely related.

### TABLE 18
TRIPEPTIDES

| Cmpd | Structure | Intensity |
|------|-----------|-----------|
| **152** | L-Asp-Gly-Gly-OMe | 0 |
| **153** | L-Asp-D-Ala-Gly-OMe | 3 (20)[a] |
| **154** | L-Asp-Gly-L-Ala-OMe | 2 |
| **155** | L-Asp-D-Ala-L-Ala-OMe | 50 |
| **156** | L-Asp-D-Ala-L-Val-OMe | 50 |

[a] ( ) Taken from Sukehiro et al.[24]

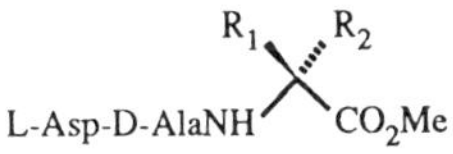

$$\text{L-Asp-D-AlaNH} \diagdown \overset{R_1 \quad R_2}{\diagup} \diagup \text{CO}_2\text{Me}$$

| Cmpd | $R_1$ | $R_2$ | Intensity |
|------|-------|-------|-----------|
| 157 | Me | Me | 20 |
| 158 | Et | Et | 20 |
| 159a-d | $(CH_2)_n$ n=2--5 | | 20 |
| 160a,b | $(CH_2)_n$ n=6,7 | | bitter |

The structure–activity trends observed for the tripeptides are consistent with those formulated for dipeptides. The small group is $R_1$ in the tripeptides while the large group is the remainder of the tripeptide. The preference for the L configuration for the third amino acid suggests that the R group may be involved in a hydrophobic interaction with the receptor.

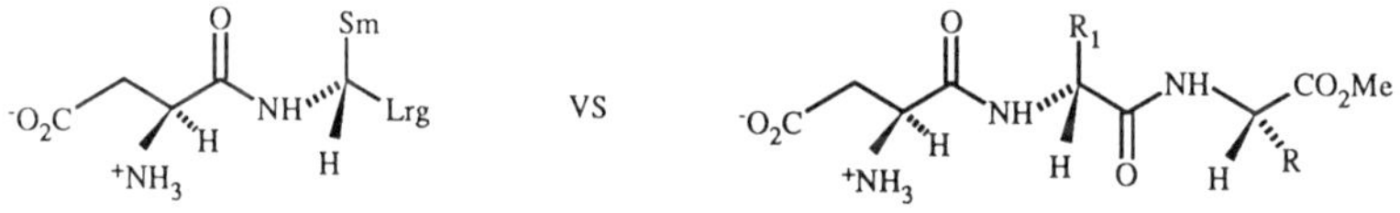

VS

Similar stereospecificity was previously observed in the Pfizer L-aspartyl-D-alanine amides as is apparent in comparing the structures shown.

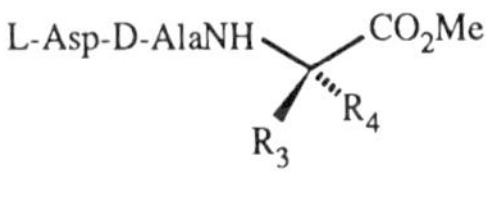

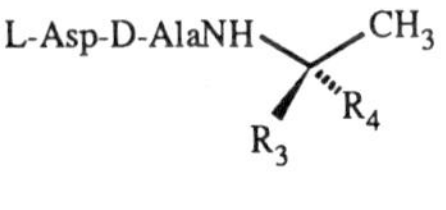

$$\text{L-Asp-D-AlaNH} \diagdown \overset{}{\diagup} \diagup \text{CO}_2\text{Me} \quad R_3, R_4$$

$$\text{L-Asp-D-AlaNH} \diagdown \overset{}{\diagup} \diagup \text{CH}_3 \quad R_3, R_4$$

| $R_3$ | $R_4$ | Intensity | $R_3$ | $R_4$ | Intensity |
|-------|-------|-----------|-------|-------|-----------|
| H | $CH_3$ | 50 | H | t-Bu | 350 |
| $CH_3$ | H | 5 | t-Bu | H | 0 |

TABLE 19

RELATIVE SWEETNESS OF DI-PENTAPEPTIDES

| Cmpd | Structure | Intensity |
|------|-----------|-----------|
| 15   | L-Asp-D-Ala-OMe | 25 |
| 155  | L-Asp-D-Ala-L-Ala-OMe | 50 |
| 161  | L-Asp-D-Ala-L-Ala-L-Ala-OMe | 0.5 |
| 162  | L-Asp-D-Ala-L-Ala-L-Ala-L-Ala-OMe | 0 |

In summary, tripeptides are less sweet than dipeptides of similar size. The loss of sweetness intensity in the tripeptides is possibly due to the increased hydrophilicity of these compounds and the restricted, and therefore non-optimal, overall size and shape.

Ariyoshi extended his investigations to tetra- and penta-peptide sweeteners.[60-62] Of 14 tetrapeptides, 3 showed sweetness intensities in the range 0·5–5 ×. None of the 7 pentapeptides was sweet (Table 19). It appears that as the oligopeptide gets larger, access to the receptor site becomes more difficult.

### 3.4 Conformationally Restricted Analogs

Aspartame is a very flexible molecule with a large number of possible conformations. The Ariyoshi model gives crude size and stereochemical requirements but says nothing about conformational requirements. Rigid dipeptides which lock the requisite functional groups in the optimal spatial arrangement for interaction with the sweet receptor would be expected to show high sweetness intensity.

Such perfect alignment of interacting groups is extremely difficult to achieve and the examples in the literature hold only a small part of the dipeptide rigid. The two rigid analogs[7] **163** and **164** are related to the L-amphetamine derivative **3** but have very different shapes. The indanyl analog **163** connects the methyl upper group to the *ortho* position of the aromatic ring. The upper and lower groups are thereby held in a planar arrangement and the resulting compound is 1/5 as sweet as the unconstrained counterpart. The cyclopropyl analog **164**, where the former methyl group is now connected to the benzyl carbon, has the aspartyl and phenyl groups held *trans*, which results in a complete loss of sweetness.

| 50x | 10x | 0x |

| 3 | 163 | 164 |

Closely related to **164** are the L-aspartyl-D,L-cyclopropylphenylalanine methyl esters **165** and **166**.[63] The *trans* compound **165** was bitter, consistent with Mazur's observation on **164**, which lacks the $CO_2Me$ group. However, one of the diastereomers of the *cis* compound **166** was sweet (100–200 ×). King and Stammer[63] concluded that the aromatic ring must be *cis* to the peptide bond for sweetness to be perceived. In contrast, the related *cis*-dehydroaspartame **167** (L-aspartyl-*cis*-dehydrophenylalanine methyl ester)[64] was not sweet. Subtle conformational differences and/or electronic effects could account for the difference between **166** and **167**.

| 165 | 166 | 167 |

Goodman and co-workers[65] prepared a series of compounds which help to probe the spatial volume requirement of the upper group (Table 20). In these compounds the carbon center shown is fully substituted. Mazur[7] and Ariyoshi[13] had demonstrated that full substitution reduced but did not eliminate sweetness (**7, 168**). Goodman found that the dimethylated methyl ester **170** was sweet while the diethyl version **171** was not. In the cycloalkyl series the cyclopropyl to cyclopentyl compounds were sweet. The cyclopentyl sweetener can be considered a constrained version of the diethyl compound formed by joining the methyl groups. The smaller volume occupied by the cyclopentyl compound allows access to the receptor. As the ring size gets larger the increased hydrophobic interaction leads to a bitter taste, which is consistent with a receptor model proposed by

## TABLE 20

$$\text{AspNH} \overset{R_1}{\underset{R_3}{\diagdown}} \!\!\! \overset{R_2}{\diagup}$$

| Cmpd | $R_1$ | $R_2$ | $R_3$ | Intensity |
|------|-------|-------|-------|-----------|
| 7 | $CH_3$ | $CH_3$ | $CH_2Ph$ | 20 |
| 168 | $CH_3$ | $CO_2Me$ | $CH_2Ph$ | 5 |
| 169 | $CH_3$ | $CH_3$ | $CO_2CH_2Ph$ | 0 |
| 170 | $CH_3$ | $CH_3$ | $CO_2Me$ | sweet |
| 171 | Et | Et | $CO_2Me$ | 0 |
| 172a-c | $(CH_2)_n$ n=2-4 | | $CO_2Me$ | sweet |
| 173a,b | $(CH_2)_n$ n=5, 6 | | $CO_2Me$ | bitter |
| 174 | $(CH_2)_n$ n=7 | | $CO_2Me$ | 0 |

Temussi and co-workers.[66] Finally, taste disappears for the cyclooctyl compound, which is too big to interact with either receptor.

In subsequent work, the lower ester group was optimized for the cyclopropyl sweeteners in Table 21.[67] Sweetness intensity reached a maximum for the n-propyl compound. The benzyl ester was not sweet,

TABLE 21

AspNH COR

| Cmpd | R | Intensity |
|------|---|-----------|
| 172a | OMe | 40 |
| 175 | OEt | 120 |
| 176 | OPr | 275 |
| 177 | OBu | 100 |
| 178 | OCH$_2$Ph | 0 |
| 179 | NHPr | 0 |

which agrees with the observation for the corresponding geminal dimethyl compound **169**. Interestingly, the $N$-propyl-amide **179** was not sweet.

The structure–activity constraints mentioned in the preceding sections offer qualitative guidance for the design of new dipeptide sweeteners. Following these guidelines, hundreds of dipeptide analogs ranging in intensity from 0 to 50 000 × were prepared. Two quantitative approaches have been taken to interpret these data in terms of the structural requirements for sweetness. The first is a classical quantitative structure–activity approach which attempts to correlate physical properties with activity. The second involves conformational analysis by calculations and/or NMR spectroscopy to define explicitly the preferred shapes for dipeptide sweeteners. The two approaches are considered separately below.

## 4 DIPEPTIDE MODELING

### 4.1 The QSAR Approach[68]

Unilever workers were the first to correlate space-filling properties with sweetness intensity.[69] They selected 28 dipeptide methyl esters related to aspartame by a variety of changes in the side-chain. Space-filling models built in the fully extended conformation were used for measuring length. The size and shape of the side-chain were determined by measuring the volume of 40% aqueous methanol displaced upon immersing the model. Plots of size and length versus intensity established that sweetness required side-chains from 4·8 to 8·8 Å in length and volumes $> 30\,Å^3$. A subsequent paper[70] with an expanded set of 40 dipeptide methyl esters used three sets of physical parameters in a multiple regression analysis: $P$, parachor, related to molecular volume; $f$, Rekker's hydrophobic fragment constant; and Verloop's STERIMOL parameters ($L$ for the length of the side-chain along the axis joining it to the rest of the molecule and the width parameters $B_1$ to $B_4$ for the width in directions perpendicular to the $L$ axis (right, left, up, down); $B_5$ is the maximum width). A subset of 31 compounds was used to obtain the equation

$$\log S \text{(sweetness)} = 0·194f + 1·472 \times 10^{-2}P - 3·357B_5 \tag{1}$$

The coefficients for each term indicate that intensity increases with increasing hydrophobicity and molecular volume, but decreases with increasing width.

van der Heijden *et al.*[71] have defined the approximate dimensions of the triangle delineated by the AH–B–$\delta$ dipeptide binding sites from the effect of side-chain length on sweetness intensity. A comparison of the dipeptide binding sites with those proposed for nitroanilines[11] and sugars[72] showed that the distance to the hydrophobic binding site was greatest for dipeptides (Fig. 2).

More recently[73] these data have been refined by the use of the STERIMOL computer program to estimate the distances of side-chain

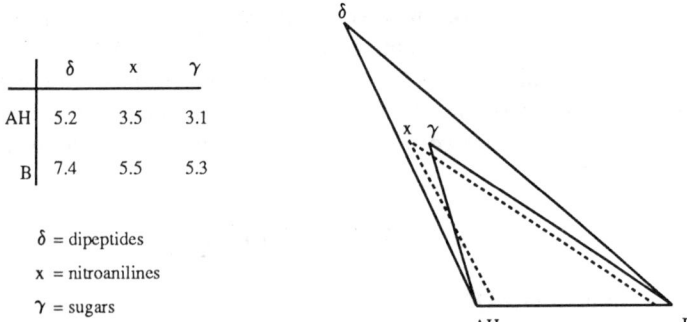

|    | $\delta$ | x | $\gamma$ |
|----|----------|-----|----------|
| AH | 5.2      | 3.5 | 3.1      |
| B  | 7.4      | 5.5 | 5.3      |

$\delta$ = dipeptides
x = nitroanilines
$\gamma$ = sugars

FIG. 2.   Hydrophobic binding site to AH–B distance (Å).

TABLE 22
OPTIMAL AH–B-$\delta$ DISTANCE (Å)

| | $F_{II}D_{II}$ | $F_{I}D_{II}$ |
|---|---|---|
| A-$\delta$ | 5·6 | 7·3 |
| B-$\delta$ | 9·4 | 10·8 |
| S | 2·2 | 4·2 |

atoms to A, H and B. The minimum, optimal and maximum length of the side-chain consistent with sweetness was estimated. The optimal distance from the hydrophobic binding site $\delta$ to the AH–B components for two assumed conformations $F_{II}D_{II}$ and $F_{I}D_{II}$ (see below) is given in Table 22. $S$ is the shortest distance from $\delta$ to the plane formed by A, H and B. In effect, $S$ defines the height of the receptor site, which for dipeptides is quite small.

A set of 217 dipeptide analogs was studied by Iwamura.[74] These were divided into 4 structural classes and each was examined separately. The multiple regression analysis considered steric and hydrophobic parameters, as did van der Heijden, but added the electronic parameter $\sigma^*$. Electronic and steric factors emerged as important in all classes, while hydrophobicity was not. The electronic factor was most important for the L-aspartylamino-acetates, which include Fujino's intensely sweet aminomalonic acid diesters (Table 10). In eqn (2), the regression equation for 56 compounds, the subscript 1 refers to the upper group ($R_1$) and the subscript 2 to the lower ester ($CO_2R$):

$$\log S(\text{sweetness}) = 1·54\sigma^* + 2·46L_2 - 0·21L_2^2 + 0·37InM + 1·49L_1$$
$$- 0·19L_1^2 - 0·74In + 1·05(W_1)_2 - 10·87 \qquad (2)$$

$$\text{AspNH} \overset{R_1}{\underset{H}{\text{'''}}}\!\!\!\!\!\!CO_2R$$

The incorporation of the squared terms reflects an optimum value for the length of $R_1$ and R (3·7 and 5·5 Å respectively). The positive coefficient for the $\sigma^*$ term reflects the increase in sweetness noted with electron-withdrawing substituents. Such groups enhance hydrogen-bonding at the amide nitrogen which may account for the increased sweetness. The extremely high sweetness intensity of the Fujino compounds can be rationalized by eqn (2). The two ester groups result in a large $\sigma^*$ value and the methyl branching in the 2,6-dimethyl cyclohexyl and fenchyl esters contributes to $(W_1)_2$.

Miyashita et al.[75] applied the SIMCA pattern recognition method to the analysis of 108 dipeptide analogs. They considered size (molar refractivity), shape (STERIMOL) and electronic ($\sigma^*$) factors in their analysis, and determined that size and shape factors were important while electronic factors were not.

The conclusions of Iwamura (steric and electronic parameters important) and van der Heijden et al. and Miyashita et al. (steric and hydrophobic parameters important) are different. Iwamura states that the discrepancy results from van der Heijden's neglect of the electronic factor. However, discrepancies of this sort are commonplace in QSAR analyses of large groups of compounds with considerable structural variability. The uniqueness of the QSAR descriptors is often an issue and can be dependent on the compounds selected for analysis. For example, the STERIMOL parameters in Iwamura's analysis correlate with hydrophobicity for certain compounds, so that log $P$ (partition coefficient) can be used instead in the analysis.[76] Additionally, the fully extended conformation was assumed to be preferred for all the dipeptides, so any dependence on conformation is ignored. Temussi et al. have recently commented on the relevance of these limitations to Iwamura's conclusions.[77]

### 4.2 Conformational Analysis

Lelj et al.[78] examined the conformational preferences for aspartame by a combination of NMR methods and molecular mechanics calculations. The side-chain conformational preferences were determined by analysis of the ABX proton NMR coupling constants of the $CHCH_2$ fragments. The relative population of each staggered conformation was then calculated. The staggered conformations for the aspartyl ($D_I$ to $D_{III}$) and phenyl-alanine ($F_I$ to $F_{III}$) residues are shown opposite. It was established that the aspartyl residue preferred the $D_{II}$ conformation. The two other staggered conformations either hold the $NH_3^+$ and $CO_2^-$ groups trans ($D_I$) and therefore unable to form a hydrogen bond simultaneously to the receptor, or crowd the bulky substituents together ($D_{III}$). For the phenylalanine residue, the $F_I$ conformation was selected over the $F_{II}$ based on the intuitive steric argument that the phenyl group would be preferentially sited gauche to the methyl ester rather than the aspartyl group. The literature proton NMR assignments for the phenylalanine $C(\beta)H_2$ protons were reversed to correlate with this conformational preference. The remaining $F_{III}$ staggered conformation again crowds all the bulky substituents.

The preferred backbone $\psi$, $\phi$ and $\chi$ angles were calculated for each of the 9 staggered conformations about the side-chain. Simplifying assumptions were made about the remaining bonds. The calculated fractional populations generally agreed with the NMR results. The preference for $F_I D_{II}$ was substantiated although many other conformations were significantly populated (Table 23).

The final distinction between the $F_I D_{II}$ and $F_{II} D_{II}$ conformations was made by invoking the Shallenberger barrier[5] which distinguishes sweet D-amino acids from their non-sweet L-enantiomers. Figure 3 shows that only the $F_I D_{II}$ conformation avoids this spatial barrier.

This conclusion was soon disputed by several workers. van der Heijden *et al.*[71] pointed out that the size and length of both R groups at the C-terminal

TABLE 23
EXPERIMENTAL VERSUS CALCULATED COMBINED
FRACTIONAL POPULATIONS[a]

|  | Experimental | Calculated |
|---|---|---|
| $F_I D_I$ | 0·136 | 0·121 |
| $F_I D_{II}$ | 0·307 | 0·184 |
| $F_I D_{III}$ | 0·148 | 0·146 |
| $F_{II} D_I$ | 0·069 | 0·073 |
| $F_{II} D_{II}$ | 0·156 | 0·141 |
| $F_{II} D_{III}$ | 0·075 | 0·088 |
| $F_{III} D_I$ | 0·025 | 0·056 |
| $F_{III} D_{II}$ | 0·057 | 0·113 |
| $F_{III} D_{III}$ | 0·027 | 0·075 |

[a] Zwitterionic form.

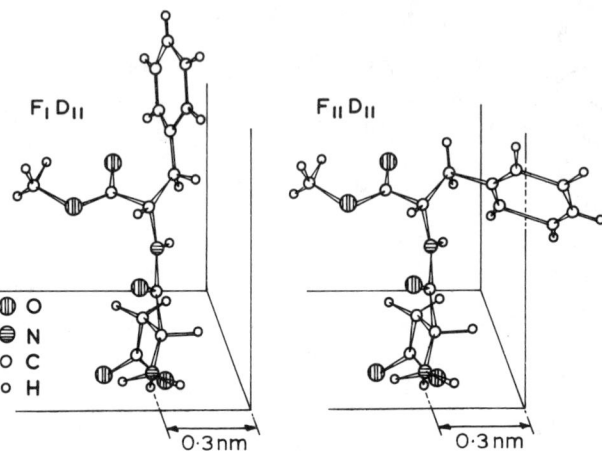

FIG. 3.    Molecular models for the $F_I D_{II}$ and $F_{II} D_{II}$ conformations of aspartame. Reprinted with permission from Ref. 78. Copyright (1976) American Chemical Society.

chiral center influence sweetness. For example, when higher esters replace the carbomethoxy group of aspartame the sweetness decreases. Likewise, when its benzyl group is replaced by groups of greater length, intensity falls. In the $F_I D_{II}$ conformation there is no apparent barrier to the length of the side-chain. (In a subsequent publication, the Italian group[79] acknowledged that the length of the side-chain is critical to sweetness; presumably a barrier roughly perpendicular to the Shallenberger barrier shown in Fig. 3 could account for the length dependence.) van der Heijden instead suggested that the $F_{II} D_{II}$ conformation interacts with the receptor. Here both R groups can interact with the receptor and, when too large, prevent access to the site. The maximum width of 15 Å was estimated from analogs with R groups of varying size.[70]

Crosby et al.[1d] objected to the criteria used in the selection of the $F_I D_{II}$ conformation. For example, the $F_I D_I$ conformation which has the aspartyl carboxylate and ammonium groups *trans* was eliminated owing to lack of ability to form a hydrogen bond at the receptor. Yet $F_I D_I$ is calculated to be only 0·2 kcal/mol higher in energy than $F_I D_{II}$ and is significantly populated despite the lack of the intramolecular H-bond calculated to be worth 1·8 kcal/mol.[76,80] In addition, the relevance of the Shallenberger barrier proposed for L-leucine to dipeptides is unclear. The hydrophobic binding site for dipeptides extends above the boundary suggested for amino acids.

Three separate NMR studies support the $F_{II} D_{II}$ conformation. Murai et al.[81] at Ajinomoto were actually the first to publish on the preferred conformation of aspartame. In a careful study, they prepared it from aspartic acid and phenylalanine which were stereospecifically deuterated at the $\beta$-$CH_2$ group. The identity of the $\beta$-$CH_2$ protons was therefore known with certainty. From the NMR coupling data they concluded that the $F_{II}D_{II}$ was preferred. More recently Takahashi et al.,[82] in a study of the inclusion complexes of aspartame with cyclodextrins, concluded that the $F_{II}D_{II}$ conformation is preferred for both the complexed and uncomplexed sweetener. Finally, Asso et al.[83] depict the $F_{II}D_{II}$ conformer as preferred for aspartame in the monodentate praseodymium and dysprosium complexes, although this conformation is called $F_{II}D_I$ for reasons that are not apparent.

Polish workers[84] examined the side-chain conformational preferences of the related sweet dipeptide L-aspartyl-L-methionine methyl ester (Table 1). Using the vicinal $CHCH_2$ couplings, they concluded that the $F_{II}D_{II}$-like conformation is preferred. They suggested that the $F_{II}D_{II}$ conformation is also preferred for aspartame, and that Lelj et al. misassigned the $H^\beta$ resonances for the phenylalanine residue.

About the same time, the Italian workers themselves briefly described[77] work not published elsewhere on the conformational preference of the methionine dipeptide. They again concluded that the $F_ID_{II}$ conformation is preferred, which is consistent with their conclusion for aspartame. Lelj et al.[85] have proposed a generalized sweet receptor site model by using the $F_ID_{II}$ conformation of aspartame along with a series of substituted saccharins as 'molecular molds' for the site. (An additional refinement, a 180° inversion of the AH–B moieties, was proposed to help rationalize the bitter taste observed for the D,D and L,D diastereomers of aspartame.[79,85]) In each case the potential sweeteners gain access to the receptor from the side opposite the Shallenberger barrier, which is postulated to be completely open.[79]

Recently Goodman et al.[86] have proposed a model for retro-inverso and dipeptide amides. The alanine amides (D-sweet (86) and L-non-sweet) and the corresponding retro-inverso amides (R-sweet (94) and S-sweet (95)) were studied by NMR. Assuming a trans peptide bond and a nearly planar shape for the zwitterionic aspartyl group, a combination of (a) coupling constant data ($J(N^1H$–$^1H)$), (b) nuclear Overhauser enhancement and (c) temperature effects on NH chemical shifts determined the conformation of these compounds. Minimum energy calculations showed that only one low-energy conformation matched the NMR data for each of the analogs. Figure 4 shows aspartame superimposed to fit the model for the preferred space for sweet taste site. This conformation of aspartame closely approximates to that found in the crystal structure.

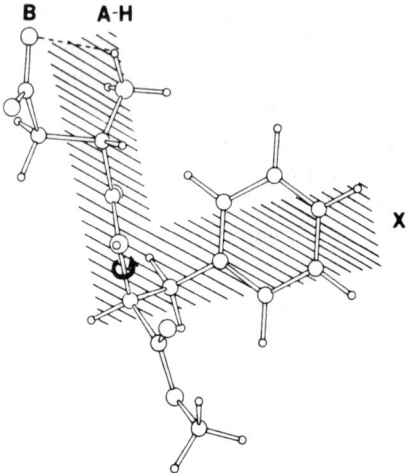

FIG. 4. Model for sweet taste with aspartame superimposed; the bond shown by the arrow has been rotated 40° from the X-ray diffraction structure. Reprinted with permission from Ref. 86. Copyright (1987) American Chemical Society.

The crystal structures of aspartame[87] and aspartame hydrochloride[88] have been determined. The bonds along the peptide backbone are all approximately *trans*. For aspartame the side-chains take up the $F_{III}D_{II}$ conformation suggested by Goodman (above) as the active one. The hydrochloride salt prefers the $F_{III}D_{I}$ conformation. The aspartyl carboxyl and ammonium group are *trans* in $F_{III}D_{I}$, and this is not the active conformation. Instead, Gorbitz[88] suggests the $F_{II}D_{III}$ as the active conformation because it best fits Kier's sweetness triangle. However, Kier's triangle was defined for nitroanilines not dipeptides.[11] van der Heijden *et al.*[71,73] have suggested that the sweetness triangle is considerably larger for dipeptides than for nitroanilines and the $F_{II}D_{II}$ conformation is in fact a better fit. Another closely related compound, the LiBr complex of aspartame, crystallized in the $F_{I}D_{III}$ conformation.[81] Thus aspartame and its HCl and LiBr salts each prefer a different conformation in the solid state.

The crystal structure of L-aspartyl-1-aminocyclopropanecarboxylic acid n-propyl ester **176** (Table 21) has also been determined.[67a] A direct comparison with aspartame is not possible because of the structural differences between these sweeteners. However, the crystal conformation of **176** is not fully extended but is clearly bent, with the alkoxy part of the ester *syn* to the aspartyl residue.

Finally, in an attempt to mimic the sweet receptor site, the NMR spectrum of aspartame hydrochloride complexed with 1 equivalent of 18-crown-6 (an 18-membered macrocyclic ring containing 6 $CH_2CH_2O$ units) in chloroform was studied.[77] The crown served to complex the $NH_3^+$ group and to promote dissolution in chloroform. Analysis of the two side-chain ABX systems showed that the $F_ID_{III}$ conformer was preferred.

The investigations outlined above allow certain conclusions to be made about the conformational requirements for sweetness. For the aspartic acid residue, it is unanimous that the $D_{II}$ conformation is preferred in solution. For the C-terminal amino acid residue, 4 out of 5 research groups favor the $F_{II}$ conformation for aspartame or the methionine analog. The weight of the evidence favors the $F_{II}D_{II}$ conformation. Goodman and co-workers suggest that $F_{III}D_{II}$ is favored by superimposing aspartame on their model for sweet taste. While there is no consensus on the conformational preference, bent conformations with the phenyl and aspartyl groups *gauche* ($F_{II}$ and $F_{III}$) are generally preferred.

The relevance of the solution conformation to the active conformation at the sweet site is an ever present point of uncertainty. Conformationally restricted analogs which maintain their shape regardless of the environment can help resolve the ambiguity. The analogs **163–167** discussed earlier have significance here. The indanyl and *trans*-substituted cyclopropyl analogs **163–165** hold the phenyl and aspartyl residues in a conformation that approximates $F_I$. The first has reduced sweetness while the latter two are not sweet. One of the diastereomers of the *cis* cyclopropyl analog **166** with a fixed conformation between $F_{II}$ and $F_{III}$ is sweet. The *cis* olefin **167** is not sweet although the relative positions of the phenyl and aspartyl groups resemble those in **166**. However, the strictly planar shape for the dehydrophenylalanine part may prevent the methyl ester group from taking part in a necessary hydrophobic interaction with the receptor. The electronic characteristics of **166** and **167** differ considerably (amide versus enamide) and may account at least in part for the difference in intensity.

## 5 CONCLUSIONS

Figure 5 summarizes the structural variations of the prototypical dipeptide sweetener aspartame. Based on the pioneering work of the Searle group, Ariyoshi proposed a model with L-aspartic acid coupled to an amine $\alpha$-substituted with a small group ($R_1$) and a large group ($R_2$) with the proper stereochemical configurations. Refinements of the model showed that rigid appropriately branched $R_2$ groups can significantly boost sweetness

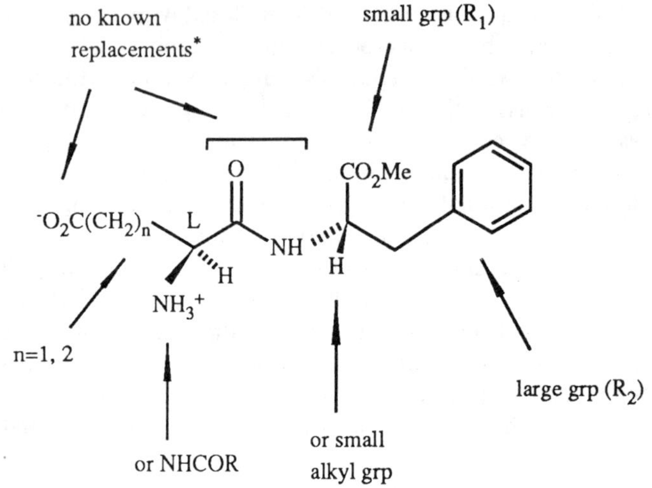

* thioamide has some sweetness

FIG. 5.   Structure–activity summary for sweet dipeptides.

intensity. It is noteworthy that virtually all high-intensity sweeteners ($>1000 \times$) bear at least one ester or amide group as $R_1$ or $R_2$. Thus far, no substitutes for the peptide bond have been found. The aspartyl residue can be modified by acylation of the amino group, and in some cases very high sweetness intensities result.

Significant progress in understanding the relationship between dipeptide structure and sweetness has been made, as witnessed by the many analogs 10 to $>100$ times as sweet as aspartame. In addition, conformational preferences, although controversial, have begun to be understood. The issues of stability and taste quality are occasionally addressed in the references cited, but the search for a high-intensity, stable, sucrose-like dipeptide sweetener remains a challenge. The continuing research activity in the field gives assurance that further progress will be made.

## REFERENCES

1. (a) Crammer, B. & Ikan, R. *Chem. Soc. Rev.*, **6** (1977) 431–65. (b) Bragg, R. W., Chow, Y., Dennis, L., Ferguson, L. N., Howell, S., Morga, G., Ogino, C., Pugh, H. & Winters, M. *J. Chem. Ed.*, **55** (1978) 281–5. (c) Crosby, G. A. & Wingard, R. E. Jr. In *Developments in Sweeteners*, Vol. 1, ed. C. A. M. Hough, K. J. Parker

& A. J. Vlitos. Applied Science Publishers, London, 1979, pp. 135–64. (d) Crosby, G. A., Dubois, G. E. & Wingard, R. E. Jr. In *Drug Design*, Vol. VIII, ed. E. J. Ariëns. Academic Press, New York, 1979, pp. 215–310. (e) Dubois, G. E. In *Ann. Reports in Med. Chem.*, Vol. 17, ed. H.-J. Hess. Academic Press, New York, 1982, pp. 323–32. (f) van der Wel, H., van der Heijden, A. & Peer, H. G. In *Food Rev. Int.*, Vol. 3, ed. R. Teranishi & I. Hornstein. Marcel Dekker, New York, 1987, pp. 193–268. (g) Lee, C.-K. *Adv. Carbohydr. Chem. Biochem.*, **45** (1987) 199–351.

2.  (a) Ariyoshi, Y. In *Food Taste Chem.*, ACS Symposium Series 115, ed. J. C. Boudreau. American Chemical Society, Washington, DC, 1979, pp. 133–48. (b) Pavlova, L. A., Komarova, T. V., Davidovich, Yu. A. & Rogozhin, S. V. *Russ. Chem. Rev.*, **50** (1981) 316–23.

3.  Mazur, R. H. *Food Sci. and Technol.*, **12** (1984) 3–9.

4.  Mazur, R. H., Schlatter, J. M. & Goldkamp, A. H. *J. Am. Chem. Soc.*, **91** (1969) 2684–91.

5.  Shallenberger, R. S. & Acree, T. E. (a) *Nature*, **216** (1967) 480–2. (b) *J. Agric. Food Chem.*, **17** (1969) 701–3.

6.  Kawai, M., Chorev, M., Marin-Rose, J. & Goodman, M. *J. Med. Chem.*, **23** (1980) 420–4.

7.  Mazur, R. H., Goldkamp, A. H., James, P. A. & Schlatter, J. M. *J. Med. Chem.*, **13** (1970) 1217–21.

8.  Mazur, R. H., Reuter, J. A., Swiatek, K. A. & Schlatter, J. M. *J. Med. Chem.*, **16** (1973) 1284–7.

9.  Ariyoshi, Y. *Agric. Biol. Chem.*, **40** (1976) 983–92.

10. Kaneko, T. *J. Chem. Soc. Japan*, **60** (1939) 531.

11. Kier, L. B. *J. Pharm. Sci.*, **61** (1972) 1394–7.

12. Goodman, M. & Gilon, G. In *Peptides 1974*, Proceedings of 13th European Peptide Symposium, ed. Y. Wolman. Wiley, New York, 1975, pp. 271–8.

13. Ariyoshi, Y., Yasuda, N. & Yamatani, T. *Bull. Chem. Soc. Japan*, **47** (1974) 326–30.

14. (a) Rizzi, G. *J. Agric. Food Chem.*, **33** (1985) 19–24. (b) Rizzi, G. P. & Echler, R. S. *Food Chem.*, **20** (1986) 165–74. (c) Brand, L. M. US patent 4 338 346 (1982).

15. Roy, G. M., Barnett, R. E. & Zanno, P. R. European patent application 203 540 (1986).

16. Yuasa, Y., Okeda, Y., Tachikawa, A., Nagakura, A. & Tsuruta, H. European patent application 255 343 (1988).

17. (a) Barnett, R. E., Zanno, P. R. & Roy, G. M. US patent 4 654 219 (1987). (b) Roy, G. M., Barnett, R. E. & Zanno, P. R. US patent 4 603 011 (1986).

18. Nagakura, A., Oketa, Y., Tachikawa, A., Yuasa, Y., Tsuruta, H. & Akutagawa, S. Japanese patent 62 198 700 (1987).

19. Miyoshi, M., Nunami, K., Sugano, H. & Fujii, T. *Bull. Chem. Soc. Japan*, **51** (1978) 1433–40.

20. (a) Fujino, M., Wakimasu, M., Tanaka, K., Aoki, H. & Nakajima, N. *Naturwiss.*, **60** (1973) 351. (b) Fujino, M., Wakimasu, M., Mano, M., Tanaka, K., Nakajima, N. & Aoki, H. *Chem. Pharm. Bull.*, **24** (1976) 2112–17.

21. (a) Liu, Y. *et al. Medical Industry (Yiyao Gongyi)*, **10** (1980) 11. (b) Liu, Y., Wang, X., Xie, W. & Hsu, T. *Acta Biochim. Biophys. Sinica*, **16** (1984) 581–5.

22. Liu, Y., Xie, H., Jian, G. & Zeng, G. *Youji Huaxue*, **1** (1982) 40–6.

23. Nagakura, A., Yuasa, Y., Tsuruta, H. & Akutagawa, S. Japanese patent 61 200 999 (1986).
24. Sukehiro, M., Minematsu, H. & Noda, K. *Science of Human Life*, **11** (1977) 9–16.
25. Brennan, T. M. & Hendrick, M. E. US patent 4 411 925 (1983).
26. Brennan, T. M. & Hendrick, M. E. US patent 4 399 163 (1983).
27. (a) Verlander, M. S., Fuller, W. D. & Goodman, M. US patent 4 571 345 (1986). (b) Fuller, W. D., Goodman, M. & Verlander, M. S. *J. Am. Chem. Soc.*, **107** (1985) 5821–2.
28. Yuasa, Y., Nagakura, A., Tsuruta, H. & Akutagawa, S. Japanese patent 62 39 599 (1987).
29. Blum, R. B. (Procter & Gamble), unpublished results.
30. Blum, R. B., Gardlik, J. M., Janusz, J. M. & Rizzi, G. P. US patent 4 692 513 (1987).
31. (a) Janusz, J. M. US patent 4 692 512 (1987). (b) Janusz, J. M. & Gardlik, J. M. US patent 4 677 126 (1987).
32. (a) Zanno, P. R., Barnett, R. E. & Roy, G. M. US patent 4 766 246 (1988). (b) Nagakura, A., Yuasa, Y., Tsuruta, H. & Akutagawa, S. Japanese patents 61 200 999 (1986), (c) 61 291 596 (1986), (d) 61 291 597 (1986).
33. Zanno, P. R., Barnett, R. E. & Roy, G. M. US patents (a) 4 571 308 (1986), (b) 4 619 834 (1986), (c) 4 634 792 (1987), (d) 4 652 457 (1987), (e) 4 652 676 (1987). (f) Roy, G. M., Barnett, R. E. & Zanno, P. R. UK patent application 2 191 775 (1987).
34. Zanno, P. R., Barnett, R. E. & Roy, G. M. US patents (a) 4 603 012 (1986), (b) 4 622 232 (1986), (c) 4 701 522 (1987).
35. Zanno, P. R., Barnett, R. E. & Roy, G. M. US patents (a) 4 572 799 (1986), (b) 4 678 674 (1987).
36. Zanno, P. R., Barnett, R. E. & Roy, G. M. US patents (a) 4 619 782 (1986), (b) 4 636 396 (1987).
37. Nelson, V. & Mastalerz, P. *J. Pharm. Sci.*, **73** (1984) 1844–6.
38. Seltzman, H. H., Hsieh, Y., Cook, C. E., Hughes, T. & Hendren, R. W. Am. Chem. Soc. National Meeting, Spring 1985, Miami Beach, Abs. 44.
39. Tinti, J. M., Durozard, D. & Nofre, C. *Naturwiss.*, **67** (1980) 193.
40. De Nardo, M. *Il Farmaco, Ed. Sc.*, **32** (1977) 522–4.
41. MacDonald, S. A., Willson, C. G., Chorev, M., Vernacchia, F. & Goodman, M. *J. Med. Chem.*, **23** (1980) 413–20.
42. Dutta, A. S. & Morley, J. S. In *Peptides 1976*, Proceedings of 14th European Peptide Symposium, ed. A. Loffet. Wiley, New York, 1976, pp. 517–22.
43. Yde, B., Thomsen, I., Thorsen, M., Clausen, K. & Lawesson, S. O. *Tetrahedron*, **39** (1983) 4121–6.
44. Briggs, M. T. & Murley, J. S. UK patent specification 1 299 265 (1972).
45. Chorev, M., Willson, C. G. & Goodman, M. In *Peptides 1977*, Proceedings of 5th American Peptide Symposium, ed. M. Goodman & J. Meienhofer. Wiley, New York, 1977, pp. 572–4.
46. (a) Seltzman, H. H. US patent 4 564 528 (1986). (b) Hendren, R. W., Hsieh, Y. A., Cook, C. E., Seltzman, H. H., Lee, Y. & Hughes, T. J. Research Triangle Institute Report, 1984, RTI/2075/00-06F.
47. Lapidus, M. & Sweeney, M. *J. Med. Chem.*, **16** (1973) 163–6.

48. De Nardo, M., Runti, C., Ulian, F. & Vio, L. *Il Farmaco, Ed. Sc.*, **31** (1976) 906–16.
49. Tinti, J. M., Nofre, C. & Durozard, D. *Naturwiss.*, **68** (1981) 143.
50. Kawai, M., Nyfeler, R., Berman, J. M. & Goodman, M. *J. Med. Chem.*, **25** (1982) 397–402.
51. Boesten, W. H. J. & Schiepers, L. A. C. (a) US patent 4 371 464 (1983); (b) European patent specification 048 051 (1984).
52. Nofre, C. & Tinti, J. M. US patent 4 645 678 (1987).
53. Nofre, C. & Tinti, J. M. US patent 4 673 582 (1987).
54. Petersen, S. & Müller, E. *Chem. Ber.*, **81** (1948) 31–5.
55. (a) Nofre, C., Tinti, J. M. & Ouar, F. European patent application 195 730 (1986). (b) Tinti, J. M., Nofre, C. & Peytavi, A. M. *Z. Lebensm.-Unters. Forsch.*, **175** (1982) 266–8.
56. Nofre, C., Tinti, J. M. & Chatzopoulos, O. F. European patent 241 395 (1987).
57. (a) Takemoto, T., Hijiya, T. & Yukawa, H. European patent application 0 186 292 (1986). (b) Ariyoshi, Y., Hasegawa, Y. & Nio, N. *Peptide Chemistry 1986*, ed. T. Miyazawa. Protein Res. Found., Osaka, 1987, pp. 251–6.
58. Ariyoshi, Y. (a) *Agric. Biol. Chem.*, **44** (1980) 943–5; (b) *Bull. Chem. Soc. Japan*, **57** (1984) 3197–202.
59. Rodriguez, M., Bland, J. M., Tsang, J. W. & Goodman, M. *J. Med. Chem.*, **28** (1985) 1527–9.
60. Ariyoshi, Y. *Bull. Chem. Soc. Japan*, **58** (1985) 1727–30.
61. Ariyoshi, Y. *Bull. Chem. Soc. Japan*, **59** (1986) 1027–30.
62. Ariyoshi, Y. In *Peptide Chemistry 1984*, ed. N. Izumiya. Protein Res. Found., Osaka, 1985, pp. 383–8.
63. (a) King, S. W., PhD dissertation. *Diss. Abs.* 8206319 (1981). (b) Stammer, C. H. World patent application WO 85/00809 (1985).
64. King, S. W. & Stammer, C. H. *J. Org. Chem.*, **46** (1981) 4780–2.
65. Tsang, J. W., Schmied, B., Nyfeler, R. & Goodman, M. *J. Med. Chem.*, **27** (1984) 1663–8.
66. Ciajolo, M. R., Lelj, F., Tancredi, T., Temussi, P. A. & Tuzi, A. *J. Med. Chem.*, **26** (1983) 1060–5.
67. (a) Mapelli, C., Newton, M. G., Ringold, C. E. & Stammer, C. H. *Int. J. Peptide Prot. Res.*, **30** (1987) 498–510. (b) Stammer, C. H. WO 87/00732 (1987). (c) Joyce, P. J. WO 86/03378 (1986).
68. (a) Cramer, R. D. III. In *Annual Reports of Medicinal Chemistry*, Vol. 11, ed. F. H. Clarke. Academic Press, New York, 1976, pp. 301–10. (b) Hansch, C. *Accounts Chem. Res.*, **2** (1969) 232–9. (c) Martin, Y. C. *J. Med. Chem.*, **24** (1981) 229–37.
69. Brussel, L. B. P., Peer, H. G. & van der Heijden, A. *Z. Lebensm.-Unters. Forsch.*, **159** (1975) 337–43.
70. van der Heijden, A., Brussel, L. B. P. & Peer, H. G. *Chem. Senses and Flavour*, **4** (1979) 141–52.
71. van der Heijden, A., Brussel, L. B. P. & Peer, H. G. *Food Chem.*, **3** (1978) 207–11.
72. Shallenberger, R. S. & Lindley, M. G. *Food Chem.*, **2** (1977) 145.
73. van der Heijden, A., van der Wel, H. & Peer, H. G. *Chem. Senses*, **10** (1985) 57–72.
74. Iwamura, H. *J. Med. Chem.*, **24** (1981) 572–83.

75. Miyashita, Y., Takahashi, Y., Takayama, C., Sumi, K., Nakatsuka, K., Ohkubo, T., Abe, H. & Sasaki, S. *J. Med. Chem.*, **29** (1986) 906–12.
76. Hopfinger, A. J. & Walters, D. E. In *Computers in Flavor and Fragrance Research*, ACS Symposium Series 261, ed. C. B. Warren & J. P. Walradt. American Chemical Society, Washington, DC, 1984, pp. 19–32.
77. Temussi, P. A., Lelj, F., Tancredi, T., Casteglione-Morelli, M. A. & Pastore, A. *Int. J. Quantum Chem.*, **XXVI** (1984) 889–906.
78. Lelj, F., Tancredi, T., Temussi, P. A. & Toniolo, C. *J. Am. Chem. Soc.*, **98** (1976) 6669–75.
79. Temussi, P. A., Lelj, F. & Tancredi, T. *J. Med. Chem.*, **21** (1978) 1154–8.
80. Hopfinger, A. J. & Jabloner, H. In *The Quality of Foods and Beverages*, ed. G. Charalambous & G. Inglett. Academic Press, New York, 1981, pp. 83–9.
81. Murai, A., Ajisaka, K., Nagashima, N., Takeuchi, Y., Kamisaku, M. & Kainosho, M. In *13th Symposium on Peptide Chemistry*, ed. S. Yamado. Protein Research Foundation, Osaka, 1975, pp. 52–6.
82. Takahashi, S., Suzuki, E. & Nagashima, N. *Bull. Chem. Soc. Japan*, **59** (1986) 1129–32.
83. Asso, M., Zineddine, H. & Benlian, D. *Int. J. Peptide Protein Res.*, **28** (1986) 437–43.
84. Siemion, I. Z. & Picur, B. *Polish J. Chem.*, **58** (1984) 475–8.
85. Lelj, F., Tancredi, T., Temussi, P. A. & Toniolo, C. *Il Farmaco, Ed. Sc.*, **35** (1980) 988–96.
86. Goodman, M., Coddington, J., Mierke, D. F. & Fuller, W. D. *J. Am. Chem. Soc.*, **109** (1987) 4712–4.
87. Hatada, M., Jancarik, J., Graves, B. & Kim, S.-H. *J. Am. Chem. Soc.*, **107** (1985) 4279–82.
88. Görbitz, C. H. *Acta Chem. Scand.*, **B41** (1987) 87–92.

*Chapter 2*

# EXPLOITATION OF GENETIC ENGINEERING TO PRODUCE NOVEL PROTEIN SWEETENERS

Joachim L. Weickmann, Jar-How Lee, Lindley C. Blair,
Pradip Ghosh-Dastidar & Raju K. Koduri

*INGENE, Santa Monica, California, USA*

## SUMMARY

*Thaumatin is a naturally occurring plant protein which tastes sweet at remarkably low concentrations. We have chemically synthesized a gene encoding thaumatin, cloned it into a recombinant vector containing a strong promoter of gene expression in yeast, and introduced it into* Saccharomyces cerevisiae *for thaumatin expression. Yeast cells transformed with this vector produce large amounts of thaumatin intracellularly. The yeast thaumatin, which is produced as a denatured, insoluble material, has been purified and then folded in* vitro *to its native conformation. By adding certain secretion signals to the recombinant vector, we have also developed yeast strains which secrete sweet protein into the growth medium. Both the folded and the secreted forms of yeast thaumatin are indistinguishable from natural plant thaumatin in sweetness intensity and immunoreactivity in an assay which recognizes the sweet form of thaumatin. Using recombinant DNA technology, we have introduced specific changes into the synthetic gene and produced the resulting variants of thaumatin by secretion under fermentation conditions. The sweet proteins were purified to homogeneity. Several of them display altered taste properties when compared to thaumatin isolated from plants.*

## 1 INTRODUCTION AND BACKGROUND

The 'rediscovery' in the mid-1960s of a class of naturally occurring plant-derived sweet proteins has stimulated considerable interest in the scientific community and in the food industry. Monellin[1] and thaumatin,[2] two

47

structurally different protein molecules, have been shown to elicit sweet taste. Both have been purified and studied in detail. A third protein, miraculin,[3] is not itself sweet but gives a sweet taste to substances that normally taste sour to human subjects.[4]

A unique feature common to thaumatin and monellin is their sweetness potency; both are several orders of magnitude sweeter than sucrose on a weight basis,[5] and one report claims that these proteins are perceived as sweet by only humans and old-world monkeys.[5] As a class of molecules that evoke a similar physiological response, these proteins attest to the diversity of structures that interact with and stimulate the human sweetness receptor(s). Monellin is comprised of two similar-sized non-covalently linked subunits with a combined molecular weight of 11 000 daltons; the thaumatin molecule is a single (unmodified) polypeptide chain of molecular weight 23 000 daltons stabilized by 8 intrachain disulfide bonds; miraculin is a glycoprotein of about 43 000 daltons molecular weight. No information is currently available about which structural features of these proteins are responsible for eliciting sweet taste. Apart from their utility for research into the physiology of sweet-taste perception, the structure, stability and taste disadvantages of monellin and miraculin have precluded their development as food additives.

Thaumatin is non-toxic, non-cariogenic, and its natural origin fits well with current dietary trends. On a weight basis, the caloric content of thaumatin (or any protein) is similar to that of sucrose. However, since gram quantities of the protein are comparable in sweetness intensity to kilogram quantities of sucrose, the caloric contribution of thaumatin as a food additive is negligible. The stability of thaumatin over a wide range of pH and temperature values makes it attractive as a food additive despite some of its undesirable taste properties, e.g. metallic taste, slow onset and a persistent aftertaste. With proper development through biotechnology, this naturally occurring intense sweetener is poised to share in the tremendous success of the low-calorie synthetic sweeteners of the 1980s. Several major companies have set out to explore the commercial potential of plant thaumatin, and a number of patents for use ranging from animal foods (the protein presumably acting as a flavor enhancer rather than as a sweetener) to tobacco applications have been issued.[6,7]

At present the main agricultural source of thaumatin is found along the edge of the rain forest belt in certain regions of West Africa. Ghana appears to be the country of choice for plantations that grow the plant *Thaumatococcus daniellii* for thaumatin harvesting. *T. daniellii* has been cultivated outside of these areas, especially on the Malaysian peninsula, but these plants produce only small fruits. This limits the production of thaumatin from plants and its use as a food additive.

Another factor resulting from cultivation is the molecular heterogeneity observed in the plant thaumatin mixture. We have observed wide variations from sample lot to lot in the relative ratios of the two major forms of thaumatin. This may reflect genetic variation in the plants or a dependence of protein production in the fruits of the plant on seasonal and/or climatic conditions.

In addition to its current use in the food industry, the intense sweetness and protein-nature of thaumatin make it an attractive molecule for further study and development. We have undertaken an extensive research program designed to characterize the thaumatin molecule and understand its properties.[8] The biological function of thaumatin in the plant is unknown; however, other plant proteins with a close sequence homology to thaumatin may be expressed in response to stress such as wounding or infection.[9,10] A knowledge of the molecular structure of thaumatin and thaumatin-like proteins may provide us with a unique tool to study the mechanism of human taste perception and an opportunity to create a better-tasting sweet molecule through the use of genetic engineering.

The arils of the fruit of the African shrub *T. daniellii* (Benth) contain the natural source of thaumatin.[11,12] Traditionally, the fruit or fruit extract has been used in West Africa as a sweetening substance for sour fruit, corn bread and palm wine. The intense sweetness of these fruits was recognized and first described in the literature in the mid-1850s.[12] More than 100 years later, Inglett and May[13] described the sweet, water-soluble nature of the extract from *T. daniellii*, but it was not until 1972 that the identity of the 'sweet principle' was confirmed to be a protein.[2] The fruits contain significant amounts of sweet protein. One report indicated that 900 mg of thaumatin could be extracted from 1 kg of fruit.[2] In another study it was shown that low concentrations of aluminum salts could increase this quantity to $6 \text{ g/kg}$.[14]

Thaumatin isolated from its natural source was found to consist of at least five variant proteins in different ratios and separable from one another depending on their charge characteristics.[15] A generalized procedure of aqueous homogenization of the arils followed by ultrafiltration and ion-exchange chromatography on SP-Sephadex C-25 (SP = sulfopropyl) was developed by van der Wel and Loeve for purification.[2] Separation of the two major variants, designated thaumatins I and II, was accomplished by ion-exchange chromatography. Additional minor variants (thaumatins a, b and c) are also separable from thaumatins I and II by ion-exchange chromatography on Whatman CM32 resin. All thaumatins were shown to have similar properties, i.e. in amino acid composition, sweetness intensity and molecular weight.

The two major molecular species of thaumatin found in plants have been characterized further. The entire 207 amino acid sequence for thaumatin I was determined directly.[16] A similar but non-identical peptide sequence was predicted by the nucleotide sequence of the complementary DNA (cDNA) synthesized from thaumatin messenger RNA.[17] This predicted sequence differs by 5 amino acids from the thaumatin I sequence determined by direct peptide sequencing. The second sequence is believed to correspond to that of thaumatin II, with the amino acid substitutions accounting in part for the charge differences between the two molecules. The cDNA-derived thaumatin II sequence was also shown to contain a 22 amino acid long N-terminal secretion signal sequence (pre-sequence) and a 6 amino acid long C-terminal pro-sequence.[17] Both the pre- and pro-sequences are removed to produce the mature protein. Southern blot analysis of *T. daniellii* genomic DNA using cloned thaumatin II cDNA as a probe indicates that there exists a family of thaumatin genes.[18] The gene coding for thaumatin II contains two small introns (DNA coding sequences).

Rabbit antibodies against thaumatin have been used as probes of 'sweet structure' for both thaumatin and monellin, and for other non-protein sweet molecules. One study showed that these reagents cross-react with monellin, suggesting that one or more identical conformational antigenic determinants exist in both molecules.[19] A tentative conclusion that this common determinant coincides with the active site responsible for the sweet-taste sensation is intriguing. Somewhat perplexing, however, is another report that rabbit antibodies against thaumatin also recognized several small molecular weight non-protein sweeteners, most notably aspartame.[20] We have been unable to confirm any of these results in our laboratory using monospecific antibodies raised against purified thaumatin. Furthermore, a partial three-dimensional structure reported for thaumatin I[21] and monellin[22] showed no obvious common determinant between these two molecules, even though they share five common tripeptides.[23]

Advances in molecular biology have provided an opportunity for producing thaumatin in heterologous hosts, thus bypassing the problems associated with agricultural production and avoiding the molecular heterogeneity inherent in preparations obtained from plant extracts. Microorganisms are particularly well suited for this since they can be engineered to produce thaumatin under fermentation conditions. This would ensure a stable source of a homogeneous preparation containing a single chosen molecular form of plant thaumatin, with a possible significant reduction in the cost of production.

Previous attempts by other workers to express the cloned thaumatin cDNA gene in yeast either in the mature form (without the pre- and pro-sequences) or in the secreted form (with the pre- and pro-sequences) have met with limited success.[24] In spite of using a strong promoter (yeast glyceraldehyde 3-phosphate dehydrogenase, GAPDH) to express the thaumatin II cDNA, the levels of thaumatin production in yeast were very low. The cause of the low expression level is unclear from the published data, and no evidence has been presented showing that a functional (sweet) protein had been expressed.

This report describes in general terms the various approaches we have taken to produce plant thaumatin in yeast. Two procedures—protein folding of denatured thaumatin and protein secretion from yeast—were successful in producing functional (sweet) thaumatin. Site-directed mutagenesis was used to introduce specific alterations into the thaumatin coding sequence. The effects of these changes in protein sequence on taste properties were analyzed and used to develop thaumatin variants with new or altered taste properties. These studies have given rise to a new class of thaumatin-related sweet proteins.

## 2 ASSAYS FOR SWEETNESS

The development of a new approach/process for the production of biologically active (i.e. sweet-tasting) thaumatin is greatly facilitated by a sensitive and quantitative assay for thaumatin. An antibody-based assay that specifically recognizes the native (sweet-tasting) form of thaumatin would provide the broadest utility for quantitating biological activity. We therefore developed a radioimmunoassay (RIA) which allows us to determine the amount of thaumatin in a test preparation, to detect the presence of thaumatin in impure starting materials (such as cell extracts where taste tests would be inappropriate), and to quantitate the amount of sweet protein in purified samples for taste-testing and food formulation.

Monospecific antibodies were raised in rabbits against purified thaumatin A (see Table 1). One of the inoculated rabbits produced a high titer specific antiserum which at dilutions of nearly 50 000-fold preferentially reacted with the native (sweet-tasting) structure of thaumatin. The data indicate that non-native or denatured thaumatin, or peptide fragments derived from thaumatin, do not effectively compete with native thaumatin to bind to the rabbit antibody.

Impure preparations of monellin were weakly cross-reactive in the RIA but a highly purified sample of monellin, kindly provided by Dr Robert Cagan, was not recognized by either immunoassay or Western blot. Non-protein sweeteners (aspartame, sucrose, etc.) were similarly undetected in the RIA, even at high concentrations of the sweetener used as the competing ligand and/or the antibody used to recognize the sweet molecules.

TABLE 1

COMPARISON OF THE AMINO ACID SEQUENCE DIFFERENCES IN FOUR
THAUMATIN VARIANTS

| *Thaumatin variant* | *Amino acid position* | | | | |
|---|---|---|---|---|---|
| | *46* | *63* | *67* | *76* | *113* |
| A | Asn | Ser | Lys | Arg | Asp |
| B | Lys | Ser | Lys | Arg | Asp |
| I | Asn | Ser | Lys | Arg | Asn |
| II | Lys | Arg | Arg | Gln | Asp |

The RIA results were correlated to the sweetness threshold of thaumatin as measured by blind taste tests. The sweet-taste threshold of native plant thaumatin was determined by taste test experiments in our laboratory to be approximately $2\,\mu g/ml$. Each sample described as sweet in taste tests was recognized in the RIA. The RIA, however, could detect the presence of native thaumatin in solutions containing as little as $1\,ng/ml$ of cross-reactive protein. Because the RIA is more objective and is approximately 1000-fold more sensitive than taste testing, it was used to quantitate sweet protein in test samples.

In addition to generating rabbit polyclonal antibodies for use in immunoassays, a selected number of hybridoma cells were cloned to propagate cell lines producing monoclonal antibodies against thaumatin. These antibodies recognized either the higher order tertiary structure of thaumatin or its primary sequence. The monoclonals also failed to recognize low molecular weight sweeteners such as aspartame and sucrose. These structural reagents are useful for immunoaffinity isolation and purification, and for quantitative immunological assays for thaumatin and thaumatin-like molecules.

## 3 PLANT THAUMATIN PURIFICATION, CHARACTERIZATION, AND SEQUENCE DETERMINATION

Since a number of distinct thaumatins exist in the plant extract, gram quantities of the major types were purified to homogeneity. These were used for biochemical characterization, amino acid composition and sequence analysis, antibody generation in rabbits, and the development of monoclonal antibodies.

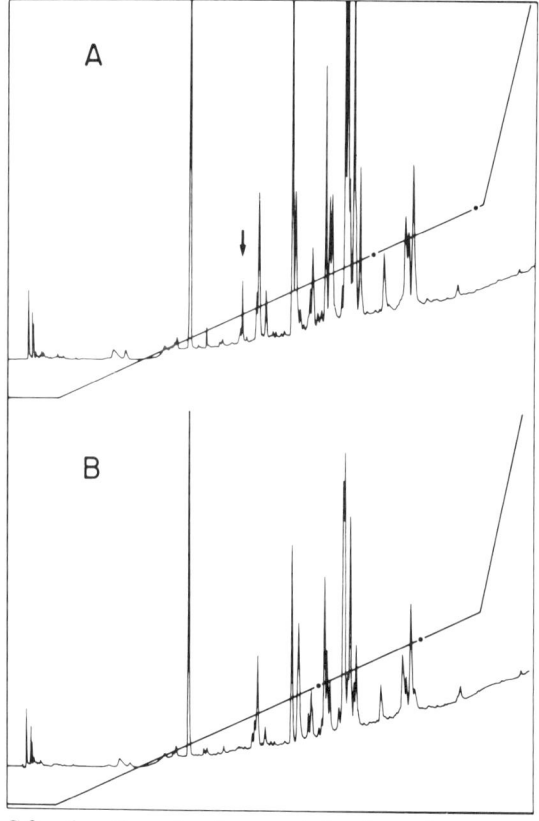

FIG. 1.   HPLC fractionation of carboxymethylated tryptic peptides of thaumatin from (A) yeast thaumatin B and (B) yeast thaumatin A. The arrow in panel A points to a peptide having an altered mobility (and not seen) in panel B. A linear acetonitrile gradient (5–50%) eluted the peptides over a period of 45 min. The ordinate axis shows the change in absorbance at 214 nm. Comparison of the two sequences enabled us to detect differences in the amino acid sequences of the two thaumatins.

Plant thaumatin, subjected to SP-Sephadex ion-exchange chromatography, was separated into two major species. Gel electrophoresis in urea under acid conditions distinguished easily between proteins differing from one another by one charge and demonstrated the completeness of the separation. The purified proteins were digested into peptides with TPCK-modified trypsin after carboxymethylation (TPCK blocks chymotryptic activity), and the fragments were separated by reverse-phase HPLC on a $C_{18}$ column. The peptide elution profiles of both thaumatins were similar but had minor differences (see arrow in Fig. 1). The peptides were collected, and the amino acid composition and sequence of each were determined. Knowledge of the primary sequence of thaumatin from previous studies[16,17] permitted placement (by similarity) of peptides in the sequence.

On comparison, only one amino acid difference between these two forms (asparagine in place of lysine) at position 46 was found. However, both sequences differed from those published for thaumatins I and II (see Table 1). To avoid confusing these new sequences with those previously published, we have designated our sequences thaumatins A and B. The thaumatin A sequence differs from the two published sequences by either 1 or 4 amino acids and the thaumatin B sequence differs by either 2 or 3 amino acids, respectively (Table 1).

## 4 THAUMATIN GENE SYNTHESIS

The starting point for our attempt to express thaumatin in yeast was to synthesize the corresponding gene. A coding sequence based on the amino acid sequence of thaumatin I[16] was designed, synthesized and cloned into the appropriate vectors (see Fig. 2). Yeast-preferred codons[25,26] were used so that the gene could be expressed efficiently in yeast. The gene was designed to contain multiple restriction enzyme sites to facilitate manipulation of the coding sequence. These sites allowed us to alter the coding sequence throughout the entire gene. The completed thaumatin I gene consists of 630 base pairs bracketed by a *BclI* restriction site at the 5′ end and an *XhoI* restriction site at the 3′ end of the DNA. This synthetic thaumatin I gene also served as the basis for construction of genes coding for thaumatins A and B.

Site-directed mutagenesis[27] and DNA fragment replacement in the thaumatin I gene were used to obtain the thaumatin A and B coding sequences. The thaumatin A sequence differs from that of thaumatin I only at residue 113, involving an asparagine to aspartic acid change. The thaumatin A gene was then used to generate a coding sequence for thaumatin B which differs from thaumatin A only at position 46. Replacement of the codon for asparagine with lysine at this position completed the construction of all three coding sequences.

## 5 VECTOR CONSTRUCTION FOR THAUMATIN EXPRESSION

In order to express the thaumatin genes in yeast, a strong yeast promoter, the 3-phosphoglycerate kinase (PGK) promoter, was joined to the 5′ end of the thaumatin I, A and B genes. Also, the PGK terminator was linked to the 3′ end of all three thaumatin variant genes to provide an efficient transcription termination and polyadenylation signal. The thaumatin gene expression cassette, containing the PGK promoter, the thaumatin gene and the PGK terminator in a linear array, was cloned into an *E. coli*–yeast shuttle vector, pJDB209.[28] This shuttle vector, which can replicate in both *E. coli* and yeast, provides the convenience of performing DNA manipulations in *E. coli* while allowing the functional testing of genes to take place in yeast. Plasmid pJDB209 contains a *leu2-d* marker which confers low-level expression of β-isopropylmalate dehydrogenase in yeast. When these thaumatin expression plasmids were transformed into a *leu2* yeast mutant, approximately 200 copies per cell[28] were apparently required to complement the *leu2* mutation of the host cell. Concomitantly the copy number of the thaumatin expression cassette also increases to approximately 200 copies per cell, resulting in an increase in the level of expression of the thaumatin gene.

## 6 THAUMATIN EXPRESSION IN YEAST CELLS

Plasmids containing the thaumatin I, A and B expression cassettes (Fig. 2) were transformed into yeast strain AH22 and BB25-1d.[29,34] A control plasmid, which is similar to the thaumatin expression vector but missing the PGK promoter and thaumatin gene, was also transformed. Cells were grown in selective media under fermentation conditions, harvested by centrifugation and lysed, and the resulting extract was analyzed. Sodium dodecyl sulfate (SDS) gel electrophoresis showed the presence of a new 23 000 daltons protein visualized by Coomassie Brilliant Blue R staining. Figure 3 shows a comparison of the quantities of thaumatin (and other yeast proteins) solubilized by SDS extraction of the insoluble fraction of yeast extract

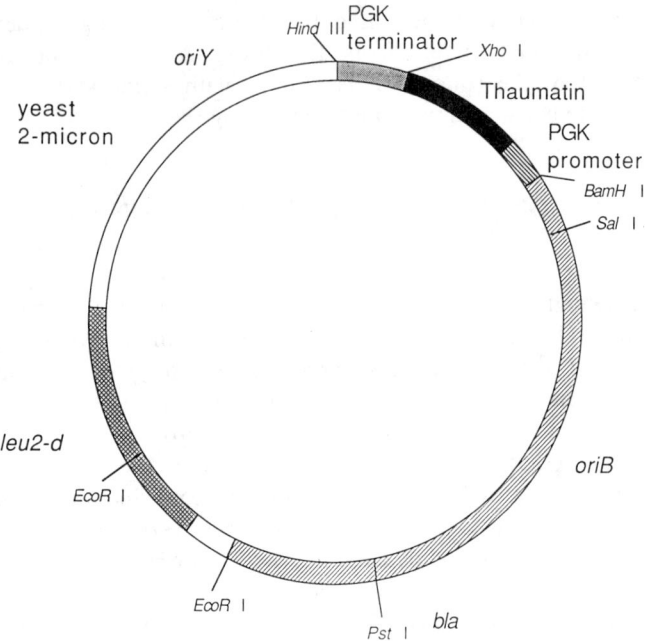

Fig. 2. Plasmid map of the basic thaumatin expression vector. The selectable markers are *bla* (in *E. coli*) and *leu2-d* (in yeast). *oriY* in the yeast 2-micron sequence and *oriB* in pAT153 provide replication functions in yeast and *E. coli*, respectively.

containing thaumatins I, A and B. The control culture grown under identical conditions but not containing the thaumatin gene did not produce any thaumatin-like protein.

The identity of yeast thaumatin was confirmed by Western blot[30] analysis of the SDS-solubilized proteins (data not shown). Rabbit antiserum to thaumatin A at a 2000-fold dilution was used to probe proteins that had been separated by SDS gel electrophoresis and transferred to nitrocellulose paper. The antiserum at this dilution recognizes denatured and aggregated thaumatin and its fragments in the SDS-solubilized cell extract.

The major protein recognized by the antiserum was localized in the insoluble portion of the yeast extract; it co-migrated with a plant thaumatin standard. No cross-reactive protein was observed in the soluble protein fraction. This is not surprising since no signal sequence was attached to the N-terminus of the protein. The cell therefore lacked the information necessary to direct the protein into the appropriate secretory pathway.

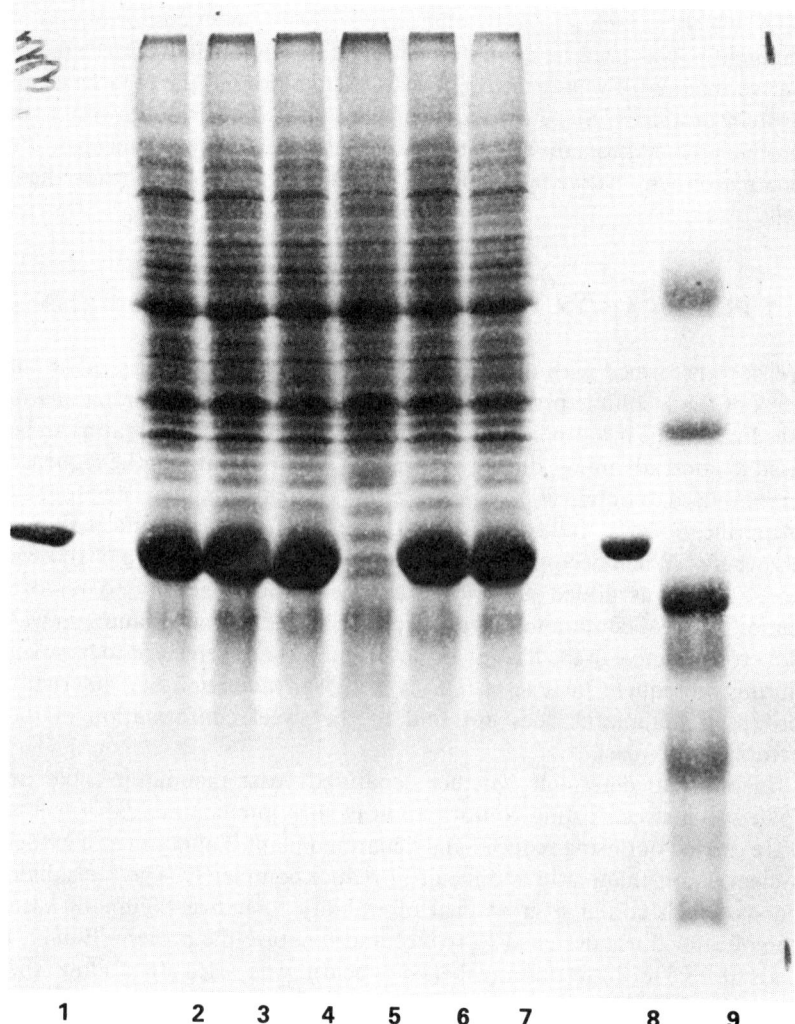

FIG. 3.   Analysis by SDS gel electrophoresis of different yeast cell extracts showing the levels of thaumatin expression for several derivatives. Samples in the gel were stained with Coomassie Brilliant Blue. No thaumatin is detected in the soluble fraction of the cell extract. Lanes 1 and 8, plant thaumatin; lane 2, thaumatin I; lane 3, thaumatin B; lane 4, thaumatin A; lane 5, no thaumatin gene; lanes 6 and 7, thaumatin B, yeast colonies a and b; lane 9, BRL protein standards (from the top: 43, 29, 18, 14, 6 and 3 kilodaltons).

No sweet protein was found inside the yeast cells. Our thaumatin-specific RIA, which, at high antibody dilution, detects the sweet conformation of thaumatin, was used to quantitate levels of immunoreactive protein. No native thaumatin was detected in yeast extract despite the high levels of protein produced. Since reduction of the thaumatin disulfides results in protein precipitation, the reducing intracellular milieu of yeast may provide one reason why native (and soluble) thaumatin is not found inside these cells.

## 7  PURIFICATION AND FOLDING OF YEAST THAUMATIN

Yeast transformed with the thaumatin expression vector produced about 20% of the insoluble proteins (or 10% of the total protein) as thaumatin (see Fig. 3). The insolubility and charge characteristics of thaumatin can be used to good advantage during purification. Figure 4 outlines the sequence of steps used to obtain 80% yields of > 95% pure thaumatin. The progress of purification was followed using SDS gel electrophoresis and is shown elsewhere.[8] When necessary, a Sephadex G-75 gel filtration step in 100 mM acetic acid was added after ion-exchange chromatography to remove minor traces of contaminating proteins. Again the purified thaumatin was not cross-reactive in the RIA at any level and was not perceived to be sweet during subsequent taste tests. This was a clear indication that internally produced thaumatin does not fold to the sweet conformation in the cytoplasm of yeast.

In order to determine whether denatured yeast thaumatin could be folded to a sweet-tasting conformation *in vitro*, preliminary experiments were carried out using reduced and denatured plant thaumatin as a model system. Plant thaumatin is difficult to reduce completely. The 8 disulfide bonds which confer a great deal of stability to native thaumatin also complicate efforts designed to reduce and renature the protein. Ellman's reagent (5,5′-dithiobis-(2-nitrobenzoic acid)) was used to follow the progress of disulfide reduction and showed that plant thaumatin contained a disulfide region that was stabilized against reduction. Two particularly resistant disulfides require 50 mM $\beta$-mercaptoethanol at 37°C for 2 h in 8M urea at pH 9 for complete reduction.

For initial folding experiments, published procedures were used[31,32] to generate a sweet molecule from denatured plant thaumatin. These procedures showed that, in addition to the difficulties encountered in denaturing thaumatin, refolding the protein would also present some

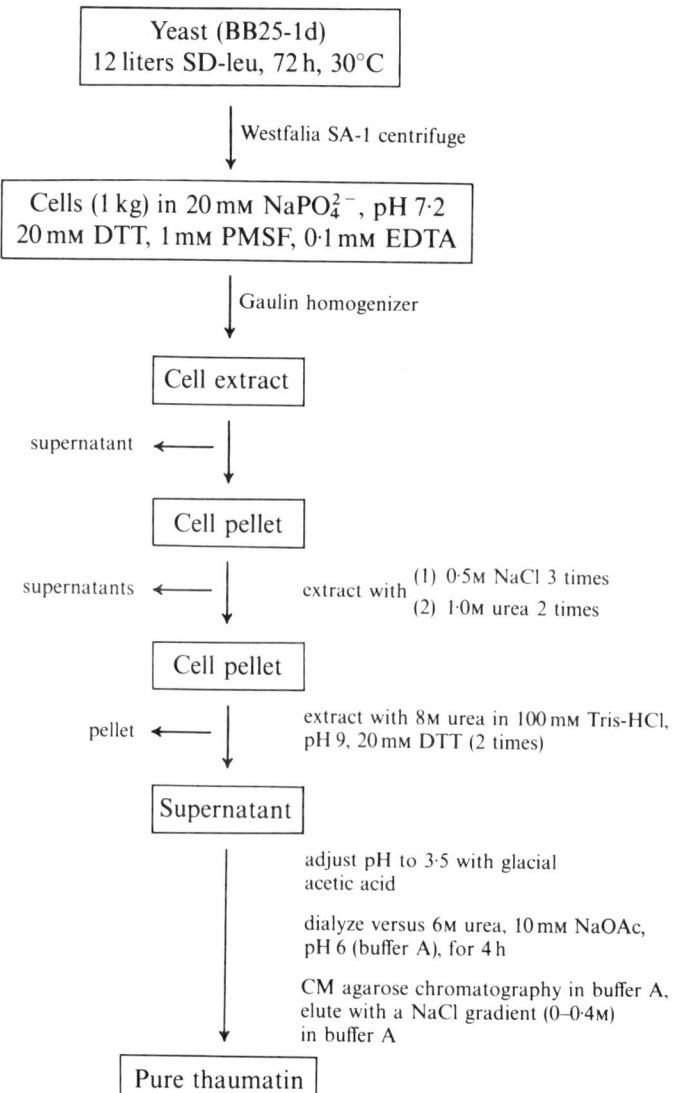

FIG. 4.   Protocol used for the isolation and purification of thaumatin from yeast cells (DTT = dithiothreitol).

Reduce thaumatin (10–20 mg/ml) in 100 mM Tris-HCl, pH 9, 6M urea, 200 mM $\beta$-mercaptoethanol, flush with nitrogen, incubate 3 h at 37°C.

REDUCTION

↓

Gel filtration through a column (2·2 × 45 cm) of Sephadex G-25 equilibrated with 50 mM Tris-HCl, pH 9, 6M urea, 100 mM cystamine.

THAUMATIN–
CYSTEAMINE
ADDUCT
FORMATION

↓

Collect fractions and locate protein using Bio-Rad protein assay. Pool thaumatin fractions, concentrate with Amicon YM-10, flush with nitrogen, incubate 3 h at 25°C.

↓

Adjust pH to 3·5 with glacial acetic acid. Gel filtration through a column (2·2 × 45 cm) of Sephadex G-25 equilibrated with 100 mM acetic acid, 0·1 mM EDTA. Locate protein by absorbance at 280 nm.

↓

Dilute peak fractions to 20 $\mu$g/ml into 50 mM Tris-HCl, pH 8 (0°C, degassed), containing 2 mM cysteine. Let stand 20 h at 4°C.

FOLDING

↓

Dialyze against 100 mM ammonium bicarbonate for 3 h, followed by five changes of distilled water in 4–5 h.

FIG. 5. Procedure *in vitro* developed to fold denatured thaumatin into its native conformation (PMSF = phenylmethane sulfonyl fluoride).

formidable problems. During protein folding, hydrophobic interactions between regions of denatured protein and improper disulfide pairings of the 16 sulfhydryl groups result in the aggregation of a large percentage of denatured protein. Plant thaumatin folding was optimized by allowing it to take place as a slow and controlled process at low protein concentration. However, even at an initial concentration of 20 $\mu$g/ml, the overall yield of native thaumatin was only about 1% or 0·2 $\mu$g/ml. These levels could only be detected by RIA or by taste testing following extensive concentration of the dilute solution. The limited success of these experiments caused us to try to develop less costly and more efficient procedures to generate the native structure from purified, denatured yeast thaumatin.

A variety of sulfhydryl reagents, e.g. oxidized glutathione[31] and cystine,[33] have been used to renature proteins. In our procedure the sulfhydryl reagent cystamine was used to form the protein–disulfide–cysteamine adduct. 'Blocking' the free protein sulfhydryl groups is advantageous because (1) the solubility of the denatured protein is greatly increased and (2) the sulfhydryls do not interchange until a weak reducing agent is added. Figure 5 outlines the new reduction/cysteamine adduct formation/folding procedure developed to produce native thaumatin from denatured protein. Use of cystamine (rather than oxidized glutathione) to form the mixed disulfide reduces the cost. The yield of sweet protein from denatured plant thaumatin is also increased to about 5 $\mu$g/ml (or about 25%), easily detected in taste tests. Slightly lower amounts were obtained from yeast thaumatins A and B. To our surprise, we were unable to fold yeast-produced thaumatin I into its native (sweet) conformation. The N-terminal amino acid of yeast-produced thaumatin (alanine) is blocked with an acetyl group.[34] Through a series of studies, we determined that this modification has no effect on protein folding or sweet taste. These results suggest that the amino acid sequence at position 113, the only other difference between thaumatins I and A, may be suspect and that, rather than an asparagine at position 113,[16] the correct amino acid is actually aspartic acid.

## 8  THAUMATIN SECRETION FROM YEAST

The development of a yeast secretion system was tested as an alternative means of producing thaumatin. This would allow us to compare the efficiency of yeast secretion to that of protein folding *in vitro*. Since plant thaumatin is produced by secretion, these experiments were designed to mimic the actual production process in the plant. The synthetic genes for thaumatins I, A and B were fused to the 3′ end of a sequence coding for the signal sequence of a secreted yeast protein, invertase. This signal sequence should direct the thaumatin into the membrane transport secretory pathway for eventual secretion into the culture medium.

When yeast transformed with a plasmid containing each of the various fusion genes is grown on selective media, immunoreactive, sweet thaumatins A and B are found in the corresponding culture medium. The molecular weight of these proteins is identical to that of mature thaumatin. Protein sequencing of the N-terminus of the secreted protein indicates a precise removal of the signal peptide. Interestingly, thaumatin I could not be detected in the growth medium when its gene was cloned into a secretion

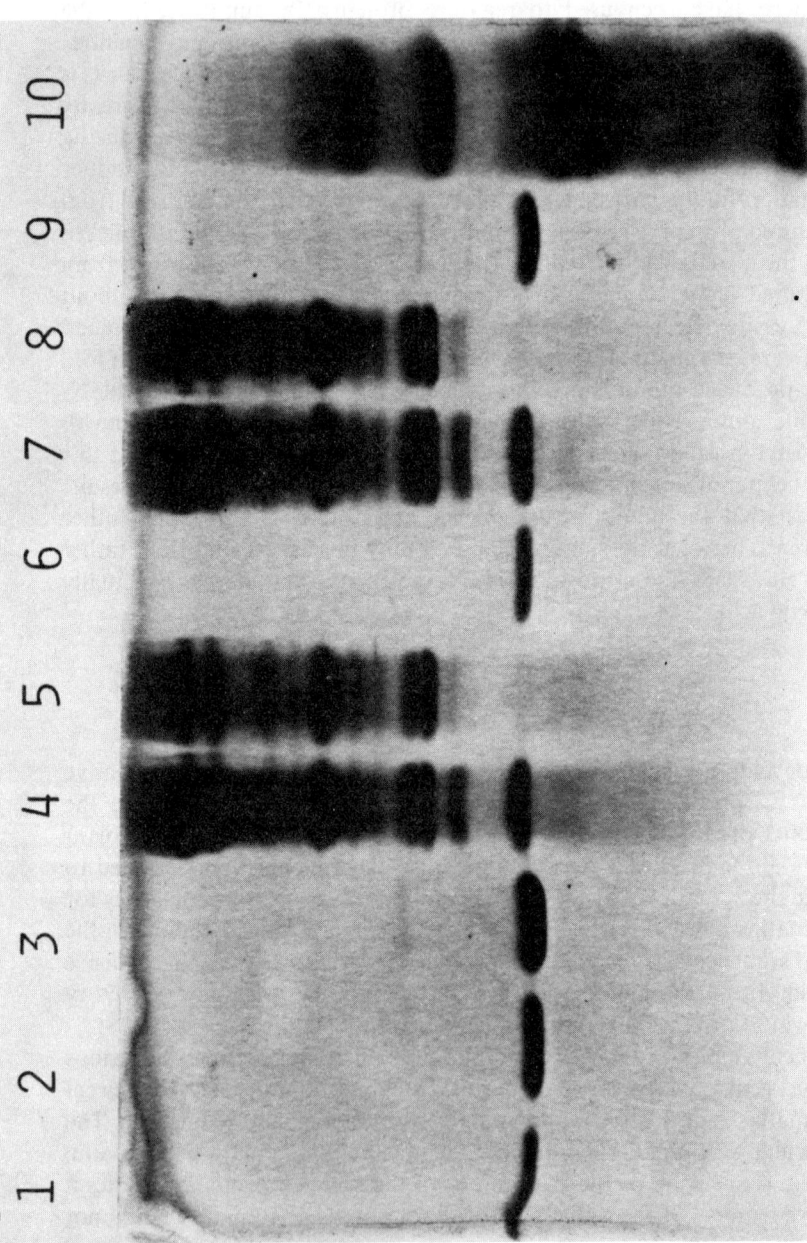

Fig. 6. Analysis by SDS gel electrophoresis of the secretion and subsequent purification of yeast thaumatin. Lanes 1–3, plant thaumatin; lanes 4 and 7, secreted yeast proteins; lanes 5 and 8, secreted yeast proteins with thaumatin variants; lanes 6 and 9, purified yeast thaumatin variants.

vector for production. This result is consistent with the in-vitro folding data for thaumatin I, suggesting that the single amino acid change at residue 113 is important for protein structure.

Not all the thaumatin produced in yeast is found in the growth medium. A large percentage of the protein remained inside the cell. This internal form has the molecular weight of mature thaumatin, suggesting that, although it has entered the secretory pathway via the endoplasmic reticulum, it has not completed its journey through the Golgi apparatus and the secretory vesicles.[35] This internal form is neither immunoreactive nor sweet. The percentage of thaumatin remaining inside these cells in relation to the total thaumatin produced is approximately 50%.

Test quantities of sweet thaumatin have been produced in fermenters of various sizes using the yeast secretion pathway. A purification scheme that allows us to process large volumes of cell-free culture media efficiently has been developed for the isolation of secreted thaumatin. Figure 6 shows an SDS gel electrophoresis pattern for all the proteins secreted into the growth medium by yeast. Also shown in Fig. 6 is the level of homogeneity achieved when secreted thaumatin is purified from this protein mixture. Improvements in the level of thaumatin secretion will ultimately allow us to use fermentation and protein purification as a means of producing thaumatin and its derivatives on a large scale.

## 9 TASTE TESTS USING YEAST AND PLANT THAUMATIN

In order to measure the sweetness thresholds for plant and yeast thaumatins, taste tests were conducted in our laboratory. The exact total protein concentration was determined by two methods: (1) absorbance measurement at 280 nm; (2) a colorimetric protein assay. These results were then checked by RIA. This final assay is important since the RIA can quantitate biological activity as it relates to protein structure rather than just giving a value for total protein. By progressive dilution and taste testing, we determined the sweetness threshold for plant thaumatin to be $2.3 \pm 0.9 \, \mu g/ml$. Similar results were obtained with thaumatin produced from yeast by either protein folding or secretion.

Another taste testing procedure that compared the aftertaste characteristics of yeast-produced thaumatin and the plant protein was a time–sweetness intensity profile. Individuals were asked to evaluate the sweetness intensity of a sample as a function of time (up to 120 s). The profiles obtained were very informative when done several times by 5 or more individuals.

In this study we generated a large number of altered thaumatin genes to make new sweet protein variants of thaumatin. These sweet proteins were initially screened by blind taste tests in our laboratory. Promising variants as judged by us were evaluated by trained personnel at Beatrice Foods Company. These taste evaluations were most helpful in our efforts to design and produce an altered form of thaumatin with improved taste properties. The next section summarizes our approach to this problem.

## 10 GENERATION OF THAUMATIN DERIVATIVES WITH ALTERED TASTE PROPERTIES

One of the limitations in the development of plant thaumatin as a widely used sweetener is its persistent aftertaste. This property may prove to be advantageous for some applications (chewing-gum, toothpaste, etc.), but for most products aftertaste is undesirable. A challenge to biotechnology is to design changes in the amino acid sequence of thaumatin to provide variants which retain the sweetness potency of the plant protein yet exhibit a reduced aftertaste.

A large number of specific changes were introduced into the thaumatin gene by mutagenesis *in vitro* or by replacing DNA sequences by insertion of synthesized oligonucleotides. These altered genes were cloned into the expression vector which produces secreted thaumatin. Most of these variant thaumatin plasmids enabled transformed yeast to secrete a variant thaumatin. Some plasmids do not enable thaumatin secretion and presumably encode secretion/folding mutations. By studying these variant thaumatins we hope to be able to designate domains in the primary sequence which may be important either for protein folding to the proper conformation or for integrity of the structures conferring sweet taste.

Figure 7 illustrates several examples of taste profiles for altered thaumatins, and shows the results of tests carried out by trained personnel who evaluated the sweetness intensity and taste characteristics of these variants, compared with plant thaumatin, sucrose and aspartame. One derivative, thaumatin 1, while retaining a high sweetness potency, has consistently demonstrated a time–intensity taste profile more sucrose-like than plant thaumatin. Several additional candidates have since been evaluated and have shown promising results. We are most interested in

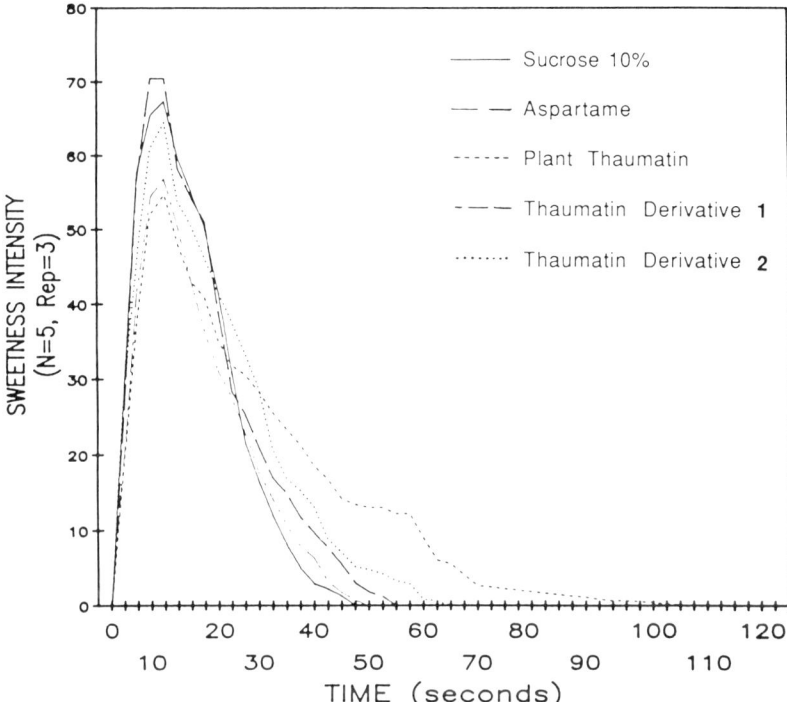

FIG. 7.   Time versus sweetness intensity profiles for sucrose, aspartame, plant thaumatin and two yeast thaumatin derivatives.

examining the effect these sequence alterations have had on protein structure.

## 11   DISCUSSION AND FUTURE SCOPE

In this chapter we have shown that a coding sequence for plant thaumatin can be expressed at high levels in yeast as an intracellular protein or, alternatively, as a secreted protein if a secretion signal sequence precedes the mature protein sequence. Intracellular thaumatin, found in the reducing environment in the cytoplasm, is not sweet presumably because it is not folded into the proper protein conformation. However, a procedure can be used *in vitro* to fold low concentrations of purified thaumatin into a sweet form. Thaumatin presents an interesting, albeit formidable, problem for protein folding *in vitro* and disulfide bond formation, and we are currently working to improve the folding efficiency. Development of an efficient folding procedure which allows correct disulfide bond formation would be useful not only for making biologically active thaumatin but also for producing many other normally secreted proteins that have disulfide bonds.

Secreted thaumatin is folded into a sweet form presumably because the environment and cellular machinery used to fold and form disulfide bonds in exported proteins are similar in plants and yeast. Secretion simplifies the process of thaumatin production. This property of secreting thaumatin has enabled us to introduce specific alterations into the coding sequence and produce many variants of thaumatin in their biologically active form. Careful taste-testing of these variants allows us to approach the problem of how the structure of thaumatin relates to its sweet taste. Are particular parts of the molecule responsible for sweetness? How do various amino acid substitutions alter the taste character?

At present we are concentrating our research efforts on three major fronts. These include (1) the economical production of thaumatin variants under fermentation conditions, (2) the development of better tasting, novel sweet proteins, and (3) structural characterization of the thaumatin molecule.

Our present system for secreting thaumatin from yeast makes it possible to produce test quantities of a number of thaumatin variants, each purified to homogeneity. Improvements in the levels of thaumatin production are being developed by optimizing our fermentation technology.

The notion that the character of the thaumatin molecule may be altered to suit consumer taste preferences is very exciting. Several of our thaumatin protein sequence changes have resulted in the secretion of variants with reduced aftertaste. These proteins can be highly purified and will be used to conduct additional research into the molecular structure of thaumatin, and how this relates to its function.

Studies by X-ray crystallography, which provide detailed structural information about thaumatin and its derivatives, combined with taste-test data and sequence information, would add to our knowledge about those regions of thaumatin that are important for taste. For instance, knowing the orientation of amino acid side-chains within the molecule could help to determine which side-chains interact with receptor molecules and/or one another through hydrophobic, ionic or hydrogen bonding. This information might allow us to design a molecule that would reduce or enhance these intramolecular interactions, thus generating a desired organoleptic effect.

The three-dimensional structure of thaumatin may also show us which amino acid side-chains face toward the outside of the molecule. It is these structures that would most likely be integral parts of the sweet site. Knowing the amino acid sequences that form the sweet site might allow us to design a smaller, less complicated protein or peptide that would have sweet properties similar to those of thaumatin. This information could also lead to an understanding of how the molecule interacts with its receptors on the tongue.

The rational selection and positioning of specific amino acid sequence changes in proteins to generate desired biological activities and properties is still in its infancy. We have begun to explore this fascinating area of research by studying the effect that single (and in some cases multiple) amino acid substitutions in the thaumatin molecule have on the sweet taste properties of the respective variants. The results to date have been encouraging and suggest that alterations in the structure of thaumatin do affect its recognition and stimulation of the sweetness receptor(s). We believe that the use of structural information and taste-test data to direct iterative rounds of modifications in the thaumatin gene will allow us to develop a new generation of novel low-calorie protein sweeteners for various food and beverage applications.

## ACKNOWLEDGEMENTS

We thank Dr Tom Parsons, Dr Arup Sen and Dr Gary Wilcox for critical reading of the manuscript and constructive comments. Meredith Alcus provided excellent and patient typing assistance while Kevin Bryan helped to prepare the figures. We thank personnel at Beatreme Food Ingredients Inc. and the Swift-Eckrich Division of Beatrice Foods for performing taste tests and data analysis. This work was supported by Beatrice Foods Co., Chicago, Illinois.

## REFERENCES

1. Morris, J. A. & Cagan, R. H., Purification of monellin, the sweet principle of *Dioscorephyllum cumminsii. Biochim. Biophys. Acta*, **261** (1972) 114–22.
2. van der Wel, H. & Loeve, K., Isolation and characterization of thaumatin I and II, the sweet-tasting proteins from *Thaumatococcus daniellii* Benth. *Eur. J. Biochem.*, **31** (1972) 221–5.
3. Brouwer, J. N., van der Wel, H., Francke, A. & Henning, G. T., Miraculin, the sweetness-inducing protein from miracle fruit. *Nature*, **220** (1968) 373–4.

4. Bartoshuk, L. M., Dateo, G. P., Vanderbelt, D. J., Buttrick, R. L. & Long, L., Effects of *Gymnema sylvestre* and *Synsepalum dulcificum* on taste in man. In *Taste and Olfaction*, III, ed. C. Pfaffmann. Rockefeller University Press, New York, 1969, pp. 436–44.

5. van der Wel, H., Physiological action and structure characteristics of the sweet-tasting proteins thaumatin and monellin. *Trends Biochem. Sci.*, **5** (1980) 122–3.

6. Higginbotham, J. D., Combination of high intensity sweetener and thaumatin—for use in rearing farm animals, to increase food conversion ratio. UK patent 2 185 674 (1987).

7. van der Loo, H. E., Wiener, C. & Higginbotham, J. D., Smoking articles containing thaumatin and monellin. UK patent 1 604 217 (1981).

8. Lee, J.-H., Weickmann, J. L., Koduri, R. K., Ghosh-Dastidar, P., Saito, K., Blair, L. C., Date, T., Lai, J. S., Hollenberg, S. M. & Kendall, R. L., Expression of synthetic thaumatin genes in yeast. *Biochemistry*, **27** (1988) 5101–7.

9. Cornelissen, B. J. C., Hooft van Huijsduijnen, R. A. M. & Bol, J. F., A tobacco mosaic virus-induced tobacco protein is homologous to the sweet-tasting protein thaumatin. *Nature* (London), **321** (1986) 531–2.

10. Richardson, M., Valdes-Rodriguez, S. & Blanco-Labra, A., A possible function for thaumatin and a TMV-induced protein suggested by homology to a maize inhibitor. *Nature* (London), **327** (1987) 432–4.

11. Higginbotham, J. D., Talin—a novel sweet protein from *T. daniellii*. In *Health and Sugar Substitutes*, Proc. ERGOB Conf., Geneva, 1978, pp. 172–7.

12. Daniell, W. F., Katemfe, or the miraculous fruit of Soudan. *Pharm. J.*, **14** (1855) 158–9.

13. Inglett, G. E. & May, J. F., Tropical fruits with unusual taste properties. *Econ. Bot.*, **22** (1968) 326–31.

14. Higginbotham, J. D., Extraction of a sweet substance from *Thaumatococcus daniellii* fruit. US patent 4 011 206 (1977).

15. van der Wel, H. & Bel, W. J., Effect of acetylation and methylation on the sweetness intensity of thaumatin I. *Chem. Senses and Flavour*, **2** (1976) 211–18.

16. Iyengar, R. B., Smits, P. S., van der Ouderaa, F., van der Wel, H., van Brouwershaven, J., Ravestein, P., Richters, G. & van Wasenaar, P. D., The complete amino acid sequence of the sweet protein thaumatin I. *Eur. J. Biochem.*, **96** (1979) 193–204.

17. Edens, L., Heslinga, L., Klok, R., Ledeboer, A. M., Maat, J., Toonen, M. Y., Visser, C. & Verrips, C. T., Cloning of cDNA encoding the sweet-tasting plant protein thaumatin and its expression in *Escherichia coli*. *Gene*, **18** (1982) 1–12.

18. Ledeboer, A. M., Verrips, C. T. & Dekker, B. M. M., Cloning of the natural gene for the sweet-tasting plant protein thaumatin. *Gene*, **30** (1984) 23–32.

19. van der Wel, H. & Bel, W. J., Structural investigation of the sweet-tasting proteins thaumatin and monellin by immunological studies. *Chem. Senses and Flavour*, **3** (1978) 99–104.

20. Hough, C. A. M. & Edwardson, J. A., Antibodies to thaumatin as a model of the sweet taste receptor. *Nature* (London), **271** (1978) 381–3.

21. De Vos, A., Hatada, M., van der Wel, H., Krabbendam, H., Peerdeman, A. & Kim, S.-H., Three-dimensional structure of thaumatin I, an intensely sweet protein. *Proc. Nat. Acad. Sci. USA*, **82** (1985) 1406–9.

22. Ogata, C., Hatada, M., Tomlinson, G., Shin, W.-C. & Kim, S.-H., Crystal

structure of the intensely sweet protein monellin. *Nature* (London), **328** (1987) 739–42.

23. Kim, S.-H., de Vos, A. & Ogata, C., Crystal structures of two intensely sweet proteins. *Trends Biochem. Sci.*, **13** (1988) 13–15.

24. Edens, L., Bom, I., Ledeboer, A. M., Maat, J., Toonen, M. Y., Visser, C. & Verrips, C. T., Synthesis and processing of the plant protein thaumatin in yeast. *Cell* (Cambridge, Mass.), **37** (1984) 629–33.

25. Hoekema, A., Kastelein, R. A., Vasser, M. & De Boer, H. A., Codon replacement in the PGK1 gene of *Saccharomyces cerevisiae*: experimental approach to study the role of biased codon usage in gene expression. *Molec. Cell Biol.*, **7** (1987) 2914–24.

26. Bennetzen, J. L. & Hall, B. D., Codon selection in yeast. *J. Biol. Chem.*, **257** (1982) 3026–31.

27. Miyada, C. G., Soberon, X., Itakura, K. & Wilcox, G., The use of synthetic oligodeoxyribonucleotides to produce specific deletions in the *araBAD* promoter of *Escherichia coli* B/r. *Gene*, **17** (1982) 167–77.

28. Beggs, J. D., Multiple-copy yeast plasmid vectors. In *Molecular Genetics in Yeast*, Alfred Benzon Symp. 16, ed. D. von Wettstein, J. Friis, M. Kielland-Brandt & A. Stenderup. Munksgaard, Copenhagen, 1981, pp. 383–90.

29. Hinnen, A., Hicks, J. B. & Fink, G. R., Transformation of yeast. *Proc. Nat. Acad. Sci. USA*, **75** (1978) 1929–33.

30. Towbin, H., Staehlin, T. & Gordon, J., Electrophoretic transfer of proteins from polyacrylamide gels to nitrocellulose sheets: procedure and some applications. *Proc. Nat. Acad. Sci. USA*, **76** (1979) 4350–4.

31. Odorzynski, T. W. & Light, A., Refolding of the mixed disulfide of bovine trypsinogen and glutathione. *J. Biol. Chem.*, **254** (1979) 4291–5.

32. Light, A., Protein solubility, protein modifications, and protein folding. *Biotechniques*, **4** (1985) 298–306.

33. Bradshaw, R. A., Kanarek, L. & Hill, R. L., The preparation, properties, and reactivation of the mixed disulfide derivative of egg white lysozyme and L-cystine. *J. Biol. Chem.*, **242** (1967) 3789–98.

34. Huang, S., Elliot, R. C., Liu, P.-S., Koduri, R. K., Weickmann, J. L., Lee, J.-H., Blair, L. C., Ghosh-Dastidar, P., Bradshaw, R. A., Bryan, K. M., Einarson, B., Kendall, R. L., Kolacz, K. H. & Saito, K., Specificity of cotranslational amino-terminal processing of proteins in yeast. *Biochemistry*, **26** (1987) 8242–6.

35. Schekman, R. & Novick, P., The secretory process and yeast cell-surface assembly. In *The Molecular Biology of the Yeast Saccharomyces: Metabolism and Gene Expression*, ed. J. Strathern, E. Jones & J. Broach. Cold Spring Harbor Laboratory, New York, 1981, pp. 361–93.

*Chapter* 3

# GLYCYRRHIZIN: THE BASIC FACTS PLUS MEDICAL AND DENTAL BENEFITS

M. N. Sela & D. Steinberg

*Department of Oral Biology, Hadassah Medical School, Jerusalem, Israel*

## DEDICATION

The authors wish to dedicate this chapter to Professor Ruth Segal, who passed away in the prime of her scientific research work.

## SUMMARY

*Glycyrrhizin is a triterpenoid saponin found in the roots of the herb liquorice. Both liquorice and glycyrrhizin possess medicinal properties. Deglycyrrhizinized liquorice is used as a drug to combat ulcers. In addition to its antiinflammatory properties, glycyrrhizin was also found to act as an inhibitor of carbohydrate metabolism in dental bacteria. The possible anti-cariogenic effect of glycyrrhizin is of interest, as it interferes in the mechanisn by which dental bacteria adhere to smooth surfaces, by enzymatic inhibition of the extracellular enzyme responsible for the production of the sticky polysaccharides. Glycyrrhizin also affects tooth enamel by raising fluoride uptake and reducing its solubility. In addition to its therapeutic effects, glycyrrhizin may reduce the toxicity of other medications, although several case reports have shown that consumption of large quantities of glycyrrhizin or liquorice may result in severe side-effects. However, they remain drugs which possess important and interesting medicinal and dental properties.*

## 1 NATURE AND CHEMISTRY OF GLYCYRRHIZIN

*Glycyrrhiza*, of Greek origin, meaning 'sweet root', is the dried rhizome and roots of various species of the plant *Glycyrrhiza*, better known as liquorice. The most abundant plants for commercial use are *Glycyrrhiza glabra* L. commonly known as 'Spanish liquorice', which is cultivated in Spain, Sicily and England; *Glycyrrhiza glabra* L. var. *gladulifera*, known in commerce as 'Russian liquorice' and grown mainly in Russia; and *Glycyrrhiza glabra* var. *β*-violacea, known as 'Persian liquorice' and grown in Iran and Iraq.

### 1.1 Cultivation and Collection
Liquorice is a wild plant that grows in deep, sandy fertilized soil near water. It can also be cultivated and propagated by cutting the rhizome and planting it in rows 2–4 feet apart. The preferred planting time is March. In October of the third or fourth year, when the leaves fall off, the plants are dug up, washed and air-dried for 4–6 months. They are then packed into bales or bundles of straight cylindrical pieces. The final product may be marked as peeled or unpeeled roots. Part of the crop might also be shipped as sticks or blocks of liquorice. The extraction of liquorice is achieved through a decoction process. After the liquid has been clarified and the excess evaporated, the resulting extract is moulded into blocks, sticks or any other desired form.

### 1.2 Constituents
The sweet taste of liquorice is due to its glycyrrhizin component, and it is about 50–100 times sweeter than cane-sugar. Glycyrrhizin is a triterpenoid saponin in which the aglycone is bound to a sugar. The aglycone, glycyrrhetinic acid, is attached to 2 molecules of glucuronic acid at the C-3 atom of the aglycone moiety. Upon hydrolysis the saponin-like glycoside converts to the aglycone and the sugar residue (Fig. 1).

Glycyrrhizic acid is found predominantly in the main roots of *Glycyrrhiza glabra* L., while the lateral roots contain lower amounts. The green parts of the plant do not contain any glycyrrhizin.[1] Different concentrations of glycyrrhizin are to be found in the various commercial products of liquorice. These differences may be attributed to the types of plants, to different ways of cultivation or to variations in the analytical methods used.

When glycyrrhizic acid was determined by thin-layer chromatography (TLC) densitometry,[2] it was found that Chinese liquorice roots contain 3·11–6·53% of glycyrrhizic acid, Russian liquorice contains 2·2% and an

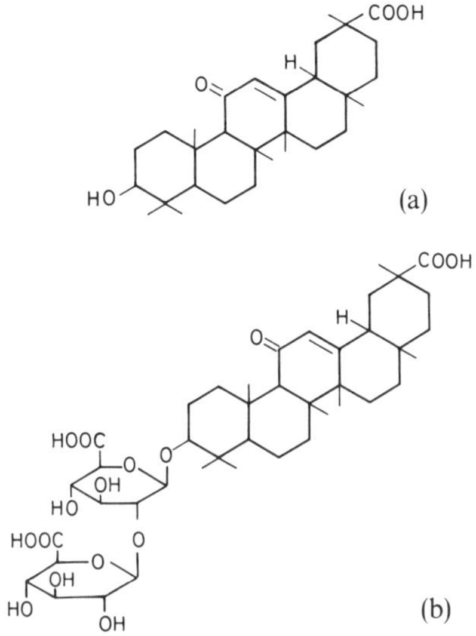

FIG. 1. (a) Glycyrrhizin (glycyrrhizic acid). (b) Glycyrrhetinic acid (glycyrrhetic acid).

Afghanistan product 5·11%. The dry extract of the Chinese liquorice contained 9·3% and a semi-dry extract contained about 7·05%. The contents of glycyrrhizic acid in the same roots and extracts were compared using a gas–liquid chromatography (GLC) method. The ratio of glycyrrhizic contents determined by the TLC densitometry to glycyrrhizic acid found by the GLC method had an average value of 0·75. In another study[3] $\beta$-glycyrrhetinic acid was determined by high-pressure liquid chromatography. It was found that in commercial liquorice products the $\beta$-glycyrrhetinic content (w/w) was as follows: Spanish liquorice, 0·88%; decortified Iraqi liquorice, 0·64%; Persian liquorice, 0·53%.

Among the triterpene-like saponins which are found in liquorice, one can also find such flavonoids as isoliqueritin (chalcone) and liquiritin. Bitter substances, such as glycyramarin, are abundant mainly in the outer tissues of the plant and are therefore removed upon peeling.

## 2 GENERAL PHARMACOLOGICAL AND MEDICINAL PROPERTIES

Liquorice, glycyrrhizin and glycyrrhetinic acid possess many pharmaceutical advantages, including anti-ulcer, anti-inflammatory, anti-viral and anti-cariogenic properties, and they have been used in traditional treatments of bronchitis and coughs. Some of their therapeutic properties will be discussed below.

## 2.1 Role of Glycyrrhizin in Gastric Ulcer

It has been reported that deglycyrrhizinized liquorice (DGL) has a potential healing effect in cases of peptic ulcers. In a double blind clinical trial,[4] 760 mg of pure DGL in a pharmaceutical tablet form [Caved-(S)® containing (per tablet) block liquorice ($\leq 3\%$ glycyrrhizic acid) 380 mg, bismuth subnitrate 100 mg, dried aluminium oxide gel 100 mg, light magnesium carbonate 200 mg, sodium bicarbonate 100 mg and frangula bark 30 mg] was given 3 times a day for 4 weeks to patients with gastric ulcer. The control group received a placebo containing caramel-coloured lactose. The efficacy of the treatment was assessed by the size of the ulcer after treatment. In 44% of the patients the ulcer niche disappeared after treatment, compared with 6% in the control group. Among treated patients the average reduction in the size of the ulcer niche was 78%, as compared with only 34% in the control group. Similar results were obtained in a different study,[5] demonstrating a 92% average reduction in ulcer size, compared with 39% in a control group given lactose as a placebo.

Other workers[6] reported different results using DGL, with 96 patients randomly divided into two groups. One group was asked to take 500 mg DGL in capsule form (Ulcedal), while the other took 200 mg of sucrose. The patients were also supplied with antacid tablets. Examinations following 4 weeks of treatment revealed no differences between the test and control groups. The same number of patients in each group exhibited healing at the end of the trial. The authors emphasized that the DGL tested in this trial was different from the pharmaceutical formulation usually employed in other clinical tests, but they did not attribute the failure of the test to the difference in the pharmaceutical preparations. However, another publication[7] suggested that the difference in the results may be attributed to biopharmacological factors, claiming that the bioavailability of DGL in capsules (Ulcedal) is poor. It has now been withdrawn from the market pending reformulation.

## 2.2 Role of Glycyrrhizin in Duodenal Ulcer

Findings on the effect of DGL on duodenal ulcers are more contradictory than the gastric ulcer results. In one clinical trial[8] a group of 32 male patients with endoscopically confirmed duodenal ulcers and chronic clinical histories were instructed to use DGL tablets (Caved-(S)). According to the manufacturer's instructions the tablets were to be taken 4 or 6 times per day after meals, lightly chewed and swallowed with water. However, in this experiment the patients were asked to chew the tablets thoroughly and to swallow 2 tablets 5 times a day on an empty stomach. Clinical results revealed that healing of ulcerations occurred in all 32 patients. In a further study[9] it was shown that DGL was superior to a placebo in treating this clinical condition. However, several other clinical tests cast doubt on the clinical efficacy of DGL. There was no significant improvement in patients suffering from duodenal ulcer who took 380 mg of DGL 6 times a day, even though the clinical impression was that the tablets had an advantage over the placebo.[5] Other studies,[10] as well as a multicentre trial,[11] showed that the effect of DGL on duodenal ulcer was equivalent to that of a placebo.

## 2.3 Mode of Action Against Ulcers

The ulcer-healing mechanism of DGL is not well established. Animal studies have shown that DGL possesses an antispasmodic action. DGL suppressed constriction of an isolated rabbit ileum induced by acetylcholine, histamine and barium chloride.[12] In man it was observed[13] that DGL caused an antispasmodic effect in patients with duodenal ulcer. It was also found that in animals the anti-ulceration protection of DGL was not due to its effect on gastric acid secretion.[14,15] As prostaglandins can play an important role in protecting gastric mucosa from damage, the effect of DGL on prostanoid synthesis by rat gastric mucosa was tested.[16] DGL had no significant effect on prostaglandin production and little effect on 6-keto-PGF 1-$\alpha$, one of the prostaglandin metabolites. Thus the authors concluded that the anti-ulcer effect of DGL is via a non-prostaglandin mechanism.

## 2.4 Metabolism of Glycyrrhizin

One of the important characteristics of a digested drug is its metabolic pathway in the gastro-intestinal tract. Metabolic studies of a drug are needed in order to establish its pharmacological properties, toxicology and pharmacokinetics. Drugs like glycyrrhizin taken by mouth in tablet, capsule, infusion or decoction form are brought into direct contact with the gastro-intestinal flora, so assessing the effect of the human intestinal flora on the metabolism of glycyrrhizin is important in understanding its pharmacological properties. An experiment *in vitro*, using an intestinal bacterial mixture, was therefore conducted in order to track the metabolites of glycyrrhizin caused by intestinal bacteria. It was found[17] that glycyrrhizin was hydrolysed by human intestinal flora to the aglycone, 18$\beta$-glycyrrhetic acid, and to its sugar moiety. The aglycone was then transformed reversibly to 3-dehydro-18$\beta$-glycyrrhetic acid, which in turn was transformed reversibly to 3-epi-18$\beta$-glycyrrhetic acid. Several attempts were made[18] to isolate the specific bacteria responsible for these reactions. Many species of bacteria were found to possess $\beta$-glucuronidase activity, which may hydrolyse glycyrrhizin to glycyrrhetic acid. Only two types of bacteria were able to reduce 3-dehydro-18$\beta$-glycyrrhetic acid to glycyrrhetic acid and/or to 3-epi-18$\beta$-glycyrrhetic acid. *Ruminococcus* isolated from fresh human faeces was found to have the ability to hydrolyse glycyrrhizin to 18$\beta$-glycyrrhtenic acid. The reduction of 3-dehydro-18$\beta$-glycyrrhetic acid to the stereomeric form of 3-epi-18$\beta$-glycyrrhetic acid was found to be brought about by *Clostridium*, also isolated from fresh human faeces. The combination of these two bacteria can isomerize glycyrrhetic acid to 3-epi-18$\beta$-glycyrrhetic acid and vice versa, possibly via the oxidized intermediate of 3-dehydro-18$\beta$-glycyrrhetic acid. Glycyrrhizin is transformed to 3-epi-18$\beta$-glycyrrhetic acid through stages in which the final isomer is a product of several bacteria. The changes are summarized in the scheme glycyrrhizin$\rightleftharpoons$18$\beta$-glycyrrhetic acid$\rightleftharpoons$3-dehydro-18$\beta$-glycyrrhetic acid$\rightleftharpoons$ 3-epi-18$\beta$-glycyrrhetic acid.

The amounts of glycyrrhizin and glycyrrhetic acid were detected in plasma, using a high-speed liquid chromatography technique.[19] The concentrations of glycyrrhizin fell rapidly after an injection into the portal vein of rats. A sharp decrease in glycyrrhizin plasma level was observed during the first 60 min. Afterwards only a slight fall in the concentration of glycyrrhizin was observed, following a 12·5 mg/0·5 ml injection, resulting in a steady level of concentration in the plasma of about 0·4 $\mu$g/ml for a further 120 min. A 5 mg/0·5 ml injection resulted in a sharp decrease in glycyrrhizin within the first 60 min, and almost none was detected after 90 min. After oral administration of glycyrrhizin, glycyrrhetic acid was detectable in the plasma. Glycyrrhetic acid reached a peak after 30 min, and then declined over more than 240 min. The amount of glycyrrhizin increased moderately. During the initial 240 min, the concentration of glycyrrhetic acid was higher than that of glycyrrhizin; only at 240 min were their concentrations in the plasma almost similar. Owing to its higher molecular weight, it seems that glycyrrhizin is converted into glycyrrhetic acid in the intestine, and is absorbed from the small intestine in that form. Following intravenous administration of glycyrrhizin and glycyrrhetic acid to rats, most of the drug was found in the plasma and blood after 60 min. Other organs (e.g. lung, heart, stomach, small intestine) contained less glycyrrhizin, and no glycyrrhizin could be detected in the brain.[20] The highest concentration of glycyrrhetic acid was found in the plasma and in the blood, while smaller amounts were found in other organs, including the brain. The cumulative amounts of glycyrrhizin excreted 24 h˙ after intravenous administration were found to be 88·5% in the bile and only 4·83% in the urine. Only 0·36% of glycyrrhetic acid was found in the bile and 0·09% in the urine. These results suggest that glycyrrhizin and glycyrrhetic acid are largely bound to plasma protein. Enterohepatic circulation (i.e. the secretion and reabsorption cycle of the bile acids) is observed in the case of glycyrrhizin, and glycyrrhetic acid is almost completely metabolized.

## 2.5 Corticoid Action

Glycyrrhizin and glycyrrhetic acid both have a chemical structure resembling steroids. Thus they possess some mineralocorticoid and glucocorticoid properties, and can influence steroid metabolism. Mineralo-corticoids affect the electrolyte metabolism and retention of sodium and potassium balance in the human fluids, thus maintaining blood pressure and volume. Glucocorticoids affect carbohydrate and protein metabolism, maintaining blood glucose level and facilitating glycogen deposition. Glucocorticoids also elicit both anti-inflammatory and immuno-suppressant effects. It has been found that glycyrrhizin and glycyrrhetic acid bind to glucocorticoid and mineralocorticoid with moderate affinity. On the other hand, they bind with very weak affinity to oestrogen receptor, sex hormone-binding globulin and corticosteroid-binding globulin, and do not bind at all to progesterone receptors.[21,22] When binding to the glucocorticoid receptors, they may influence the corticosteroid activity via a receptor mechanism. They also may release free corticosteroids from corticoid-binding globulins. This, plus the inhibitory effect of cortisone on metabolic activity in the liver, may cause the effect of glycyrrhizin on corticoid activity.

## 2.6 Anti-inflammatory Effects

Although it has been reported that glycyrrhizin possesses anti-inflam-matory properties, the mechanism has not been fully established. Glycyrrhetinic acid, administered as a single dose 60 min prior to carrageenan, reduced the inflammatory effect of 0·1 ml of a 1% carrageenan suspension injected into the paw of a rat.[23] Glycyrrhetinic acid prevented the migration of leukocytes into a pleural space induced by dextran. However, the migration of polymorphs was hardly altered. Glycyrrhetinic acid did not suppress the release of prostaglandins and the biosynthesis of prostaglandins from rat peritoneal leukocytes. It also did not modify the contractions induced by prostaglandins on a guinea-pig isolated ileum. Several derivatives of glycyrrhizin were tested for their influence on the activity of cyclo-oxygenase, a compound which leads to the formation of prostaglandins.[24] Glycyrrhizin showed no significant inhibition, while glycyrrhetic acid demonstrated a moderate inhibitory effect of 26% in the cyclo-oxygenase activity. A glycyrrhizin derivative, the disodium salt of olean-11,13(18)-diene-3$\beta$,30-diol 3$\beta$,30-di-$O$-hemiphthalate, exhibited 71% inhibition. The possible effects of glycyrrhizin and glycyrrhetic acid were tested on the activity of 5-lipoxygenase, which is the first enzyme involved in the biosynthesis of several leukotrienes related to asthma, allergic diseases and inflammation. While glycyrrhizin had a minor effect on 5-lipoxygenase activity (14% inhibition), glycyrrhetic acid had a 43% inhibition. However, the disodium salt of olean-12-ene-3$\beta$,30-diol 3$\beta$,30-di-$O$-hemiphthalate completely inhibited the activity of 5-lipoxygenase. It appears that the hemiphthalate derivative of glycyrrhizin may have an important role in the inhibition of inflammatory processes.

## 2.7 Cell Biology

Glycyrrhetic acid was also found to have an effect on intercellular junctions, such as 'gap' junctions, which are aqueous channels that link between cells. Through these junctions a small molecular exchange can be conducted between two neighbouring cells. It was found that 18$\alpha$-glycyrrhetic acid was capable of inhibiting 95% of the intercellular junction communication.[25] The inhibition was reversible after washes with an albumin-containing medium. The rapid rate of junction recovery was attributed to the reopening of the existing junction but not to the formation of new junctions. It was possible to maintain the inhibition for 20 days. Related compounds of glycyrrhetic acid, such as 18$\beta$-glycyrrhetinic acid and carbenoxolone, also demonstrated an inhibition at the intercellular junctions. Although glycyrrhetic acid possesses a similar activity to mineralocorticoids and glucocorticoids, aldosterone spirolactase and glucocorticoids do not have the same inhibitory effect at the intercellular junctions.

In addition to its direct therapeutic properties, glycyrrhizin might also be used as an adjunct medicament to reduce the toxicity of other drugs. Although saponins possess great therapeutic advantages, they are poorly absorbed from the intestines, and when given intravenously they are toxic due to their haemolytic characteristics. Studies of saponin haemolysis *in vitro* in the presence of glycyrrhizin revealed that glycyrrhizin can inhibit saponin and sapogenin haemolysis.[26] The best anti-haemolysis results were obtained when glycyrrhizin was pre-incubated with erythrocytes prior to the addition of the saponins. The anti-haemolytic action of glycyrrhizin can be attributed to its absorption by the surface of erythrocytes, thus preventing access of the haemolysin to its receptor. The use *in vivo* of glycyrrhizin as an anti-haemolytic agent is of doubtful value because of the large amounts necessary in order to achieve the same results as those shown *in vitro*.

## 2.8 Toxic Effects

Several studies and case reports have been published on the side-effects of liquorice and glycyrrhizin.

Hypokalemia myopathy with myoglobinuria developed in a woman who consumed 30–40 g per day of liquorice for a period of 9 months.[27] Her serum potassium concentration was 1·6 mmol/litre and chloride concentration 68 mmol/litre. Although the patient was also receiving 50 mg of hydrochlorothiazide 3 times weekly, it is unlikely that it caused the potassium depletion. Her clinical treatment, which included an intravenous infusion of potassium chloride, resulted in a clear improvement. There is a case report of a 70-year-old woman who had been using a liquorice mixture as a laxative for 2–3 years.[28] It was calculated that she ingested about 94–141 mg of the calcium and potassium salts of glycyrrhizic acid. Potassium concentration in her serum was 1·1 mmol/litre and the electrocardiogram showed a change typical of severe hypokalaemia. The treatment consisted of intravenous potassium supplemented with spironolactone. The authors attributed the hypokalaemia to her unusual sensitivity to liquorice.

Hypertension was observed in a 53-year-old man after consuming 700 g of liquorice candy over 9 days.[29] His electrocardiogram indicated hypokalaemia; 2 g of sodium diet without any other medication noticeably improved his clinical condition. Hyperprolactinaemia was observed in a 22-year-old female who admitted to consuming excessive amounts of liquorice in the preceding few years;[30] 6 months after the liquorice was withdrawn from her diet her hormone levels returned to normal. Diarrhoea and unpleasant taste in the mouth have been reported in clinical trials of patients with gastric ulcer treated with DGL.[31] In other clinical tests using DGL in patients suffering from ulcers, no side-effects were observed even after administration of the drug for several weeks.

## 3 EFFECT OF GLYCYRRHIZIN ON DENTAL CARIES AND CARIOGENIC BACTERIA

### 3.1 Role of Oral Bacteria in Dental Caries

Epidemiological studies have shown that after human populations introduce sucrose into their diet they exhibit a significant increase in the incidence and prevalence of dental caries.[32,33] A restriction in sucrose availability, like that experienced during World War II, led to a drop in caries prevalence among European and Japanese populations.[34,35] It was through animal experimentation that the cariogenicity of sucrose compared with other dietary components was confirmed. Such techniques also led to the discovery that dental caries result from an endogenic infection caused by *Streptococcus mutans* and other oral bacteria, and it is sucrose-dependent.[36] The pathogenic properties of *S. mutans* in relation to sucrose are determined by acid production and bacterial adherence.

### 3.2 Acid Production

As early as 1890 Miller[37] and others suggested that *in vivo* bacterial acid production could be responsible for dental decay. Recent bacteriological findings suggest that bacterial deposits taken from caries-active areas of the teeth have higher proportions of *S. mutans* than caries-free areas.[38] Furthermore, other studies have shown that such caries-active bacterial plaques are significantly more acidogenic than caries-free plaques. It has also been shown that *S. mutans* is capable of converting sucrose into acids much more efficiently than other oral bacteria.[39]

### 3.3 Bacterial Adherence

A major pathogenic property of these bacteria is their ability to accumulate and adhere to tooth surfaces. Several possible mechanisms for this important initial phase of caries production might include: (a) the presence of a bacteria-free film, namely the acquired pellicle on tooth surfaces, which may initiate bacterial adherence; (b) reversible adherence through non-specific properties of bacterial envelope structure and chemical composition; (c) non-reversible adherence moderated by extracellular polysaccharides produced enzymatically from sucrose by specific cariogenic bacteria.

There are two stages involved in the adherence of *S. mutans* to surfaces: an initial period of reversible non-specific binding and a second irreversible stage. Several characteristics of *S. mutans* plus many additional mediators play a role in the adherence and accumulation process. Although the role of extracellular polysaccharides in this respect is not fully understood,[40-44] insoluble glucans were found to facilitate the formation of plaque, and thus promote cariogenic activity.

Glucosyltransferases (GFT) are extracellular enzymes of *S. mutans* which produce various glucans. It has been suggested that the formation of the insoluble glucans is a result of a sequence of enzymic activities.[45] One stage in this is the polymerization of the glucosyl moiety of the sucrose to a soluble glucan (SG) by the enzyme GTF-SG. A further stage is the formation of insoluble sticky glucans (ISG) produced by another enzyme (GTF-ISG).

Although the addition of preformed soluble glucans to a GTF-ISG preparation does not cause the formation of sticky glucans, it has been suggested that the activity of GTF-SG is essential for the production of the insoluble glucan.[46] Inhibition of any one of these enzymic processes will therefore interfere with the production and accumulation of glucans, and may thus partially prevent dental plaque formation.

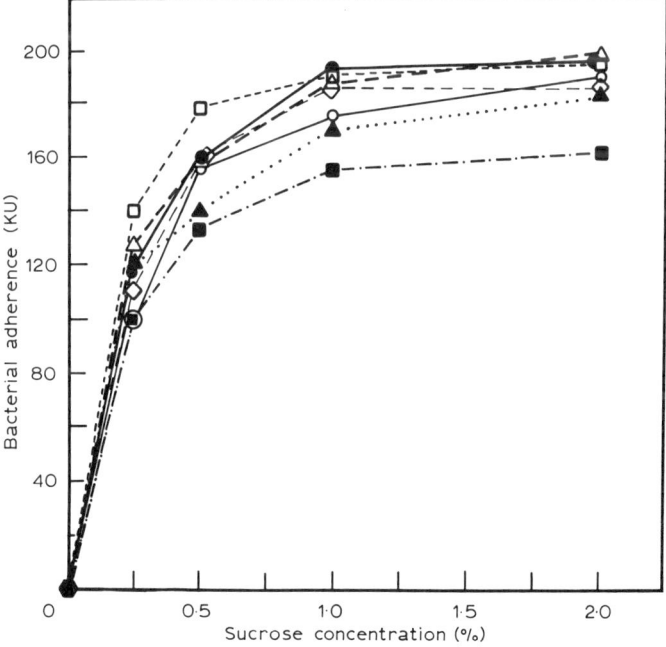

FIG. 2. Effect of sucrose concentration on adherence of *S. mutans*, measured in Klett units (KU): ●, *S. mutans* 6715; ▲, *S. mutans* 5608; □, *S. mutans* 1895; ○, *S. mutans* ZAHT; △, *S. mutans* SB25; ◇, *S. mutans* SL-1; ■, *S. mutans* NS-XIII.

### 3.4 Effect of Glycyrrhizin on Bacterial Growth and Adherence

An examination of the effect of glycyrrhizin as well as liquorice powder, *in vitro*, on three oral microorganisms—*Streptococci* (*mutans* and *sanguis*), *Actinomyces* (*viscosus* and *naeslundii*) and *Bacteroides* (*matruchotii*)—has demonstrated a significant reduction both in growth and in acid production by each strain metabolizing liquorice or glycyrrhizin alone as compared with sucrose, glucose or fructose alone.[47] Glycyrrhizin compounds were found to reduce the metabolism of sucrose, glucose and fructose effectively. The authors conclude that glycyrrhizin and liquorice powder are minimally fermentable and act as potent inhibitors of carbohydrate metabolism.

The role of glycyrrhizin in growth and plaque formation by *S. mutans* has been examined *in vitro*.[48] Seven strains of *S. mutans* which showed a strong tendency to adhere to glass in the presence of sucrose were studied (Fig. 2). While growth of *S. mutans* was hardly affected by glycyrrhizin in concentrations as high as 1%, the adherence of the bacteria (in the presence

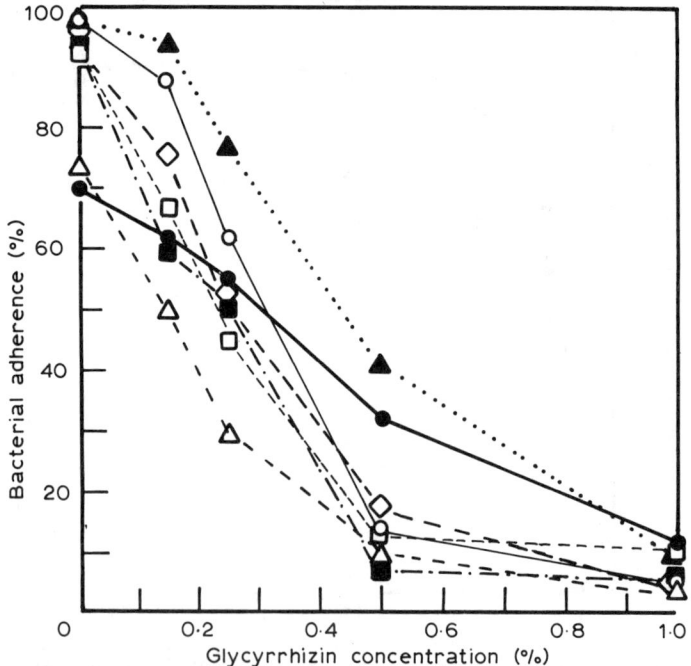

FIG. 3.    Effect of glycyrrhizin on adherence of *S. mutans* in presence of 2% sucrose. Key as Fig. 2.

of 2% sucrose) was greatly inhibited. At high glycyrrhizin concentrations (0·5–1%) adherence was almost completely lost (Fig. 3).

It was found that glycyrrhizin did not enhance the growth of S. *mutans*, nor did it induce plaque formation. However, the inhibition of bacterial adherence to a smooth surface in a sucrose-containing medium is outstanding. The results suggest that, when used orally, glycyrrhizin will not affect the delicate balance of the oral bacterial flora. As the adherence of S. *mutans* to teeth is essential for caries development, the fact that glycyrrhizin inhibits dental plaque formation of cariogenic bacteria in the presence of sucrose is of great interest. The anti-adherent effect may be due to an enzymic inhibition of GTF. This possibility was tested by examining the direct effect of glycyrrhizin on sticky polysaccharides produced by GTF prepared from a cariogenic strain of S. *mutans* (6715).[49]

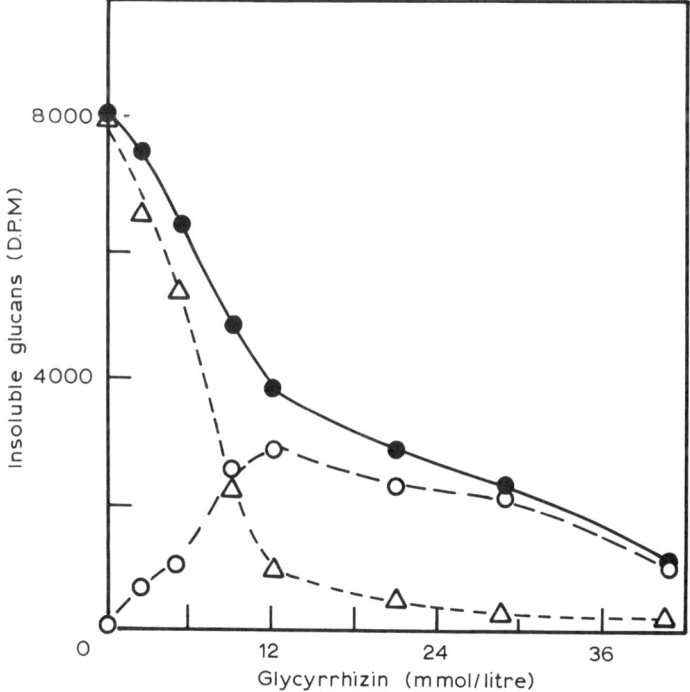

FIG. 4.    Effect of glycyrrhizin on the formation of insoluble glucans by crude GTF. The amount of glucans was determined by measuring incorporated [14]C-glucose: ●, total glucans; △, adhered glucans; ○, non-adhered glucans.

### 3.5 Effect of Glycyrrhizin on the Production of Insoluble Glucans

The addition of glycyrrhizin to the incubation mixture of GTF supplemented with sucrose resulted in three major effects, summarized in Fig. 4. (1) The amount of total 'adherent glucans' decreased as the glycyrrhizin concentrations increased; at concentrations exceeding 12 mmol/litre no 'adhered glucans' could be detected. (2) Depending on the glycyrrhizin concentration used, an insoluble glucan which lacked the ability to adhere to the test-tube was obtained, appearing as a flocculated suspension. The amount of the 'non-adherent glucans' produced rose with increasing glycyrrhizin concentrations, reached a maximum level at 12 mmol/litre glycyrrhizin, and then declined. (3) The total amount of the insoluble glucans (adherent and non-adherent) produced by the crude GTF was reduced in the presence of glycyrrhizin. The extent of reduction was 85% at a glycyrrhizin concentration of 40 mmol/litre. The inhibitory effect

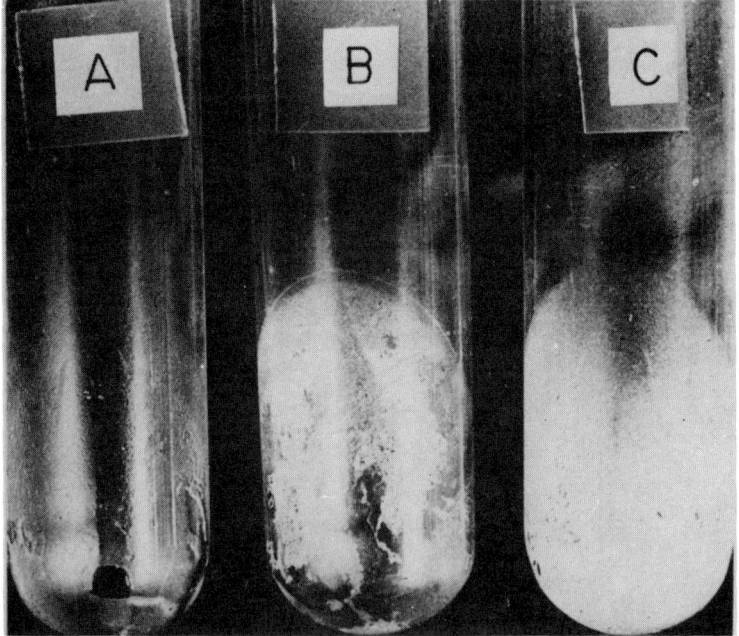

FIG. 5.    Effect of increasing concentrations of glycyrrhizin on the formation of adherent glucan: A, 5·2 mmol/litre; B, 2·6 mmol/litre; C, control without glycyrrhizin.

of two different concentrations of glycyrrhizin (5·2 and 2·6 mmol/litre) compared with a control (no glycyrrhizin added) can be seen in Fig. 5.

### 3.6 Inhibition of Soluble Glucan Production by Glycyrrhizin

In order to examine whether glycyrrhizin can affect the formation of soluble glucans (SG) as well, the supernatant of the incubation mixture, which contained crude GTF, sucrose and glycyrrhizin, was examined using paper chromatography. Since the rate of migration of the glucan is higher the lower the molecular weight of the oligosaccharides, the appearance of the spots on the chromatogram may indicate their degree of polymerization. In Fig. 6 a definite spot of stained carbohydrate may be detected at the starting line of the chromatogram in the absence of glycyrrhizin and also at low concentrations, while no such spots could be detected when

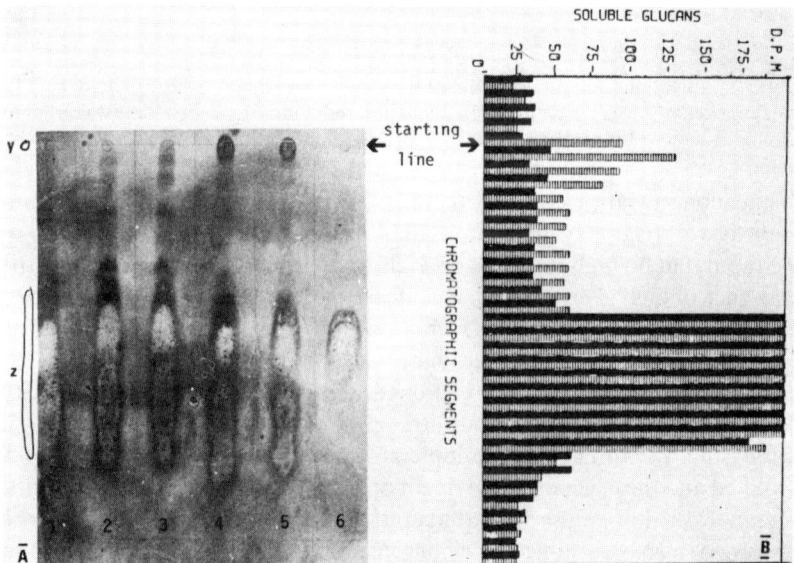

FIG. 6. (A) Migration of soluble glucans, labelled with $^{14}C$, on paper chromatography. The soluble glucans were obtained after incubation of crude GTF for 24 h with sucrose in the presence of the following concentrations (mmol/litre) of glycyrrhizin: (1) 40, (2) 20·1, (3) 9, (4) 2·6, (5) no glycyrrhizin, (6) control (no crude GTF). The presence of high molecular weight glucans was observed as a dot on the starting line of the paper (y = high-molecular-weight polymer; z = low-molecular-weight oligomers). (B) Distribution of radioactive spots of two incubation mixtures, 40 mmol/litre glycyrrhizin (dark bands) and no glycyrrhizin (light bands), measured at each chromatographic segment.

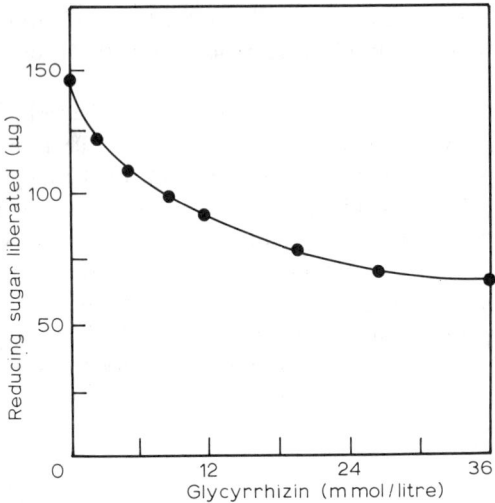

Fig. 7.    Effect of different concentrations of glycyrrhizin on GTF-SG activity. The enzyme activity was determined by liberating reducing sugar. Fructose was used as standard.

higher concentrations were used. Since low mobility on the chromatogram indicates a relatively high molecular weight of the glucan, it may be deduced that no high-molecular-weight soluble glucans were formed in the presence of high glycyrrhizin concentrations. These results were confirmed by examining the distribution of radioactivity on the chromatogram. The starting line of the chromatogram of the supernatant from the incubation mixture without glycyrrhizin exhibited about 3 times more radioactivity than that containing 40 mmol/litre glycyrrhizin. The enzyme fraction catalysing the formation of soluble glucans from sucrose (GTF-SG) was isolated and the effect of glycyrrhizin on its activity was tested. The results, summarized in Fig. 7, show that liberation of reducing sugars from sucrose by glycyrrhizin was subject to a concentration-dependent inhibition (up to 60 mmol/litre). Higher glycyrrhizin concentrations had only a moderate additional effect.

### 3.7  Effect of Surfactants on Crude GTF Activity

Since glycyrrhizin possesses strong surface activity, the effects of two other surfactants, cetomacrogol-1000 and gypsophila saponin, were tested on glucan production by crude GTF. The addition of different concentrations of these did not have much inhibitory effect on the amount of 'adherent glucans' produced. Furthermore, there was no change in the nature of the soluble glucans on examining the paper chromatograms of the reaction mixtures.

The mechanism by which glycyrrhizin may inhibit bacterial adherence is not fully understood. One possible explanation is that its surface activity is involved, as surfactants are known to affect bacterial growth and adherence.[50] However, this is unlikely since the response curves for adherence do not show any correlation to the critical micelle concentration (CMC) of this material. The CMC of glycyrrhizin is between $1.5 \times 10^{-4}$M and $4 \times 10^{-4}$M ($0.012$–$0.05\%$) in the pH range $4.2$–$5.5$, i.e. inhibition occurs only at concentrations far above the CMC. Furthermore, preliminary experiments indicate that glycyrrhizin does not dissolve plaque which has already adhered to a smooth surface.

An experimental salted liquorice with a low sugar alcohol content has been developed, its mono/disaccharide content being less than $0.1\%$.[51] As salted liquorice does not have an intensely sweet taste, small amounts of sodium-saccharin were added. Gum arabic, potato starch derivatives, soy flour, animal protein derivatives and sorbitol were added as well. The acidogenic properties of this experimental non-sugar liquorice were tested *in vitro* by means of *S. mutans* fermentation, in pooled plaque saliva mixtures, and *in vivo* by means of pH telemetry. The sugarless liquorice showed a 54% fermentability compared with a glucose control in the *S. mutans* cultures. In the plaque saliva suspension, its fermentability reached 68% compared with sugar liquorice. The authors concluded that acid production during and following consumption of their experimental liquorice was too high to be safe for teeth, probably owing to the starch content of the mixture.

The effect of several antibacterial and anti-inflammatory agents on ligature-induced gingival inflammation in beagle dogs has been studied.[52] A mixture of glycyrrhetinic acid with a hydrophilic emulsion ointment at concentrations of $0.1\%$, $0.3\%$ and $0.5\%$ (w/w) was applied to the gingiva following the induction of gingival inflammation. Acute inflammation was significantly reduced following application of the prepared mixture for 10 days.

The literature contains few studies on salivary and plaque pH following liquorice and liquorice preparation consumption. Workers who examined the effect of chewing liquorice root ('sweet wood') on salivary and plaque pH suggested that substrates for glycolysis were in a form that did not lead to a sharp fall in plaque pH.[53] Chewing liquorice root after a sucrose rinse led to a rise in plaque pH, attributed by the authors to salivary secretion, but it may also result from the buffering and anti-glycolytic actions of glycyrrhizinic acid.

The use of various chemicals for the control of dental plaque accumulation has been extensively studied.[54–56] Glycyrrhizin, the sweet component of liquorice, is the sweetest substance on the US Food and Drug Administration list of natural flavours generally accepted as safe, and is about 50–100 times sweeter than sucrose.[57]

The facts that this natural compound has a palatable taste and an ability to form stable gels in aqueous solutions support the potential application of glycyrrhizin as a suitable anti-cariogenic agent.

### 3.8 Effect of Glycyrrhizin on Plaque Formation *in vivo*

A preliminary clinical trial has been conducted on the effect of glycyrrhizin on plaque formation in humans.[58] The experiment, conducted in the clinics of the Hebrew University Hadassah School of Dental Medicine, involved 21 dental school students (approx. 1:3 male:female ratio). At the beginning of the experiment (morning of day 1) participants brushed and flossed their teeth thoroughly under supervision, until an examination by erythrosine disclosing solution (Red-Cote Butler, USA) confirmed that all teeth had plaque-free surfaces. The mouth was divided randomly into test and control sides. Subjects were directed to discontinue all oral hygiene procedures during the experimental period, and no dietary restrictions were imposed. All 21 students participated in the first 3 days of the study and 5 continued for an additional fourth day.

The experimental procedure was repeated every morning and noon of the study period. The teeth were dried with an air syringe, then cotton rolls and a saliva ejector were carefully placed in the mouth (without disturbing the plaque), and the upper and lower central incisors were isolated from each other using a celluloid strip. The test compound, 1 ml glycyrrhizin gel (Fluka AE, Buchs, Switzerland; 2% glycyrrhizin in 0·01 M acetate buffer, pH 5·0), was applied using a 1-ml syringe, on the buccal, lingual and occlusal surfaces of teeth on the experimental side of the mouth. The same procedure was repeated on the control side using 0·01 M acetate buffer alone. Subjects were instructed and supervised to remain seated for 10 min with the substances in their mouths, and then to expectorate and thoroughly rinse their mouths with water.

The two sides of the mouths were compared by two independent examiners at the end of the third day. The double blind procedure, using erythrosine solution, was similar to that previously employed.[59] Buccal surfaces of the upper and lower central incisors, canines and second

## TABLE 1

PLAQUE SCORES OF TEETH ON EXPERIMENTAL AND CONTROL SIDES OF MOUTHS IN 21
SUBJECTS AFTER 3 DAYS OF GLYCYRRHIZIN GEL APPLICATION

|  | Experimental side | | Control side | |
|---|---|---|---|---|
|  | Upper arch | Lower arch | Upper arch | Lower arch |
| Teeth number | 1  3  5 | 1  3  5 | 1  3  5 | 1  3  5 |
| Scores | 1[a]  9  9 | 5  8  2 | 11[a]  7  6 | 6  9  7 |
| Total arch | 19 | 15 | 24 | 22 |
| Total side | 34[b] | | 46[b] | |

[a] Significance of difference $p = 0.0019$.
[b] Significance of difference $p = 0.089$.

premolars were examined. The amount and thickness of plaque were
compared for each two contralateral teeth. Five subjects who volunteered
to continue the experiment for an extra day were examined on the fourth
afternoon.

For each contralateral pair, the tooth with the larger amount of plaque
was scored at 1. When both teeth had the same plaque level they were
assigned a score of 0 each. Average scores of the two examiners were
calculated. The Wilcoxon matched-pairs signed-ranks test[60] was employed
to compare the two sides of the mouth. When individual test and control
teeth were compared, the McNemar sign test[60] was used.

Table 1 demonstrates the results obtained after 3 days of glycyrrhizin
application. When comparing individual teeth, a highly significant ($p =
0.0019$) difference was found between upper central incisors (teeth 11 and
21) on the experimental side (plaque score = 1) versus the control side
(plaque score = 11). It should be noted that these teeth benefited from
optimal retention of the gel during the application. Comparisons between
other pairs of teeth failed to reveal significant differences. Comparing all
upper teeth examined, a score of 19 was found on the test side versus 24 on
the control side. A similar trend was revealed for lower teeth with a score of
15 on the test side and 22 on the control side. For all teeth on the
experimental side the total score was 34, compared with 46 on the control,
with a tendency towards statistical significance at a level of $p = 0.089$.

The results after 4 days are given in Table 2. The quantitative difference
between scores for all teeth on the experimental versus the control sides (4
and 18 respectively) was greater than that found in the 3-day experiment.
This suggests the importance of the time factor in glycyrrhizin application

TABLE 2

PLAQUE SCORES OF TEETH ON EXPERIMENTAL AND CONTROL SIDES OF MOUTHS IN 5
SUBJECTS AFTER 4 DAYS OF GLYCYRRHIZIN GEL APPLICATION

|  | Experimental side | | Control side | |
|  | Upper arch | Lower arch | Upper arch | Lower arch |
| --- | --- | --- | --- | --- |
| Teeth number | 1  3  5 | 1  3  5 | 1  3  5 | 1  3  5 |
| Scores | 2  1  0 | 0  1  0 | 3  3  2 | 4  4  2 |
| Total arch | 3 | 1 | 8 | 10 |
| Total side | | 4[a] | | 18[a] |

[a] Significance of difference $p = 0.094$.

on the reduction of dental plaque accumulation. Owing to the small group of subjects ($N = 5$), the level of significance was $p = 0.094$, but this preliminary study does indicate the potential of glycyrrhizin in preventive dentistry in combating plaque formation.

### 3.9  Effect of Glycyrrhizin on Enamel

The physicochemical characteristics of glycyrrhizin gel, together with its inhibitory effect on plaque formation by *S. mutans*, stimulated a study of the effect of this surfactant gel on fluoride uptake by tooth enamel.[61] This followed earlier studies which demonstrated that surface-active agents may affect fluoride penetration into enamel.[62]

The effect of fluoride application was evaluated *in vitro* by measuring fluoride uptake by the surface of human enamel. Fluoride uptake in the outer enamel layer was highest in specimens treated with acidulated NaF solution containing 0·5% glycyrrhizin. The authors also examined the effect of glycyrrhizin and fluoride on enamel decalcification. They suggest that a mixture of 0·5% NaF, 0·1M $H_3PO_4$ and 0·5% glycyrrhizin preserved significantly greater areas of enamel than controls (NaF alone). These results, in which enamel was exposed to fluoride mixtures for 10 days, complement work[63] (see above) which also showed that glycyrrhizin alone is capable of protecting enamel from continued demineralization owing to its anti-glycolytic and buffering capacities. However, it has been suggested that the surface enamel remains relatively intact owing to a coating layer produced by glycyrrhizin, thus reducing the diffusion of calcium and phosphate ions during incubation in acid media.[61] The same authors also point out that glycyrrhizin increased fluoride uptake to some extent but did not inhibit fluoride release in the decalcifying media. They suggest, therefore, that the demineralization protection demonstrated in their study is due to an additive effect of glycyrrhizin and fluoride.

The relative cariogenicity of different types of sweets has been studied by tests of enamel solubility *in vitro*,[64] and by the fall in plaque pH *in vivo*.[65] These studies and others have shown that liquorice-containing sweets cause less enamel dissolution and a smaller decline in plaque pH than most other types of sweets.

The reduction in enamel dissolution by a major water-soluble constituent of liquorice—glycyrrhizinic acid (obtained as the hydrated ammonium salt)—has been studied.[63] An aqueous extract of liquorice root was also tested, as well as extracts of liquorice confections, and they were found to reduce the dissolution of enamel in an acidic buffer system and in incubation with saliva and glucose. These effects may be attributed to the glycyrrhizinic acid, which was found to possess solubility-reducing, glycolysis-inhibiting and buffering properties.

## 4 EFFECT OF GLYCYRRHIZIN ON ORAL ULCERS AND VIRAL INFECTIONS

Several other studies demonstrate additional properties of glycyrrhizin which may indicate further oral use. The role of glycyrrhizin as a vehicle for the application of triamcylolone in the treatment of recurrent aphthous stomatitis was studied in 134 patients.[66] Preparations of glycyrrhizin alone and of 0·1% triamcylolone in glycyrrhizin were applied topically to the ulcers, and clinical signs and symptoms were monitored continually in comparison with untreated ulcers. The application of triamcylolone in glycyrrhizin caused a marked improvement in lesion healing. Furthermore, treatment by glycyrrhizin alone caused a reduction in symptoms of the ulcers. The authors suggest that the effect of glycyrrhizin may be attributed to its mild adrenocorticoid and anti-inflammatory properties, due to the $\alpha\beta$-unsaturated carbonyl function present in the molecule. The high level of effectiveness of the triamcylolone and glycyrrhizin mixture is thought to be (a) a result of synergism and (b) due to the saponin properties of glycyrrhizin, which may promote penetration of triamcylolone into the oral mucosa.

The clinical effect of glycyrrhizin gel as a vehicle for idoxuridine-1 (IDu) was also examined.[67] The rationale is based on the acceptance by the FDA in the USA of glycyrrhizin for food and drug use. Aqueous glycyrrhizin solutions form stable gels adequate for the topical delivery of drugs.

Furthermore, glycyrrhizin was found to inhibit *in vitro* growth of viruses.[68,69] A hydrophilic preparation containing 0·2% w/w IDu in glycyrrhizin was prepared and applied to herpes labialis and herpes nasalis lesions, both in the vesiculation and the prodromal phases, and the time to crust formation was recorded. Glycyrrhizin alone and commercial IDu ointment served as controls. While the duration of untreated lesions crust formation was 8–14 days, glycyrrhizin itself reduced crust formation to 8–10 days and 6–8 days in the vesicle and prodromal stages respectively. Application of the mixture of glycyrrhizin and IDu was even more effective, and the time to crust formation was significantly shorter than that experienced without any treatment. All patients reported relief from pain within 1 h of application of the preparation, irrespective of the stage or severity of their illness. The authors suggest that the vehicle itself, i.e. glycyrrhizin, is responsible for the fast relief from pain following application of the formulation.

## 5 CONCLUDING REMARKS

Liquorice and glycyrrhizin have been used for centuries as medicaments for different purposes. In the modern pharmaceutical industry glycyrrhizin has been used mainly on account of its palatable taste, as a sweet substitute for sugars. However, several interesting and important medicinal properties attributed to glycyrrhizin have been demonstrated. It has become clear that glycyrrhizin is effective in treating gastric and duodenal ulcers, but its mode of action as an anti-ulcer or anti-inflammatory agent has not been well established. Recent publications suggest that liquorice and its derivatives can act as anti-cariogenic agents by interfering with the adhesion of cariogenic bacteria to the tooth, and through a remineralization action. Glycyrrhizin may also intervene with the enzymic metabolic pathways of bacteria. The toxic side-effects of liquorice are due to its mineralocorticoid-like activity, and the consumption of large amounts of liquorice as a candy or as a medicament may lead to hypokalaemia, hypertension or other mineralocorticoid side-effects. Despite its disadvantages, however, glycyrrhizin possesses several promising medicinal properties which are under active investigation.

## REFERENCES

1. Fuggersberger-Heinz, R. & Franz, G., Formation of glycyrrhizinic acid in *Glycyrrhiza glabra* var. *typica*. *Planta Medica*, **50** (1984) 409–13.

2. Takino, Y., Koshioka, M., Shiokawa, M., Ishii, Y., Maruyama, S., Higashino, M. & Hayashi, T., Quantitative determination of glycyrrhizic acid in liquorice roots and extracts by TLC densitometry. *Planta Medica*, **36** (1979) 74–8.

3. Killacky, J., Ross, M. S. F. & Turner, T. D., The determination of glycyrrhetinic acid in liquorice by high-pressure liquid chromatography. *Planta Medica*, **30** (1976) 310–16.

4. Turpie, A. G. G., Runcie, J. & Thomson, T. J., Clinical trial of deglycyrrhizinized liquorice in gastric ulcer. *Gut*, **10** (1969) 299–302.

5. Russell, R. I. & Dickie, J. E. N., Clinical trial of a deglycyrrhizinised liquorice preparation in peptic ulcer. *J. Therap. Clin. Res.*, **2** (1968) 2–5.

6. Bardhan, K. D., Cumberland, D. C., Dixon, R. A. & Holdsworth, C. D., Clinical trial of deglycyrrhizinised liquorice in gastric ulcer. *Gut*, **19** (1978) 779–82.

7. Glick, L., Deglycyrrhizinated liquorice for peptic ulcer. *Lancet*, **ii** (1982) 817.

8. Larkworthy, W. & Holgate, P. F. L., Deglycyrrhizinized liquorice in the treatment of chronic duodenal ulcer. *Practitioner*, **215** (1975) 787–92.

9. Mills, D. H. & Damrau, F., Deglycyrrhizinized glycyrrhiza in treatment of peptic ulcer. *Med. Digest.*, **4** (1969) 36–44.

10. Feldman, H. & Gilat, T., A trial of deglycyrrhizinated liquorice in the treatment of duodenal ulcer. *Gut*, **12** (1971) 499–51.

11. Multicentre Trial, Treatment of duodenal ulcer with glycyrrhizinic acid-reduced liquorice. *Br. Med. J.*, **3** (1971) 501–3.

12. Aarsen, P. N. & von Noordwij, K. J. Comparison of the spasmolytic activity of succus liquiritiae in treated and untreated form. *Therapeutisch Umschau*, **20** (1963) 302–5.

13. Tewari, S. N. & Trembalowicz, F. C., Some experience with deglycyrrhizinated liquorice in the treatment of gastric and duodenal ulcers with special reference to its spasmolytic effect. *Gut*, **9** (1968) 48–51.

14. Aarsen, P. N., Standardization method of deglycyrrhizinized liquorice on experimental gastric ulcers in rats. *Arzneimittel-Forschung*, **23** (1973) 1346–8.

15. Andersson, S., Barany, F., Caboclo, J. L. F. & Mizuno, T., Protective action of deglycyrrhizinized liquorice on the occurrence of stomach ulcers in pylorusligated rats. *Scand. J. Gastroenterol.*, **6** (1971) 683–6.

16. Bennet, A., Melhuish, P. B. & Stamford, I. F., Carbenoxolone and deglycyrrhized liquorice have little or no effect on prostanoid synthesis by rat gastric mucosa *ex vivo*. *Int. J. Pharmac.*, **86** (1985) 693–5.

17. Hattori, M., Sakamoto, T., Kobashi, K. & Namba, T., Metabolism of glycyrrhizin by human intestinal flora. *Planta Medica*, **48** (1983) 38–42.

18. Hattori, M., Sakamoto, T., Yamagishi, T., Sakamoto, K., Konishi, K., Kobashi, K. & Namba, T., Metabolism of glycyrrhizin by human intestinal flora. II. Isolation and characterization of human intestinal bacteria capable of metabolizing glycyrrhizin and related compounds. *Chem. Pharm. Bull.*, **33** (1985) 210–17.

19. Sakiya, Y., Adada, Y., Kawano, S. & Miyauchi, Y., Rapid estimation of glycyrrhizin and glycyrrhetinic acid in plasma by high-speed liquid chromatography. *Chem. Pharm. Bull.*, **27** (1979) 1125–9.

20. Ichikawa, T., Ishida, S., Sakiya, Y. & Akada, Y., High-performance liquid chromatographic determination of glycyrrhizin and glycyrrhetinic acid in biological materials. *Chem. Pharm. Bull.*, **32** (1984) 3734–8.

21. Tamaya, T., Sato, S. & Okada, H. H., Possible mechanism of steroid action of the plant herb extracts glycyrrhizin, glycyrrhetinic acid, and paeoniflorin: inhibition by plant herb extracts of steroid protein binding in the rabbit. *Am. J. Obstet. Gynecol.*, **155** (1986) 1134–9.

22. Tamaya, T., Sato, S. & Okada, H., Inhibition by plant herb extracts of steroid bindings in uterus, liver and serum of the rabbit. *Acta Obstet. Gynecol. Scand.*, **65** (1986) 829–42.

23. Capasso, F., Mascolo, N., Autore, G. & Duraccio, M. R., Glycyrrhetinic acid, leukocytes and prostaglandins. *J. Pharm. Pharmacol.*, **35** (1983) 332–5.

24. Inoue, E., Saito, H., Koshihara, Y. & Murota, S., Inhibitory effect of glycyrrhetinic acid derivatives on lipoxygenase and prostaglandin synthetase. *Chem. Pharm. Bull.*, **34** (1986) 897–901.

25. Davidson, J. S., Baumgarten, I. M. & Harley, E. H., Reversible inhibition of intercellular junctional communication by glycyrrhetinic acid. *Biochem. Biophys. Res. Comm.*, **134** (1986) 29–36.

26. Segal, R., Milo-Goldzweig, I., Kaplan, G. & Weisenberg, E., The protective action of glycyrrhizin against saponin toxicity. *Biochem. Pharm.*, **26** (1977) 643–5.

27. Gross, E. G., Dexter, J. D. & Roth, R. G., Hypokalemic myopathy with myoglobinuria associated with licorice ingestion. *New Eng. J. Med.*, **274** (1966) 602–6.

28. Cumming, A. M. M., Brown, J. J., Lever, A. F., Boddy, K., Fraser, R., Padfield, P. L. & Robertson, J. I. S., Severe hypokalaemia with paralysis induced by small doses of liquorice. *Postgrad. Med. J.*, **56** (1980) 526–9.

29. Chamberlain, T. J., Licorice poisoning, pseudoaldosteronism, and heart failure. *J. American Medical Assoc.*, **213** (1970) 1343.

30. Werner, S., Brismar, K. & Olsson, S., Hyperprolactinaemia and liquorice. *Lancet*, **i** (1979) 1319.

31. Wilson, J. A. C., A comparison of carbenoxolone sodium and deglycyrrhizinated liquorice in the treatment of gastric ulcer in the ambulant patient. *Br. J. Clin. Pract.*, **26** (1972) 563–6.

32. Hardwick, J. L., The incidence and distribution of caries throughout the ages in relation to the Englishman's diet. *Br. Dent. J.*, **108** (1960) 9–17.

33. Keene, H. J., History of dental caries in human populations: the first million years. In *Symposium on Animal Models in Cariology*, ed. J. M. Tanzer (Sp. Suppl. *Microbiology Abstracts*), 1981, p. 23.

34. Takenchi, M., Epidemiological study on dental caries in Japanese children before, during and after World War II. *Int. Dent. J.*, **11** (1961) 443.

35. Toverud, G., Rubal, L. & Wiehl, D. G., The influence of war and post-war conditions on the teeth of Norwegian school children. IV. Caries in specific surfaces of the permanent teeth. *Millbank Mem. Fund Qr*, **39** (1961) 489.

36. Keyes, P. H., Research in dental caries. *J. American Dental Assoc.*, **76** (1968) 1356–73.

37. Miller, W. D., *The Microorganisms of the Human Mouth*. S. S. White Manufacturing Co., Philadelphia, 1890.

38. Loesche, W. J., The rationale for caries prevention through the use of sugar substitutes. *Int. Dent. J.*, **35** (1985) 1–8.

39. Minah, G. E. & Loesche, W. J., Sucrose metabolism by prominent members of the flora isolated from cariogenic and non-cariogenic dental plaques. *Infect. Immun.*, **17** (1977) 55–61.

40. Gibbons, R. J. & Van Houte, J., Bacterial adherence in oral microbial ecology. *Ann. Rev. Microbiol.*, **29** (1975) 19–44.

41. Gibbons, R. J., Microbial ecology adherent interactions which may affect microbial ecology in the mouth. *J. Dent. Res.*, **63** (1984) 378–85.

42. Mukasa, H. & Slade, H. D., Mechanism of adherence of *Streptococcus mutans* to smooth surfaces. III. Purification and properties of the enzyme complex responsible for adherence. *Infect. Immun.*, **10** (1974) 1135–40.

43. Torii, M. & Hamada, S., *In vitro* inhibition of glucosyltransferase activity by surfactants. *IADR 59*, No. 316 (1980).

44. Hamada, S., Role of glucosyltransferase and glucan in bacterial aggregation and adherence to smooth surfaces. In *Proceedings Glucosyltransferases, Glucans, Sucrose and Dental Caries*, ed. R. S. Doyle & J. E. Ciardi (Sp. Suppl. *Chemical Senses*), 1983, pp. 31–8.

45. Hamada, S. & Slade, H. D., Biology, immunology and cariogenicity of *Streptococcus mutans*. *Microbiol. Rev.*, **44** (1980) 331–84.

46. Hamada, S., Koga, T. & Ooshima, T., Virulence factors of *Streptococcus mutans* and dental caries prevention. *J. Dent. Res.*, **63** (1984) 407–11.

47. Berry, C. W. & Henry, C. A., Influence of glycyrrhizin on the metabolic activity of oral bacteria. *IADR Abstracts 63*, Special Issue No. 373 (1984).

48. Segal, R., Pisanty, S., Wormser, R., Azaz, E. & Sela, M. N., Anticariogenic activity of licorice and glycyrrhizin. I. Inhibition of *in vitro* plaque formation by *Streptococcus mutans*. *J. Pharm. Sci.*, **74** (1984) 79–81.

49. Sela, M. N., Steinberg, D. & Segal, R., Inhibition of the activity of glucosyltransferase from *Streptococcus mutans* by glycyrrhizin. *Oral Microbiol. Immunol.*, **2** (1987) 125–8.

50. Wernette, C. M., San Clemente, C. L. & Kabara, J. J., The effect of surfactants upon the activity and distribution of glycosyltransferase in *Streptococcus mutans* 6715. *Pharm. Therap. Dent.*, **6** (1981) 99–107.

51. Toors, F. A. & Herczog, J. I. B., Acid production from a nonsugar liquorice and different sugar substitutes in *Streptococcus mutans* monoculture and pooled plaque–saliva mixtures. *Caries Res.*, **12** (1978) 60–8.

52. Tetsuji, N., Yamashita, S., Yamashita, T., Yamazaki, K., Sasaki, S. & Hara, K., The effects of *Carthamus tinctorius* L.—glycyrrhetinic acid, carbazochrome, and chlorhexidine in inflamed gingiva induced by activating plaque accumulation in beagle dogs. *Jap. J. Oral Biol.*, **27** (1985) 299–305.

53. Silvera, R. S. & Mühlemann, H. R., Interdental plaque pH and sweet wood chewing. *Helv. Odontol. Acta*, **17** (1973) 96–8.

54. Attström, R., Mattson, L., Edwardsson, S., Willard, L. & Klinge, B., The effect of octapinol and dento-gingival plaque and development of gingivitis. *J. Periodont. Res.*, **18** (1983) 445–51.

55. Birkhed, D., Edwardsson, S., Ahlden, M. & Frostell, G., Effects of 3 months frequent consumption of hydrogenated starch hydrolysate (lycasin), maltitol, sorbitol and xylitol on human dental plaque. *Acta Odontol. Scand.*, **37** (1979) 103–15.

56. Southard, G. L., Boulwasre, R. T., Walborn, D. R., Groznik, W. J., Thorne, E. E. & Yankell, S. L., Sanguinarine, a new antiplaque agent: retention and plaque specificity. *J. American Dental Assoc.*, **108** (1984) 338–41.
57. Shaw, J. H., Sweeteners—an overview, I (Special Article). *Dent. Abst.*, **26** (1981) 116–20.
58. Steinberg, D., Sgan-Cohen, H. D., Stabholz, A., Pisanty, S., Segal, R. & Sela, M. N., The anti-cariogenic activity of glycyrrhizin: preliminary clinical trials. *Isr. J. Dent. Sci.* (1989) in press.
59. Carlsson, J. & Egelberg, J., Effect of diet on early plaque formation in man. *Odontologisk. Rev.*, **16** (1965) 112–25.
60. Chilton, N. W., *Design and Analysis in Dental and Oral Research.* Praeger, New York, 1982, pp. 327–55.
61. Gedalia, I., Stabholtz, A., Lavie, A., Shapira, L., Pisanty, S. & Segal, R., The effect of glycyrrhizin on *in vitro* fluoride uptake by tooth enamel and subsequent demineralization. *Clinical Preventive Dentistry*, **8** (1986) 5–9.
62. Caslavska, V. & Gron, P., Effect of surface-active agents on fluoride–enamel interactions, 1. *Caries Res.*, **17** (1983) 221–5.
63. Edgar, W. M., Reduction in enamel dissolution by liquorice and glycyrrhizinic acid. *J. Dent. Res.*, **57** (1978) 59–64.
64. Bibby, B. G. & Mundorff, S. A., Enamel demineralization by snack foods. *J. Dent. Res.*, **54** (1975) 461–70.
65. Edgar, W. M., Bibby, B. G., Mundorff, S. A. & Rowley, J., Acid production in plaque after eating snacks: modifying factors in foods. *J. American Dental Assoc.*, **90** (1975) 418–25.
66. Pisanty, S., Azaz, E. & Segal, R., Glycyrrhizin as a vehicle for the application of triamcynolone in the treatment of recurrent aphtous stomatitis. *Pharm. Acta*, **59** (1984) 341–4.
67. Segal, R. & Pisanty, S., Glycyrrhizin gel as a vehicle for idoxuridine. I. Clinical investigations. *J. Clin. Pharm. Therapeut.*, **12** (1987) 1–7.
68. Pompei, R., Flore, O., Pani, R., Marcialis, M. A. & Marongin, M. E., On the antiviral action of glycyrrhizic acid. *Rivistla Farmacologia Theapia*, **10** (1979) 355–9.
69. Otsuka, A., Yonezawa, Y., Iba, K., Tatsumi, T. & Sumada, H., Physicochemical properties of glycyrrhizic acid in aqueous media. I. Surface-active properties and formation of molecular aggregates. *Yakugaku Zasshi*, **96** (1976) 203–5.

Chapter 4

# ENHANCEMENT OF THE SWEETNESS OF SUCROSE BY CONVERSION INTO CHLORO-DEOXY DERIVATIVES

Leslie Hough

*Department of Chemistry, King's College London, UK*

&

Riaz Khan

*Tate & Lyle Research & Technology, Philip Lyle Memorial Research Laboratory, Reading, Berkshire, UK*

## SUMMARY

*The sweet taste of sucrose appears to be attributable to the combination of hydroxyl groups at C-2 of the glucosyl unit and at C-3', and possibly C-1', of the fructoside unit. Any increase in the size of the group at C-6 of the glucosyl unit results in loss of sweetness, as does inversion of the configuration of the 3- and 4-hydroxyl groups in* allo-*sucrose and* galacto-*sucrose respectively. Substitution of the 3'-hydroxyl group in the fructoside unit as in 3'-O-acetyl-sucrose also destroys its sweet taste. Enhancement of the natural sweetness of sucrose to levels exceeding those of synthetic high-intensity sweeteners such as aspartame, cyclamate, acesulpham-K and saccharin has been achieved by increasing the lipophilicity of the molecule by replacing specific hydrophilic hydroxyls with chloro groups. Synthesis of a wide range of chloro-deoxy sucrose derivatives for taste and toxicological evaluation has led to the commercial development of 4,1',6'-trichloro-4,1',6'-trideoxy-*galacto-*sucrose (sucralose) as a high-intensity sweetener. The introduction of hydrophobic chloro groups in sucrose at the axial 4-position of glucose and at the 1',4' and 6' positions of fructose all enhanced its natural sweetness. Assessment of the sweetness of a range of chloro-deoxy sucroses suggests that the chloro groups should be on the upper face of the molecule and that the AH/B of the glucophore is located at the 2-OH of the glucosyl unit and the 3'-OH of the fructoside unit.*

97

# 1 INTRODUCTION

Sucrose is produced from sugar cane and sugar beet, with a combined annual worldwide production in excess of 100 million metric tons, making it a vast regenerable resource and the most abundant of all crystalline organic substances. In the light of its high purity, low cost and ease of storage and transportation (it does not require refrigeration and will keep indefinitely), sucrose is an attractive commodity for chemical and microbiological exploitation for use in the chemical, pharmaceutical, food and related industries. Many of the chemical properties[1] of this unique, non-reducing disaccharide (Fig. 1) result from the multiplicity of hydroxyl

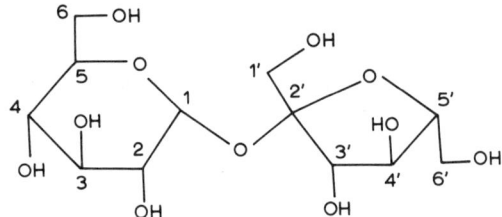

FIG. 1.   Sucrose (α-D-glucopyranosyl β-D-fructofuranoside).

groups, eight in all, from the α-D-glucopyranosyl and β-D-fructofuranosyl units, which are joined by an acid-labile inter-glycosidic bond.

Until recently, understanding the sweetness of sucrose in relation to its molecular structure was a complete mystery, although it was generally thought that certain hydroxyl groups were involved. Rapid development in the chemistry of sucrose over the past decade has led to derivatives in which specific hydroxyls are blocked or substituted, and evaluation of their taste has pin-pointed the essential hydroxyl groups, with the emergence of a structural relationship to sweetness activity. Certain specific hydroxyls on both the glucose and fructose units participate jointly in the glucophore. Furthermore, the sweetness of sucrose can be greatly enhanced when certain hydrophilic hydroxyl groups are replaced by hydrophobic halogeno substituents, and this will be the focal point of this review.

# 2 THE MOLECULAR BASIS OF SWEETNESS

The first syntheses of sucrose, using enzymes[2] and chemical methods,[3] led subsequently to a similar synthesis[4] of the unnatural L-sucrose (Fig. 2)

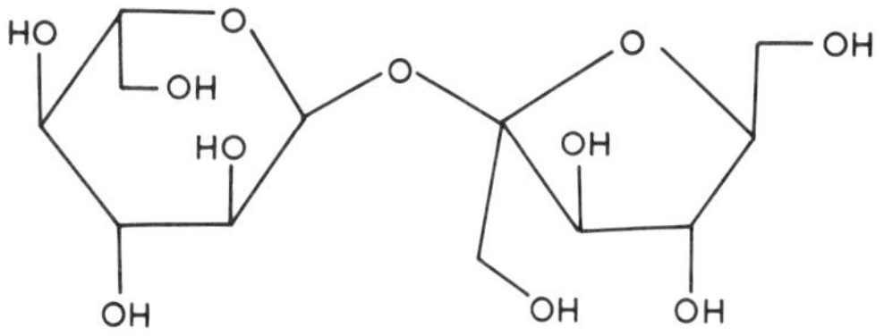

FIG. 2.   L-Sucrose, the mirror image of natural sugar.

which, surprisingly, was as sweet as natural sucrose. L-Sucrose, the mirror image of natural D-sucrose, was synthesised from L-glucose and L-fructose derivatives. It is significant that L-glucose and L-fructose are also sweet.[5] These sweet L-sugars, including L-sucrose, are of considerable interest since they are not metabolised, thus raising their possible use as non-nutritive sweeteners if they can be made economically.[6] However, the absorption of L-glucose and L-fructose and their retention in the bloodstream for long periods could prevent their use in humans.[6]

The sweetness characteristics of D- or L-sucrose are in sharp contrast to those of D- and L-amino acids, where in most cases[7] the D-forms are sweet whereas the L-forms are not. Thus D-tryptophan is 35 times sweeter than sucrose whereas L-tryptophan is bitter; D-phenylalanine is 7 times sweeter and again the L-isomer is bitter. The discovery of the sweetness of D-amino acids by Emil Fischer,[8] during his classical studies on their chemistry, led Louis Pasteur to postulate that the taste bud receptors are asymmetric or chiral, so mirror image compounds should give different responses.[9] It follows that the similar sweetness levels of D- and L-sugars seem to be anomalous.

In addition to sugars and amino acids, sweetness is induced by a range of such diverse chemical structures (Fig. 3) as saccharin, cyclamate, acesulpham-K and neohesperidin dihydrochalcone. In 1966 Deutsch and Hansch[10] suggested that the generation of sweet taste required an area of hydrophobic bonding on the molecule coupled with another area for electronic bonding, in the same way as active sites are suggested for drugs. Then in 1967 Shallenberger and Acree[11,12] noted that all sweet compounds had a structural feature in common, namely two electronegative atoms, A and B, separated by 2·5–4·0 Å (260–300 nm), and a hydrogen atom covalently linked to A (Fig. 4).

Ethylene glycol (ethane-1,2-diol) is sweet whereas ethanol is not. Hence the glycol group is the minimum requirement for sweetness in sugars. For carbohydrates, a pair of hydroxyls on adjacent carbon atoms (a glycol group) was assigned as the AH/B unit, with one hydroxyl acting as the AH

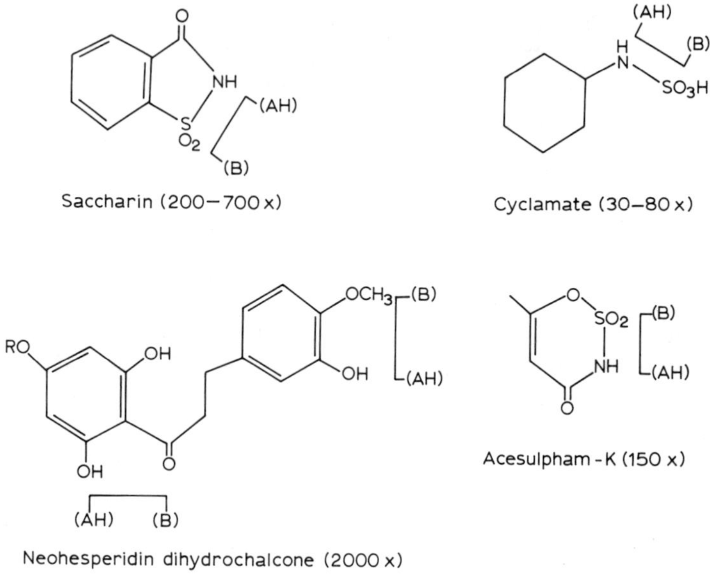

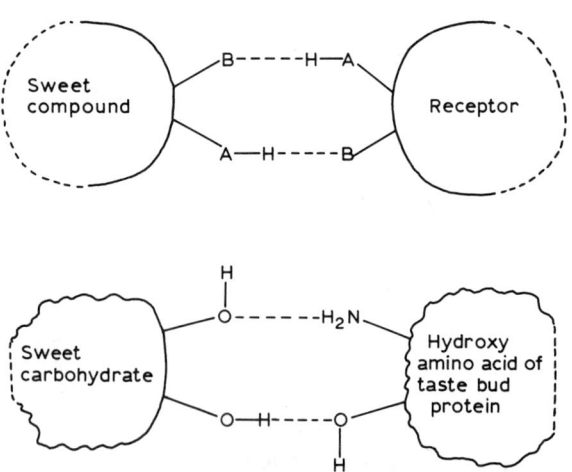

Saccharin (200–700 x)

Cyclamate (30–80 x)

Acesulpham - K (150 x)

Neohesperidin dihydrochalcone (2000 x)

Fig. 3.   Some high-intensity sweeteners and their AH/B units.

Fig. 4.   Hydrogen-bond interaction between AH/B units of sweet compounds and the receptor site.

sub-unit and the oxygen atom of the other hydroxyl as the B sub-unit (Fig. 4). Shallenberger and Acree[11] suggested that the sweetness sensation was initiated by the formation of a pair of hydrogen bonds from this AH/B unit to the proteinaceous receptor site. This site is ideally suited to bond in this way since it contains groups to act as reciprocal AH/B units, such as amide (N–H) and carbonyl (C=O) structures, and hydroxy amino acids. Thus Suami[13] has suggested that either L-serine or L-threonine units could fulfil this receptor role as N-terminal residues of an α-helix of the sweet-sensitive protein, with $NH_2 = AH$ and the oxygen of OH as the B (Fig. 4). It should be noted that, in the case of carbohydrates, either one of the hydroxyls of the glycol group can act as the AH or B (i.e. they are interchangeable), but this is not so in all other types of sweet compounds, including the amino acids, which accounts for the different sweetness of the D and L forms of amino acids and sugars.[5]

In the case of aspartame (Nutrasweet or L-aspartyl-L-phenylalanine methyl ester)[14] (Fig. 5), its sweetening power requires the ester linkage; the free acid is tasteless and only the L,L optical isomer is sweet; the D,D, D,L and L,D isomers are tasteless; and the essential unit for sweetness has been shown to be L-isoasparagine.[15] Aspartame is 100–200 times sweeter than sucrose, and increasing its lipophilicity by the introduction of a fenchyl group[16] increases the intensity of sweetness by over 100-fold. Whilst the Shallenberger theory[11,12] accounts for the sweetness of all known sweet compounds, there are many other organic compounds that are not sweet[17] but which have an identifiable AH/B unit, so there must be additional criteria.

In 1972 Kier[7] introduced another molecular feature from a study of a

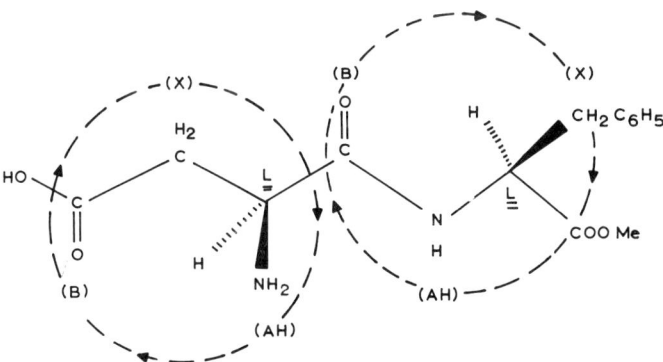

FIG. 5.  Suggested AH/B/X glucophore systems in aspartame (L-aspartyl-L-phenylalanine methyl ester).

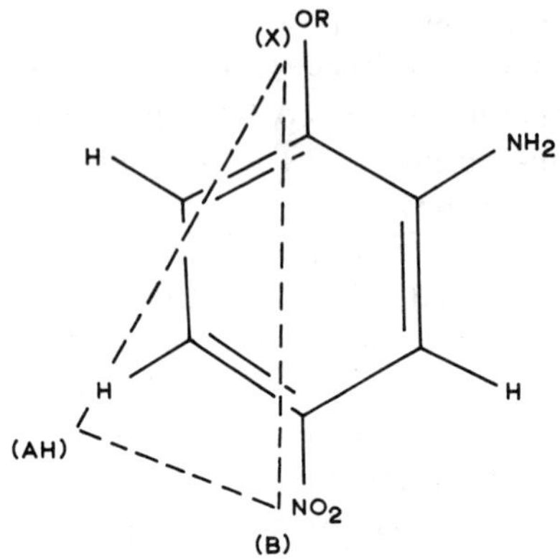

FIG. 6.    1-Alkoxy-2-amino-4-nitrobenzenes.

series of 1-alkoxy-2-amino-4-nitrobenzenes (Fig. 6), namely a third site X
that is hydrophobic (lipophilic) and binds the sweet compound to the
receptor site, thus giving a triangle of groups important in conferring
sweetness (X, AH and B),[18,19] known as the glucophore. The corresponding
*n*-propoxy derivative (Fig. 6) was 4000 times sweeter than sucrose, and
clearly the lipophile–hydrophile balance is a critical factor in sweetness
intensification. Deutsch and Hansch[10] have related the sweetness of 2-
amino-4-nitrobenzene derivatives to their partition coefficients between
water and octanol. It is significant that all substances that have a high
intensity of sweetness are substantially more hydrophobic than sucrose; all
unsubstituted sugar molecules are hydrophilic, not very sweet and weakly
absorbed by the taste buds.


## 3 THE SWEETNESS OF CARBOHYDRATES[20]

Slight variations in the structures of monosaccharides, such as configuration at individual chiral carbons, can alter sweetness. Thus D-galactose is
less sweet than D-glucose, its C-4 epimer, and L-sorbose, the 5-epimer of D-
fructose, is only one-fifth as sweet.[5] D-Fructose is sweeter than D-glucose
and sucrose, due largely if not entirely to β-D-fructopyranose (Table 1).
Only one tripartite glucophore is possible for β-D-fructopyranose, namely
AH = 1-OH, B = 2-O and X = 6-H's in its chair conformation,[21–24] and the
respective groups are arranged clockwise when viewed with the AH, B and
X groups as they would interact with the receptor, i.e. projecting towards

TABLE 1

THE SWEETNESS OF CARBOHYDRATES AND
DERIVATIVES RELATED TO SUCROSE

| Carbohydrate | Sweetness |
|---|---|
| Sucrose | 1·0 |
| D-Glucose[a] | 0·6–0·75 |
| Sorbitol (glucitol) | 0·5–0·6 |
| Xylitol | 1·0 |
| D-Galactose[a] | 0·4–0·5 |
| β-D-Fructopyranose | 1·8 |
| D-Fructose[a] | 1·1 |
| Methyl α-D-glucopyranoside | 0·1 |
| Methyl β-D-fructofuranoside | 0 |
| Trehalose | 0·1 |

[a] Mutarotation equilibrated.

the reader (Fig. 7(a)). The receptor groups (NH/CO/R) will also be disposed clockwise in order to interact with the clockwise glucophore (Fig. 8). In support, methyl β-D-fructopyranoside is much less sweet than β-D-fructopyranose. In L-fructose the role of the hydroxyls can reverse to give the same configuration (AH = 2-OH, B = 1-O and X = 6-H's). β-D-Glucopyranose is significantly sweeter than α-D-glucopyranose and β-lactose is twice as sweet as α-lactose;[20] this, coupled with the fact that methyl α-D-glucopyranoside and α,α-trehalose are only about one-eighth as sweet,[14] suggests that the anomeric hydroxyl is involved in the sweet taste.

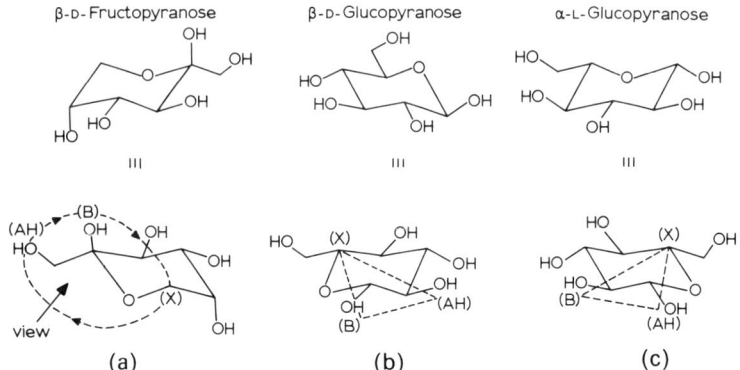

FIG. 7.   Suggested AH/B/X glucophore systems in their preferred chair conformations.

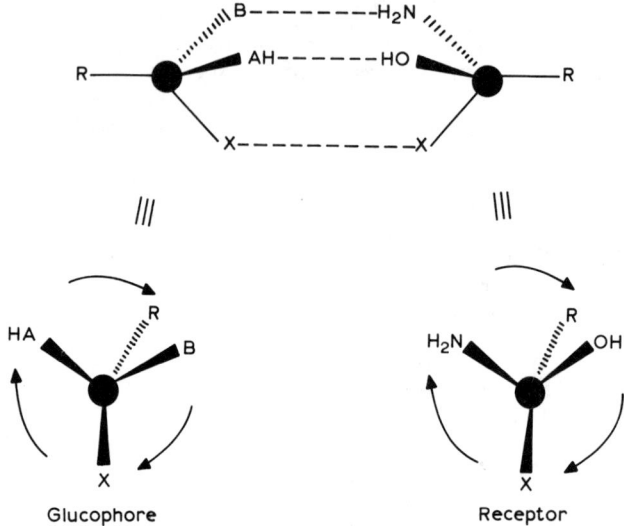

FIG. 8.    Both glucophore and receptor have clockwise AH/B/X systems.

The X = 5-H, AH = 2-OH and B = 1-O tripartite glucophore in $\beta$-D-glucopyranose is possible and arranged clockwise when viewed from below the plane of the molecule as depicted in Fig. 7(b). This is contrary to the view of Birch and his co-workers,[24,25] who favour the hydroxyls at carbons 3 and 4.

The sweetness of L-glucose can be accounted for by interchanging the roles of the hydroxyls 1 and 2 from AH/B in D-glucose to B/AH in L-glucose (Fig. 7(c)), a clockwise arrangement, when viewed from the

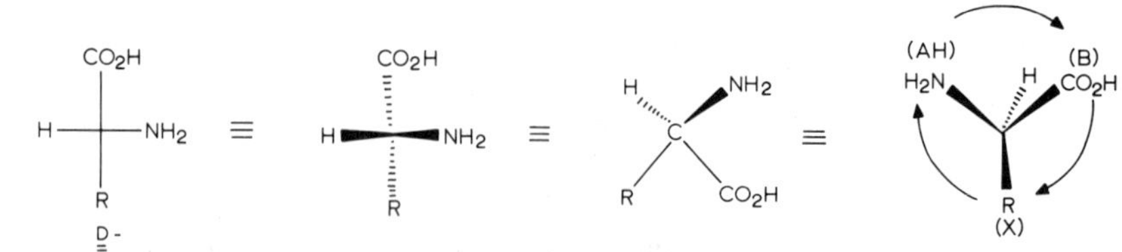

FIG. 9.    D- and L-Amino acid configurations and glucophores.

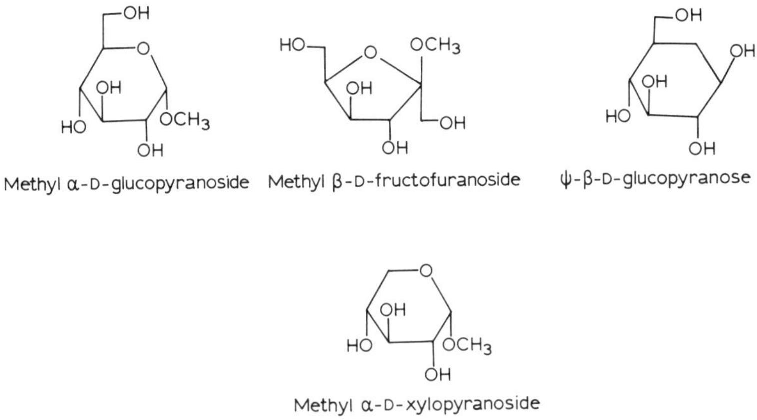

Methyl α-D-glucopyranoside   Methyl β-D-fructofuranoside   ψ-β-D-glucopyranose

Methyl α-D-xylopyranoside

FIG. 10.   Methyl pyranoside, furanoside and pseudo-sugar.

top face of the molecule (c) as required if this receptor is chiral, and conforms with that allocated to D-glucose and D-fructose. The arrangement in D-amino acids is similar but with $AH = NH_2$, $B = CO_2H$ and $X = R$, a clockwise triad (Fig. 9). This would be anticlockwise in the L forms. The glucophores assigned to aspartame (Fig. 5) are similar to those of the sweet sugars and the D-amino acids (Fig. 9) in that the $AH/B/X$ triads again have a clockwise orientation.

Pseudo-sugars are compounds in which the ring-oxygen of a pyranoid or furanoid sugar has been replaced by a methylene group, such as ψ-β-D-fructopyranose. Their conformations closely resemble those of the normal sugars, and ψ-β-DL-glucopyranose (Fig. 10), ψ-α-DL-galactopyranose and ψ-β-DL-fructopyranose are similar in sweetness to the parent sugars.[13] Consequently the ring-oxygen is of little consequence in determining the sweetness of sugars.

## 4 THE SWEETNESS OF SUCROSE[26]

In relation to sucrose, it is significant that the glucosyl and fructosyl units of this disaccharide, as represented by methyl α-D-glucopyranoside or trehalose[20] on the one hand and methyl β-D-fructofuranoside[23] on the other, have one-tenth the sweetness of sucrose and zero sweetness respectively (Table 1), suggesting a special or unique arrangement of the triad of groups for sweetness.[27] Sucrose, both in crystalline form and in solution, adopts a conformation[28] in which the α-D-glucopyranosyl unit

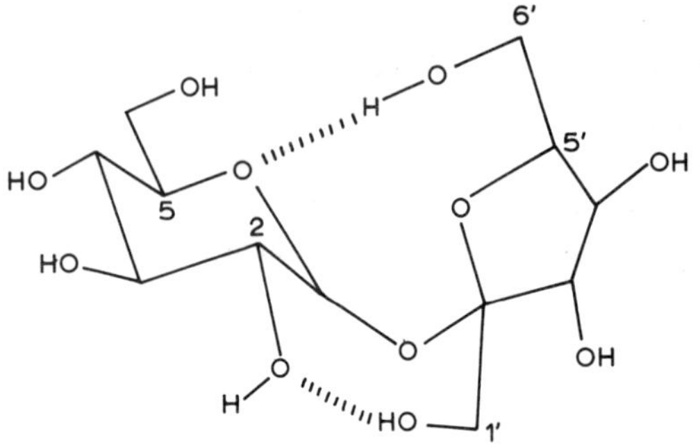

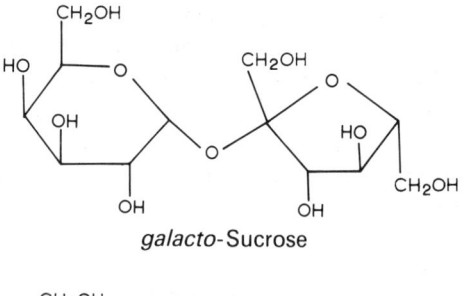

FIG. 11. Conformation of sucrose and intramolecular H-bonds.

is a chair form ($^4C_1$), and the β-D-fructofuranoside unit is a twist shape ($_3T^4$), with the furanose ring held at right-angles to the plane of the pyranose ring by two bridging, intramolecular hydrogen bonds from O-6′ to O-5 and O-1′ to O-2 (Fig. 11).[28] (The unprimed and primed numbers are used to show the carbon and attached oxygen atoms in the glycosyl and fructoside units respectively.)

Consideration of the approximate dimensions of the Kier–Shallenberger template,[20] and its application to the structure of sucrose, leads to the conclusion that two glucophores are possible: X = 4-H, AH = 1′-OH and

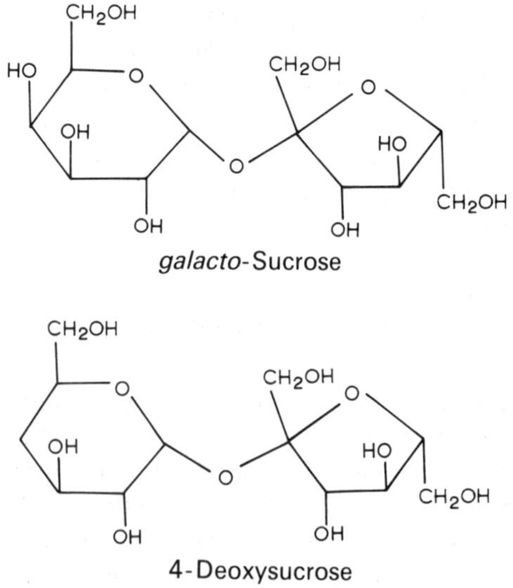

galacto-Sucrose

4-Deoxysucrose

FIG. 12. Derivatives of sucrose.

B = 2-O, and X = 4-H, AH = 3'-OH and B = 2-O, both of which are clockwise arrangements. In support of this hypothesis,[26,27] inversion of the chirality at C-4 of sucrose, to give *galacto*-sucrose (Fig. 12), results in loss of sweetness,[29] whilst 4-deoxy-sucrose (Fig. 12) is sweet; hence the configuration of the hydrophilic 4-hydroxyl of sucrose is critical. The configuration at C-3 is also important in sucrose since the C-3 epimer, namely *allo*-sucrose,[30] is tasteless; the reason for this is obscure although it may well be that an axial hydroxyl group at C-3 cannot be accommodated within the dimensions of the receptor site of the taste bud.

## 5 ESTER, ETHER AND DEOXY DERIVATIVES OF SUCROSE[1,20]

Chemical modification of mono- and di-saccharides by selective etherification, esterification or substitution of one or more hydroxyl groups has been used, notably by Birch and his colleagues,[31] to seek out those hydroxyls involved in the glucophore, namely X/AH/B systems, in these molecules. Thus, in considering glucose, the primary 6-hydroxyl and the anomeric 1-hydroxyl were discounted, since methyl D-xylopyranosides (Fig. 10) are sweet. The 4,6-O-methyl and similar derivatives of methyl α-D-glucopyranoside (Fig. 10) are not sweet.[22] Hence the 2,3-glycol [vicinal synclinal (gauche) hydroxyls] was eliminated, thereby highlighting the role of 3- and 4-hydroxyls as AH and B units. Investigation of 3-substituted gluco- and xylo-pyranosides suggested that the 3-hydroxyl was the B group.[22,20] Selective methods have been devised for blocking, modifying or replacing specific hydroxyls in sucrose;[20,26] since certain hydroxyls must be responsible for its sweet taste, the availability of these derivatives permitted a study of structure versus sweetness activity.[32]

Esterification of sucrose can be controlled to give a range of products, from the mono- to the fully substituted octa-ester, containing from 7 to zero free hydroxyls respectively. It should be appreciated that, whilst there are 8 possible monoesters and only one octaester, 28 di- and 56 tri-esters are possible in theory.[26] The preferential reactivity of the primary hydroxyls at C-6, C-1' and C-6' simplifies the selectivity observed in these reactions, the order of reactivity being HO-6, HO-6' > HO-1' > HO-2.

Sweetness falls off dramatically upon esterification of sucrose. Its 6-monoacetate[33,34] is only slightly sweet, the 6-O-benzoate and 6-phosphate are not sweet, neither are any of the 6,6'-diesters and 1',6'-diesters, whilst the octa-acetate is a well known bittering agent or denaturant. Thus the size of the groups at C-6 in particular, and also those at C-1' and C-6', would

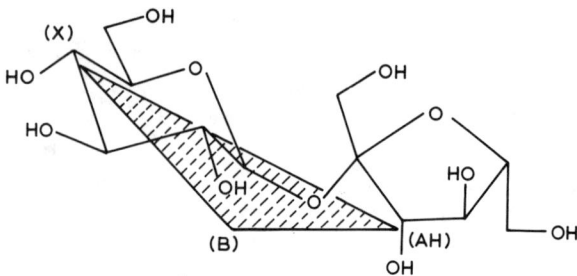

FIG. 13.    The glucophore in sucrose at C-4 (X), C-2 (B) and C-3′ (AH).

appear to be critical, any significant increase resulting in loss of sweetness due to a misfit of the enlarged molecule on the receptor site of the taste buds of the tongue. This is supported by the sweetness of 6-deoxy- and 6-O-methyl-sucrose,[35,36] which have smaller groups at C-6, whilst the larger 6-O-benzyl ether is not sweet. Like the 4-deoxy derivative, 4-O-methylsucrose is sweet and so are the 1′-deoxy and the 1′-methyl ether. These results are in accord with the location of the sweetness triangle of groups in sucrose at C-4 (X), C-2 (B) and C-3′ (AH) (Fig. 13), and this is supported by the loss of sweetness when the 3′-hydroxyl is esterified in 3′-O-acetylsucrose,[37] thus blocking out the AH of the glucophore.

Any enhancement of the natural sweetness of sucrose would reduce calorie intake from this source; should derivatisation inhibit the hydrolytic action of invertase or α-glucosidase, this would further limit its metabolism. To enhance the sweetness of sucrose, it is clear from the conclusions above that derivatisation must make the molecule more lipophilic especially at the critical axial 4-position,[32] and possibly at the 1′-position, leaving the hydroxyls free at C-2 and C-3′ if they are to serve as the AH,B of the glucophore.

The transformation of D-tryptophan (35 × the sweetness of sucrose) into its 6-chloro derivative (1000 ×) increased its sweetness 30-fold,[38] suggesting the possibility of studying chloro derivatives of sucrose already available. However, a related chlorination of saccharin (600–700 ×) to give 6-chlorosaccharin (100–350 ×) did not improve its sweetness.[38]

## 6 CHLORO DERIVATIVES FROM SUCROSE[26]

Considerable interest in deoxy-halogeno sugars has centred not only on their utility as intermediates in the synthesis of deoxy and amino-deoxy sugars, rare sugars of biological importance, but also on their potential in

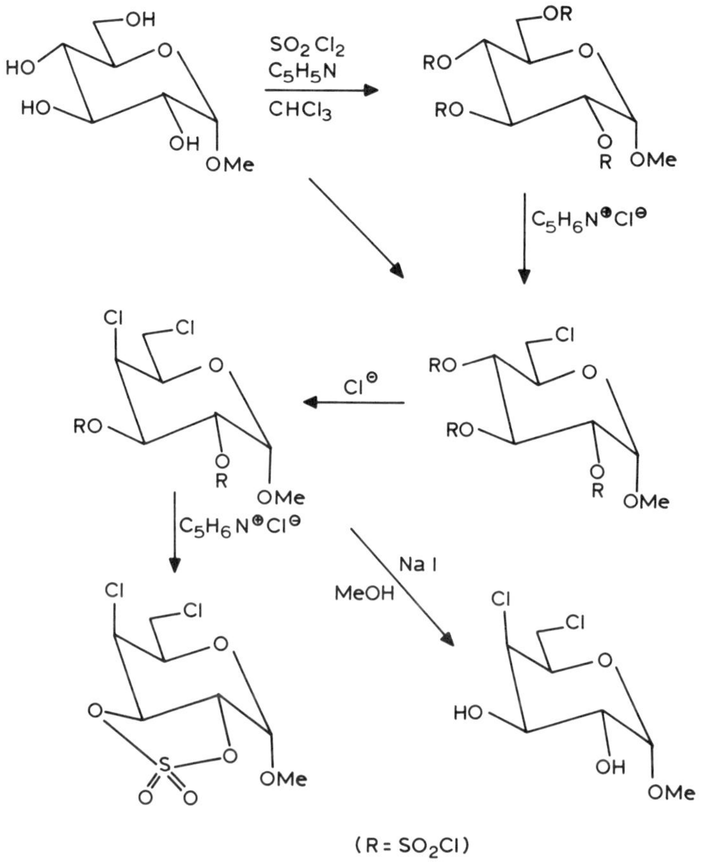

(R = SO$_2$Cl)

FIG. 14. Reactions of sulphuryl chloride with methyl α-D-glucopyranoside.

biochemistry and pharmacology.[39] Thus halogenation of the carbohydrate components of antibiotic substances has yielded products with enhanced activity and superior chemotherapeutic behaviour.[39]

Helferich and his colleagues[40-42] discovered that the reaction of sulphuryl chloride (SO$_2$Cl$_2$) in pyridine–chloroform at 5°C with carbohydrates led to the replacement of specific hydroxyl groups by chloro substituents and to the formation of cyclic sulphates. In 1925 they observed that methyl α-D-glucopyranoside gave a 4,6-dichloro-4,6-dideoxy-2,3-cyclic sulphate derivative, shown by Jones and his co-workers[43,44] to proceed with inversion of chirality at C-4 to the methyl 4,6-dichloro-4,6-dideoxy-α-D-galactopyranoside 2,3-cyclic sulphate (Fig. 14).

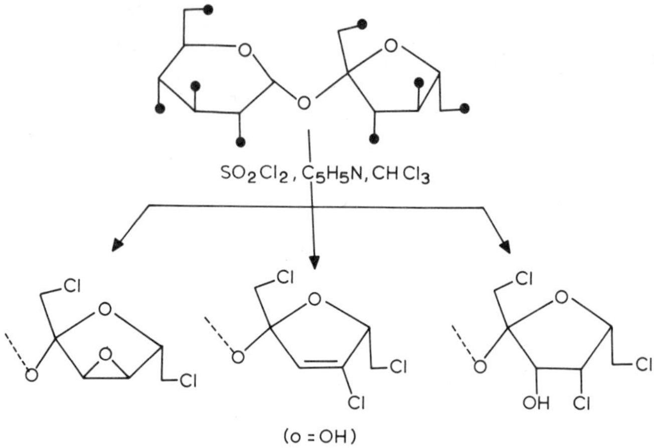

(o = OH)

FIG. 15.  Modifications of the fructofuranoside unit of sucrose on reaction with
SO$_2$Cl$_2$ in pyridine.

Application of this reaction to sucrose[43] gave a complex mixture of
products, in which the glucosyl residue had been similarly transformed into
a 4,6-dichloro galactosyl residue. Repetition of this reaction with sucrose at
50°C, followed by careful separation of the products by chromatography,
gave low yields of a tetrachloro and two pentachloro derivatives resulting
from variations in the reactions in the fructofuranoside unit, which were
identified[45] by spectroscopic methods, notably proton nuclear magnetic
resonance ($^1$H NMR), as a 3',4'-epoxide, a 3'-ene and a 1',4',6'-trichloride
(Fig. 15).

Jennings and Jones[46] showed that cyclic sulphate formation could be
avoided by minimising the quantity of pyridine used in the sulphuryl
chloride reaction; instead chlorosulphate esters are formed which can be
removed subsequently with methanolic sodium iodide solution, liberating
the hydroxyl groups. Under these conditions methyl α-D-glucopyranoside
gave the 2,3-dichlorosulphate of methyl 4,6-dichloro-4,6-dideoxy-α-D-
galactopyranoside (Fig. 14). Study of this reaction at low temperature
revealed[47] that it proceeds via the 2,3,4,6-tetrachlorosulphate, which then
undergoes nucleophilic bimolecular substitution by chloride anions, first at
the primary C-6 as expected, giving the 6-chloride, followed by slower
substitution at the secondary C-4, with inversion of configuration to yield
the 4,6-dichloro-galactoside 2,3-chlorosulphate (Fig. 14); substitution
reactions at C-2 and C-3 are inhibited by unfavourable transition states.[39]

Careful control of the multi-centred reactions of sucrose with sulphuryl

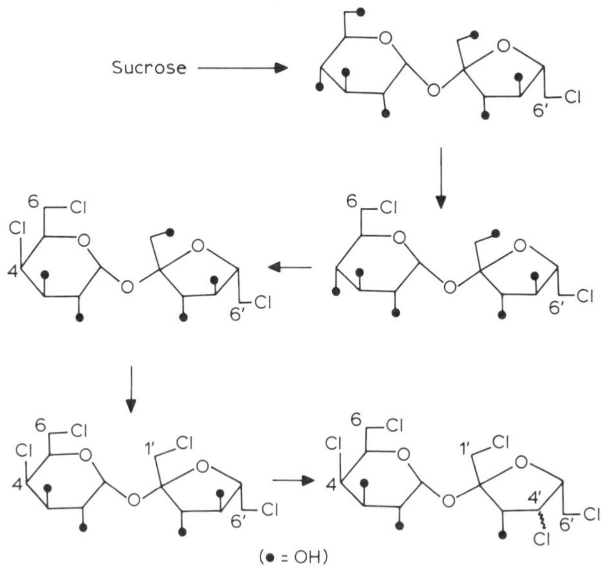

($\bullet$ = OH)

FIG. 16.   Progressive reactions of sucrose with $SO_2Cl_2$ in pyridine and chloroform.

chloride revealed the pattern of substitution by the progressive formation of mono- to penta-chloro derivatives.[26] The major pathway appeared to progress initially to the 6'-chloride (29% yield), followed by the 6,6'-dichloride (29% yield)[45] and then the 4,6,6'-trichloro (50% yield),[48,49] the 4,6,1',6'-tetrachloro (45% yield)[50] and the 4,6,1',4',6'-pentachloro derivatives of *galacto*-sucrose (Fig. 16).[50,51] Steric factors appear to prevent direct substitution of the 4'-chlorosulphate, and the introduction of the 4'-chloro substituent probably proceeds via a 3',4'-epoxide.[50] The isolation and characterisation of these chloro products suggested a sequence of stereoselective reactions where the order of reactivity was HO-6' > HO-6 > HO-4 > HO-1' > HO-4'. The chlorination at HO-1' is slow because it is a hindered neopentyl type of primary hydroxyl and adjacent to the 2'-anomeric group. The 6,6'-dichloride can now be more conveniently synthesised[52] in higher yield (> 70%) by selective reaction of sucrose with triphenylphosphine in carbon tetrachloride and pyridine. The 4,6,1',6'-tetrachloro-4,6,1',6'-tetradeoxy-*galacto*-sucrose was originally prepared[53] from sucrose 6,1',6'-trimesitylene sulphonate by substitution with lithium chloride, giving the 6,1',6'-trichloride, followed by a selective reaction with sulphuryl chloride at C-4, as anticipated by the above order of reactivity (Fig. 17). This compound was tasted by Dr Shashi P. Phadnis in 1975 and it

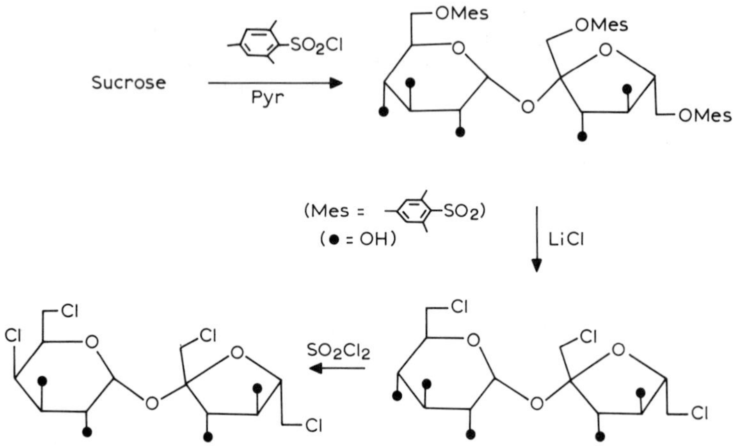

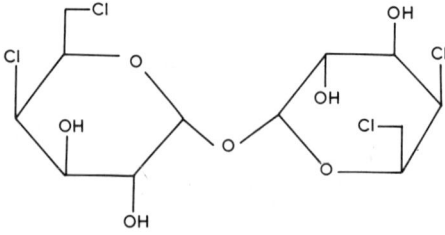

Fig. 17.   Synthesis of the 4,6,1',6'-tetrachloro derivative of *galacto*-sucrose.

was 200 times sweeter than the parent sucrose; it was the first time that a derivative of a carbohydrate had shown an enhancement of natural sweetness.[53] Not only did this product have a good taste profile, but it was not metabolised and hence a potential non-caloric high-intensity sweetener. In terms of structure and its relationship to sweetness, it is interesting that previous taste studies had revealed that an analogous trehalose derivative, 4,6,4',6'-tetrachloro-4,6,4',6'-tetra-deoxy-*galacto*-trehalose (Fig. 18), was not sweet but as bitter as quinine.[54]

This important discovery led to an extensive collaboration[55] between the carbohydrate groups at the Philip Lyle Research Laboratory (Tate & Lyle), located at Reading University, and at Queen Elizabeth College (now King's), to synthesise a wide range of chloro and related derivatives of sucrose for pharmacological and toxicological evaluation and the subsequent assessment of sweetness and taste quality, with the objective of obtaining a safe, new high-intensity sweetener.

Fig. 18.   4,6,4',6'-Tetrachloro derivative of *galacto*-trehalose.

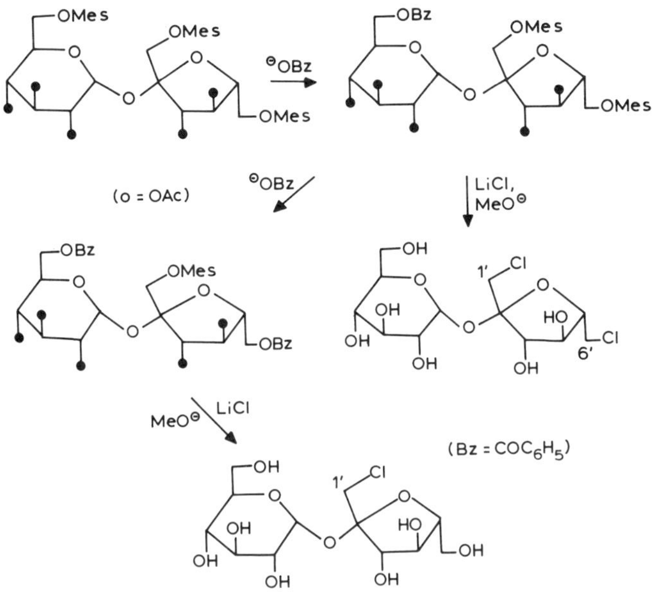

FIG. 19.    Synthesis of 1',6'-dichloro and 1'-chloro derivatives of sucrose.

6,1',6'-Trisulphonate esters of sucrose are readily available from the use of the bulky, hence more selective, 2,4,6-trimethylbenzene (mesitylene or trimsyl) sulphonyl chloride[56,57] or 2,4,6-tri-isopropylbenzene (tripsyl) sulphonyl chloride,[58] since the trisulphonates crystallise without recourse to chromatography in $> 50\%$ yield. Selective nucleophilic mono-substitution of the 6,1',6'-tri-$O$-trimsylate with benzoate anion yields the 6-benzoate 1',6'-disulphonate, which in turn affords the 1',6'-dichloride on reaction with chloride anions.[59] Under more forcing conditions disubstitution occurs and the 6,6'-dibenzoate 1'-sulphonate is obtained, from which the 1'-chloride is then available (Fig. 19). Guthrie and Watters[60] showed that 1'-chloro-1'-deoxysucrose was stable to invertase and its rate of acid hydrolysis was 10 times slower than that of sucrose. It was 20 times sweeter than sucrose, as was the 6'-chloro derivative, whereas 6-chloro-6-deoxysucrose was not sweet but bitter. The 4-chloro-*galacto*-sucrose derivative, prepared[61] from the readily available hepta-pivalate (4-OH free), was 5 times sweeter than sucrose (Fig. 20).

6,6'-Dichloro-6,6'-dideoxysucrose was not sweet,[27] the 6-chloro substituent clearly having an adverse effect due either to the increased size of the substituent at C-6, which, as described earlier, is critical, or to competition

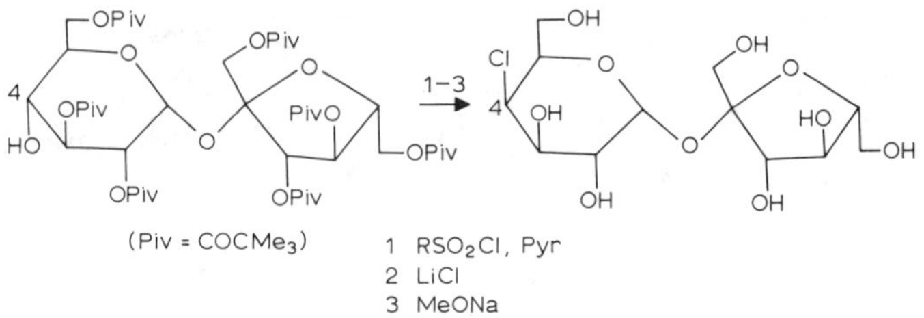

        (Piv = COCMe₃)      1   RSO₂Cl, Pyr
                            2   LiCl
                            3   MeONa

FIG. 20.    Synthesis of 4-chloro-4-deoxy-*galacto*-sucrose.

for the hydrophobic locking site on the receptor, resulting in a misfit.[27] On the other hand, in the 1',6'-dichloro derivative, the effect is synergistic, enhancing the sweetness 76 times, and with even greater effect in 4,1'-dichloro-4,1'-dideoxy-*galacto*-sucrose (120 × sweetness). This can be synthesised by direct chlorination of a 6,6'-diester of sucrose with sulphuryl chloride (Fig. 21). Addition of a further chloro substituent at the 6' position raised the sweetness by yet another order of magnitude, and at this stage 4,1',6'-trichloro-4,1',6'-trideoxy-*galacto*-sucrose (sucralose) was the sweetest derivative encountered, 650 times sweeter than sucrose.[55]

Since sucralose was non-toxic, not metabolised and 60 times more stable than sucrose to acid hydrolysis than sucrose, it was the product of choice for joint development by Tate & Lyle plc and Johnson and Johnson (USA) as a high-quality intense sweetener.[62] Sucralose is synthesised[63] from the penta-acetate of its 6,1',6'-tri-*O*-trityl ether, which on detritylation rearranges to the 2,3,6,3',4'-penta-acetate, by a 4→6 acetyl migration; subsequent chlorination with SO₂Cl₂, followed by de-esterification, yields the 4,1',6'-trichloro derivative as a white, water-soluble crystalline solid (Fig. 22).

Sucrose derivatives with a 4-chloro substituent on the fructofuranoside ring were synthesised for sweetness evaluation.[50,64] A novel reaction of sucrose with triphenylphosphine-diethylazodicarboxylate to give the 3',4'-*lyxo*-epoxide, first described in 1983 by Guthrie and his co-workers,[65] was applied to 4,1',6'-trichloro-4,1',6'-trideoxy-*galacto*-sucrose, and the resulting *lyxo*-epoxide opened stereospecifically at C-4' by nucleophilic attack with chloride anion, to revert to the *fructo* configuration, giving exclusively

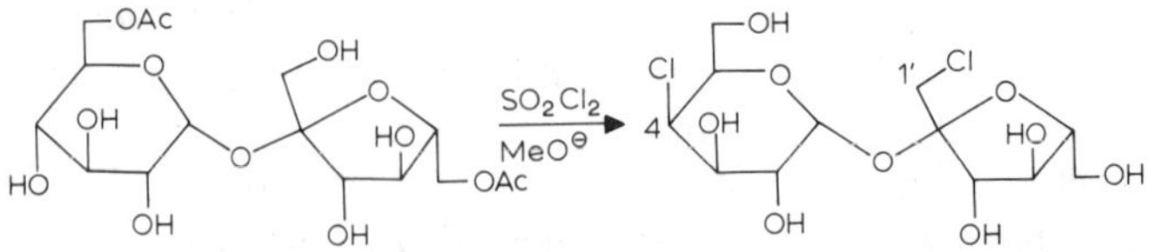

FIG. 21.    Synthesis of 4,1'-dichloro-4,1'-dideoxy-*galacto*-sucrose.

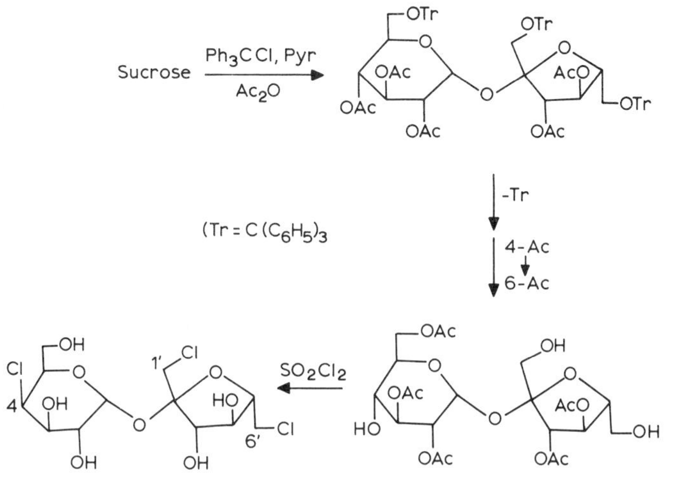

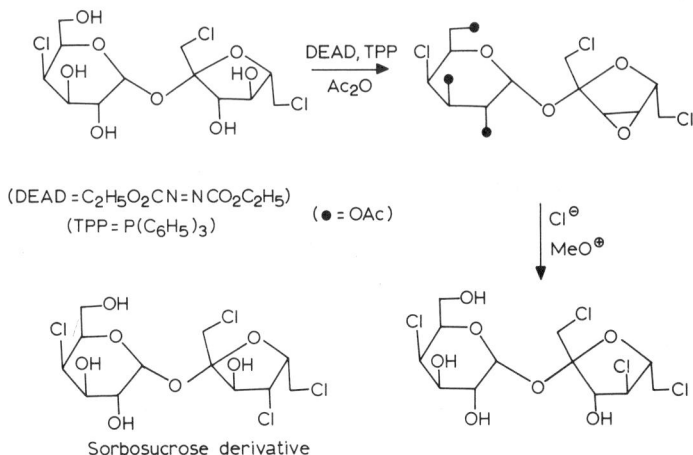

FIG. 22. Synthesis of sucralose.

the required 4,1',4',6'-tetrachloride (Fig. 23). This product was 2200 times sweeter than sucrose, a 4-fold increase in the sweetness of sucralose by replacement of the 4'-hydroxyl by chloride.[50,66] It may be significant that all the chloro substituents are on the upper face of the molecule, since the corresponding *sorbo*-tetrachloride, with the opposite configuration at C-3',4', was only 200 times sweeter; this *sorbo*-tetrachloride (Fig. 23) was synthesised from the 3',4'-*ribo*-epoxide by stereospecific attack at C-4' with chloride anions.[50]

FIG. 23. Synthesis of 4,1',4',6'-tetrachloro-4,1',4',6'-tetradeoxy-*galacto*-sucrose.

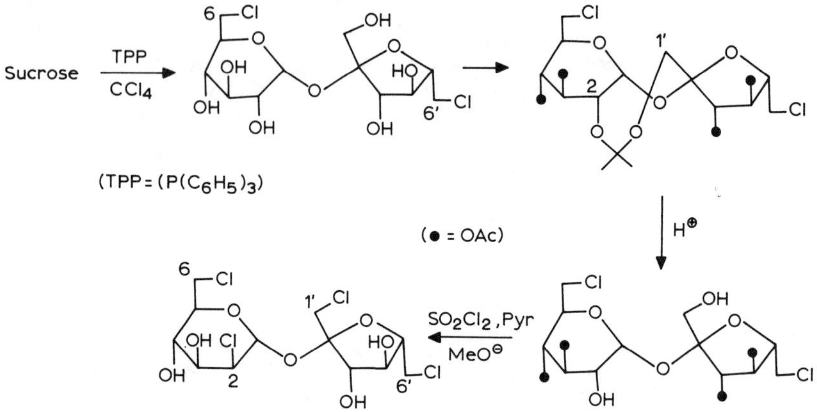

FIG. 24.    Synthesis of 2,6,1′,6′-tetrachloro-2,6,1′,6′-tetradeoxy-*manno*-sucrose.

The 6,1′,6′-trichloro derivative of sucrose is 25 times sweeter than sucrose but when the 2-hydroxyl is replaced by chloride (Fig. 24), with inversion of configuration to give the *manno* isomer, the resulting 2,6,1′,6′-tetrachloro derivative[67] is as bitter as quinine, thus supporting the theory that the 2-hydroxyl is essential for sweetness and emphasising that the triangular array of saporophoric groups has strict steric requirements for sweetness.[27]

Attempts to enhance the sweetness of D-glucose, maltose and trehalose by the introduction of chloro substituents, to give compounds such as 6-chloro-6-deoxy-D-glucose, 4′,6′-dichloro-4′,6′-dideoxymaltose and 4,6-dichloro-4,6-dideoxytrehalose, were unsuccessful, the majority being less sweet than the parent sugars and many of them bitter.[68] The taste properties of 4-chloro-D-galactose and simple glycoside derivatives have been examined, and none is sweet,[69] emphasising the importance of the interplay of hydroxyl groups across the sucrose molecule.

The observations that 6-chloro-6-deoxy-D-fructofuranose was sweet ($<1 \times$) and the 1,6-dichloro derivative even sweeter ($1 \times$) were surprising since the parent fructofuranose is believed to lack sweetness. Although not in the same league as sucralose or 1′,6′-dichlorosucrose ($80 \times$), it was suggested that this 1,6-dichlorofructofuranosyl unit may contribute to the glucophore(s) in sucralose. Interpretation of the Fourier transformation infrared spectra of sweet polyhydroxy compounds, including sucralose, has revealed that they have a sharp absorption characteristic of a free hydroxyl which is not engaged in hydrogen bonding.[25] Kanters *et al.*[70] have examined the conformation of sucralose in crystalline form using X-ray analysis, and the result demonstrated an intramolecular bond between the

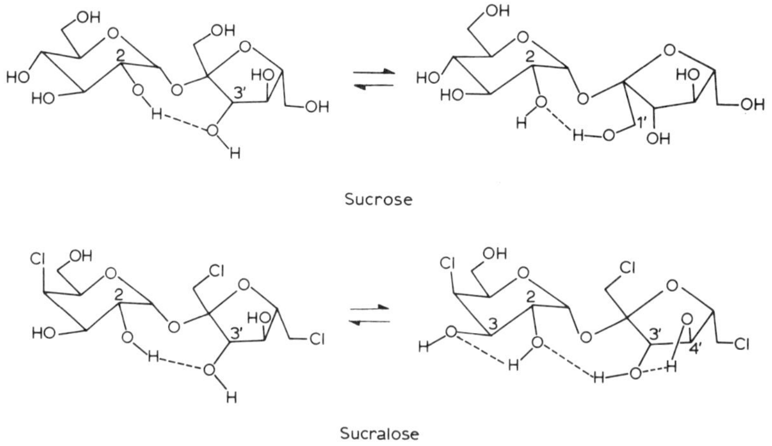

Sucrose

Sucralose

FIG. 25. The 2-OH $\cdots$ O-3′ hydrogen bond and co-operative H-bonding as revealed by X-ray analysis and SIMPLE NMR.

2-hydroxyl and the 3′-hydroxyl (Fig. 25). Consequently the overall conformation of sucralose differs from that of sucrose, which is hydrogen-bonded from the 2-hydroxyl to the 1′-hydroxyl (Fig. 11), because of rotation about the C-1–O–C-2′ interglycosidic bond. Secondary isotope multiplet partially labelled entities (SIMPLE) NMR spectroscopy[71] revealed the existence of an intramolecular 3′-OH $\cdots$ O-2 hydrogen bond, in which the 3′-OH is the donor and 2-OH is the acceptor hydroxyl group, in four 1′-chloro-1′-deoxysucrose derivatives, including sucralose, in dimethyl sulphoxide (Fig. 25). This bond stabilises co-operative bonding in both glucosyl and fructoside units (3-OH $\cdots$ 2-OH $\cdots$ 3′-OH $\cdots$ 4′-OH), and the strength of the inter-unit H-bond intensifies as the network becomes more extensive, i.e. the whole process is co-operative. However, in dilute aqueous solution the intramolecular hydrogen bonds are probably broken, allowing rotation around the interglycosidic bond.

Indeed, in the case of a solution of sucrose, SIMPLE NMR studies revealed the presence of two different intramolecular hydrogen-bonded conformations in competitive equilibrium in deuterated dimethyl sulphoxide, with the 2-OH as an acceptor for either the 1′-OH or 3′-OH of the fructofuranoside (Fig. 25).[72] These hydrogen-bonded conformations of sucrose and sucralose (Fig. 25) are obviously closely related to the glucophores suggested (Fig. 13) for these molecules since the 2-hydroxyl and 3′-hydroxyl groups are involved in both structures, and they have similar structural requirements. Clearly little structural change will take place when sucrose and sucralose are transformed from their internally hydrogen-bonded forms to the externally hydrogen-bonded complex (Fig. 4) that links them to the protein of the taste buds, thereby minimising the conformational energy required to initiate the sweetness sensation.

## 7 CONCLUSION

The increased lipophilicity of sucralose compared with sucrose, due to the introduction of chloro groups at C-1', C-4 and C-6' on the upper face of the disaccharide, results in a stronger binding of the sweetener to the protein membranes of the taste bud, thereby producing a greater, prolonged sweet response when the AH/B groups of the C-2 and C-3' hydroxyls interact by hydrogen bonding with their receptors on the taste protein.

## REFERENCES

1. Jenner, M. R. In *Developments in Food Carbohydrates*, Vol. 2, ed. C. K. Lee. Applied Science Publishers, London, 1980, p. 91.
2. Hassid, W. Z. & Doudoroff, M. *Adv. Carbohydr. Chem.*, **5** (1950) 29.
3. Lemieux, R. W. & Huber, G. *J. Amer. Chem. Soc.*, **75** (1953) 4118.
4. Szarek, W. A. & Jones, J. K. N. Queen's University, Kingston, Ontario, Canada. Canadian patent 1 556 007 (1979).
5. Shallenberger, R. S. In *Carbohydrates in Solution*, ed. R. F. Gould, Advances in Chemistry Series 117, 1973, p. 262.
6. Bakai, A. I. US patent 4 459 316 (1984).
7. Kier, B. K. *J. Pharm. Sci.*, **61** (1972) 1394.
8. Fischer, E. *Ber.*, **23** (1890) 2611.
9. Shallenberger, R. S. In *Sweetness and Sweeteners*, ed. G. G. Birch. Applied Science Publishers, London, 1971, p. 47.
10. Deutsch, E. W. & Hansch, C. *Nature*, **211** (1966) 75.
11. Shallenberger, R. S. & Acree, T. E. *Nature*, **216** (1967) 480.
12. Shallenberger, R. S. & Acree, T. E. *J. Agric. Food Chem.*, **17** (1969) 701.
13. Suami, T. *Pure Appl. Chem.*, **59** (1987) 1509.
14. Nishimura, T. & Hiromichi, K. *Food*, **4** (1988) 175.
15. Mazur, R. H. *J. Toxicol. Environ. Health*, **2** (1976) 243.
16. Fukino, M., Wakimasu, M., Tanaka, K. & Aoki, H. *Naturwiss.*, **60** (1973) 351.
17. Bragg, R. W., Chow, Y., Denis, L., Ferguson, L. N., Howell, S., Morga, G., Ogino, C., Pugh, H. & Winters, M. *J. Chem. Ed.*, **55** (1978) 281.
18. Shallenberger, R. S. & Lindley, M. G. *Food Chem.*, **2** (1977) 145.
19. van der Heijden, A., Brussel, L. P. B. & Peer, H. G. *Food Chem.*, **3** (1978) 207.
20. Lee, C. K. *Adv. Carbohydr. Chem. Biochem.*, **45** (1987) 199.
21. Shallenberger, R. S. & Lindley, M. G. *Food Chem.*, **2** (1977) 145.

22. Lindley, M. G. & Birch, G. G. *J. Sci. Food Agric.*, **26** (1975) 117.
23. Shallenberger, R. S. *Pure Appl. Chem.*, **50** (1978) 1409.
24. Birch, G. G. & Mylvaganam, A. R. *Nature*, **260** (1976) 632.
25. Mathlouthi, M., Seuvre, A.-M. & Birch, G. G. *Carbohydr. Res.*, **152** (1986) 47.
26. Hough, L. *Chem. Soc. Rev.*, **14** (1985) 357.
27. Hough, L. & Khan, R. *Trends Biochem. Sci.*, **3** (1978) 61.
28. Bock, K. & Lemieux, R. U. *Carbohydr. Res.*, **100** (1982) 63.
29. Lindley, M. G., Birch, G. G. & Khan, R. *J. Sci. Food Agric.*, **27** (1976) 140.
30. Hough, L. & O'Brien, E. *Carbohydr. Res.*, **84** (1980) 95.
31. Birch, G. G. & Lee, C. K. *J. Food Sci.*, **41** (1976) 1403.
32. Khan, R., Mufti, K. S. & Patel, G. European patent 0 103 479 (1984).
33. Kononenko, O. K. & Kestenbaum, I. L. *J. Appl. Chem.*, **11** (1961) 7.
34. Khan, R. & Mufti, K. S. UK patent 2 079 749B (1982).
35. Lindley, M. G., Birch, G. G. & Khan, R. *J. Sci. Food Agric.*, **27** (1976) 140.
36. Lindley, M. G., Birch, G. G. & Khan, R. *Carbohydr. Res.*, **43** (1975) 360.
37. James, C. E., unpublished results (1988).
38. Crammer, B. & Ikan, R. *Chem. Soc. Rev.*, **6** (1977) 455.
39. Szarek, W. A. *Adv. Carbohydr. Chem. Biochem.*, **28** (1973) 225.
40. Helferich, B. *Ber.*, **54** (1921) 1082.
41. Helferich, B., Löwa, A., Nippe, W. & Riedel, H. *Ber.*, **56** (1923) 1083.
42. Helferich, B., Sprock, G. & Besler, E. *Ber.*, **58** (1925) 886.
43. Bragg, P. D., Jones, J. K. N. & Turner, J. C. *Can. J. Chem.*, **37** (1959) 1412.
44. Jones, J. K. N., Perry, M. B. & Turner, J. C. *Can. J. Chem.*, **38** (1960) 1122.
45. Ballard, J. M., Hough, L., Richardson, A. C. & Fairclough, P. H. *J. Chem. Soc.*, *Perkin Trans. I* (1973) 1524.
46. Jennings, H. J. & Jones, J. K. N. *Can. J. Chem.*, **41** (1963) 1151.
47. Jennings, H. J. & Jones, J. K. N. *Can. J. Chem.*, **43** (1965) 2372.
48. Hough, L., Phadnis, S. P. & Tarelli, E. *Carbohydr. Res.*, **44** (1975) 37.
49. Parolis, H. *Carbohydr. Res.*, **48** (1976) 132.
50. Lee, C. K. *Carbohydr. Res.*, **162** (1987) 53.
51. Phadnis, S. P., unpublished results.
52. Anisuzzaman, A. K. & Whistler, R. L. *Carbohydr. Res.*, **61** (1978) 511; **78** (1980) 185.
53. Hough, L. & Phadnis, S. P. *Nature*, **263** (1976) 800.
54. Birch, G. G. *Olfaction and Taste*, **VI** (1977) 27.
55. Hough, L., Phadnis, S. P., Khan, R. & Jenner, M. R. UK patents 1 543 167 and 1 543 168 (1979).
56. Creasey, S. E. & Guthrie, R. D. *J. Chem. Soc., Perkin Trans. I* (1974) 1373.
57. Hough, L., Phadnis, S. P. & Tarelli, E. *Carbohydr. Res.*, **44** (1975) C12.
58. Almquist, R. G. & Reist, E. J. *Carbohydr. Res.*, **46** (1976) 33; *J. Carbohydr., Nucleosides, Nucleotides*, **3** (1976) 261.
59. Gurjar, M. R., PhD thesis, University of London, 1980.
60. Guthrie, R. D. & Watters, J. J. *Aust. J. Chem.*, **33** (1980) 2487.
61. Hough, L., Chowdhary, M. S. & Richardson, A. C. *J. Chem. Soc., Perkin Trans. I* (1984) 419; *Carbohydr. Res.*, **147** (1986) 49.
62. Jenner, M. R. This volume, Chapter 5, pp. 121–41.
63. Fairclough, P. H., Hough, L. & Richardson, A. C. *Carbohydr. Res.*, **40** (1975) 285.

64. Khan, R. *Pure Appl. Chem.*, **56** (1984) 833.
65. Guthrie, R. D., Jenkins, I. D., Thang, S. & Yamaski, R. *Carbohydr. Res.*, **85** (1980) C5; **121** (1983) 109.
66. Lee, C. K. UK patent 2 088 855A (1982).
67. Khan, R. & Jenner, M. R. UK patent 2 037 561A (1980).
68. Dziedzic, S. Z. & Birch, G. G. *J. Sci. Food Agric.*, **32** (1981) 283.
69. Thelwall, L. A. W., unpublished results.
70. Kanters, J. A., Scherrenberg, R. L., Leeflang, B. R., Kroon, J. & Mathlouthi, M. *Carbohydr. Res.*, **180** (1988) 175.
71. Christofides, J. C., Davies, D. B., Martin, J. A. & Rathbone, E. B. *J. Amer. Chem. Soc.*, **108** (1986) 5738.
72. Christofides, J. C. & Davies, D. B. *J. Chem. Soc., Chem. Commun.* (1985) 1533; *Carbohydr. Res.*, **163** (1987) 269.

*Chapter 5*

# SUCRALOSE: UNVEILING ITS PROPERTIES AND APPLICATIONS

M. R. JENNER

*Tate & Lyle Speciality Sweeteners, Reading, Berkshire, UK*

## SUMMARY

*Since its discovery in 1976 the development of the high-intensity sweetener sucralose has followed the conventional path for a major new food ingredient. After the structure–activity relationship studies had confirmed sucralose as the target compound, safety evaluation studies were initiated to ascertain its behaviour in model systems and in man. Parallel studies encompassed process research and process development followed by manufacture, applications research and product development. This chapter covers primarily the functions, performance and advantages of sucralose as a food ingredient. Progress so far suggests that sucralose is destined to be a commercial success as a result of a unique combination of features which ensure that upon approval it will rapidly become the sweetener of choice in industry.*

## 1 INTRODUCTION

The search for new high-intensity sweeteners has absorbed prodigious amounts of time and money over the last 20 years. The imagination of scientists from the academic and commercial worlds has been fired in the search for new products which not only demonstrate the curious effect of intense sweetness but also hold out the promise of significant financial rewards for a successful product.

Hundreds, if not thousands, of new high-intensity sweeteners have been discovered during this period, as shown by the proliferation of patents in this area. However, very few of these products will ever achieve commercial

121

viability owing to the size and complexity of the hurdles to be crossed in their development.

Sucralose has taken more than 12 years of intensive effort by a large team from two major international companies to bring it to the market. Very few new products show sufficient early promise to justify the expenditure necessary to carry out a successful development programme and achieve regulatory approval on a worldwide basis. A further factor inhibiting the development of new sweeteners is the existence of the current range. Many regulatory agencies require that new food ingredients are demonstrably 'needed' by the food industry before they will consider devoting their time and resources to an evaluation of safety. The advent of sucralose, which provides a high quality of sweetness in combination with good stability under acidic conditions, could make it even more difficult for manufacturers of new sweeteners to argue a case for further extending the range. An opportunity remains for a wholly natural product with a sweetness quality as good as that of sucrose. At the time of writing, this goal would appear to have a very low probability of success.

The story of the discovery of sucralose and the family of compounds to which it belongs has been covered in Chapter 4 by Professor Hough and Dr Khan. The objective of this chapter is to present the properties, functions and applications of sucralose, and to discuss its advantages to the food industry in products needing a high-quality high-intensity sweetener.


## 2 PHYSICO-CHEMICAL CHARACTERISTICS OF SUCRALOSE

Before attempting to use any new food ingredient an assessment of its physico-chemical characteristics should be made. These can have a profound influence on the way in which an ingredient is used, the materials it can be combined with and the final products.

### 2.1 Structure and Nomenclature

Sucralose (Fig. 1) is produced from ordinary table sugar, sucrose, by a process involving selective chlorination at the 4, 1' and 6' positions of the sugar molecule. The correct chemical nomenclature is 1,6-dichloro-1,6-dideoxy-$\beta$-D-fructofuranosyl 4-chloro-4-deoxy-$\alpha$-D-galactopyranoside. This is sometimes abbreviated to 4,1',6'-trichloro-*galacto*-sucrose, giving rise to the abbreviation TGS which was used in some of the earlier papers and patents.

The structure of crystalline sucralose was determined by X-ray

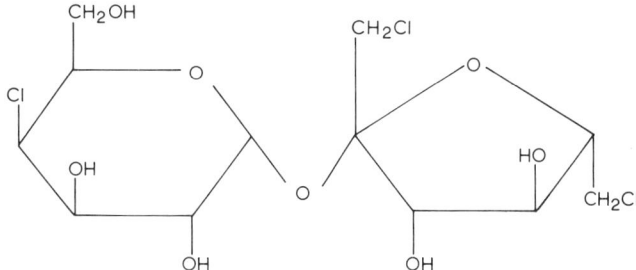

FIG. 1. Structure of sucralose.

diffraction,[1] which showed that the sucralose crystal is orthorhombic with the space-group $P2_12_12_1$. The unit cell contains 4 sucralose molecules, and there is one intramolecular hydrogen bond between OH-2 and O-3', which inhibits rotation around the glycosidic linkage between the two ring systems. There is some evidence from NMR spectroscopy studies[2] that this overall conformation is maintained in solution although, in the dimethyl sulphoxide solution used, the predominant hydrogen bond was between OH-3' and O-2. It is likely that one or more of these hydrogen bonds persists in aqueous solutions, and may make an important contribution to the exceptional stability of sucralose in aqueous acidic conditions.

## 2.2 Properties

The main properties of sucralose are listed in Table 1. Sucralose is typically produced in the form of a white crystalline powder obtained by micronisation of larger crystals. It is odourless, with an intense sweet taste approximately 600 times that of sugar (compared with a 5% sugar solution).

The high solubility of sucralose in water, for example 28·2 g/100 ml at 20°C, means that it can readily be incorporated in aqueous products. If desired, concentrated aqueous solutions can be pumped or metered into a food manufacturing process. An octanol/water partition coefficient of 0·32 implies a very low fat solubility, reflected in the finding that sucralose is virtually insoluble in corn oil. The implication for the food technologist is that in multiphase food systems such as emulsions involving a fat or oil component, sucralose will partition into the aqueous phase and thus will behave in a similar fashion to sugar.

A surface tension of 71·8 mN/m for a 0·1 g/100 ml aqueous solution at 20°C indicates a negligible lowering of the surface tension with respect to water. This has a particular relevance to the use of sucralose in carbonated

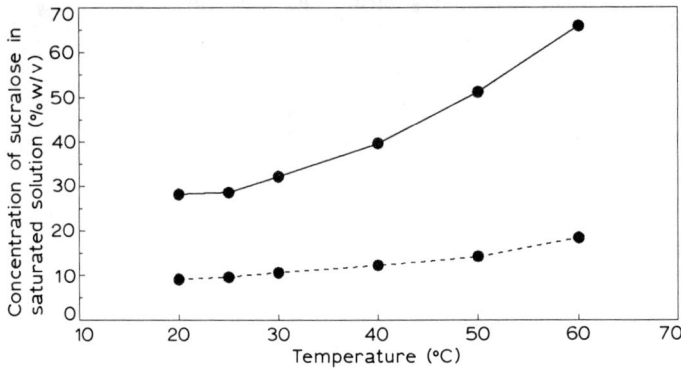

Fig. 2. Solubility of sucralose in water and ethanol. ——, Water; – – –, ethanol.

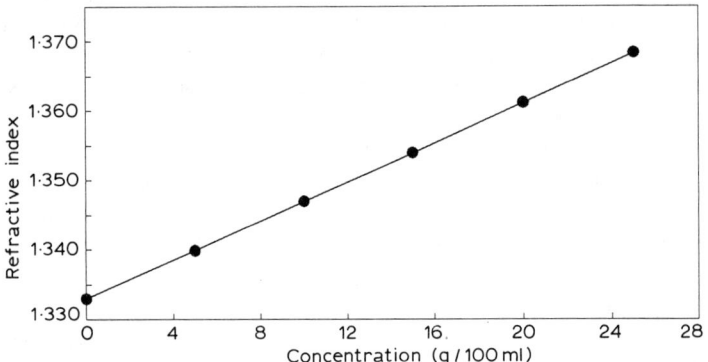

Fig. 3. Refractive index of aqueous sucralose solutions.

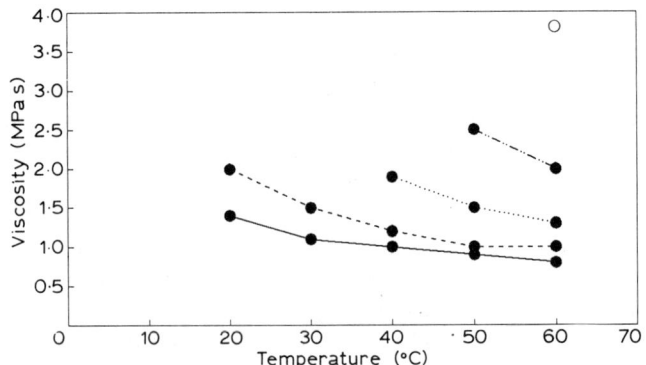

Fig. 4. Dynamic viscosity of aqueous sucralose. ○, 50 g/100 g; – ·· –, 40 g/100 g; ···, 30 g/100 g; – – –, 20 g/100 g; ——, 10 g/100 g.

## TABLE 1
PROPERTIES OF SUCRALOSE

| | |
|---|---|
| Physical form | White crystalline powder |
| Odour | Odourless |
| Taste | Intensely sweet |
| Taste intensity | 400–800 times sweeter than sugar |
| Caloric content | Zero |
| Cariogenicity | Non-cariogenic |
| Solubility | Freely soluble in water and ethanol (see Fig. 2) |
| | Insoluble in corn oil ($<0.1$ g/100 g at 20°C) |
| Specific optical rotation | $[\alpha]_D^{20} + 85.8°$ (C10, aqueous) |
| Octanol/water partition coefficient | 0·32 (20°C) |
| Surface tension of aqueous solutions | 71·8 mN/m (20°C, 0·1 g/100 ml) |
| Melting (decomposition) point | 125°C (when heated from 115°C at 5°/min) |
| Specific gravity (10% aq. solution) | 1·04 (20°C) |
| Specific gravity (crystals) | 1·66 (20°C) |
| Refractive index | Linear correlation with concentration (see Fig. 3) |
| Molecular weight | 397·64 |
| Viscosity of aqueous solutions | Newtonian behaviour (see Fig. 4) |
| pH of 10% aqueous solution | 5–8 |

soft drinks, in which any significant lowering of surface tension can result in excessive generation of foam. The lack of surfactant activity means that sucralose can be used in high-speed bottling and canning lines.

When the refractive index of a series of sucralose solutions was measured using an Abbé refractometer, a linear correlation with concentration was observed (see Fig. 3). This means that measurement of refractive index provides a rapid and accurate method for determining the concentration of aqueous sucralose solutions and will be particularly useful in food manufacturing where concentrates are being handled.

### 2.3 Stability in Aqueous Solutions

Sucralose is exceptionally stable in acidic aqueous systems such as soft drinks, a major factor in deciding which sweetener to use from the range available.

Sucralose has two potential breakdown mechanisms in aqueous solution. At low pH sucralose will slowly hydrolyse to its component monosaccharide derivatives at a rate dependent on pH and temperature. At high pH it can undergo a base-catalysed elimination of hydrogen chloride

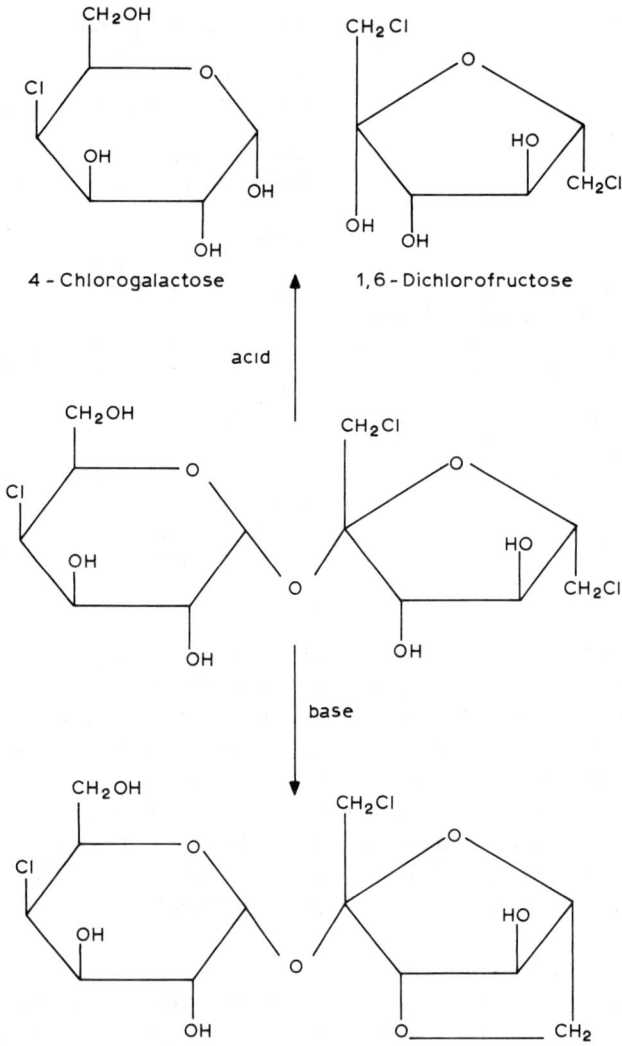

3', 6'- anhydro - 4, 1'- dichlorogalactosucrose

FIG. 5.   Breakdown of sucralose in aqueous solutions.

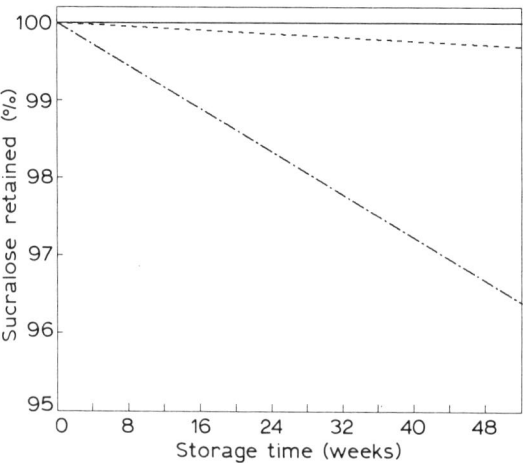

FIG. 6.    Aqueous stability of sucralose; effect of pH, 20°C. – – –, pH 3·0; ——, pH 4·0, 6·0, 7·5; –·–·–, pH 8·6.

from the 3′ and 6′ positions, resulting in the formation of a 3′,6′-anhydride. Figure 5 shows these two possible pathways although, in practice, only the acid catalysed mechanism is relevant for sucralose in food products.

The stability of sucralose in solution is a function of pH, time and temperature. The maximum stability of sucralose is found at approximately pH 5. Figure 6 shows the stability of sucralose at different pH values

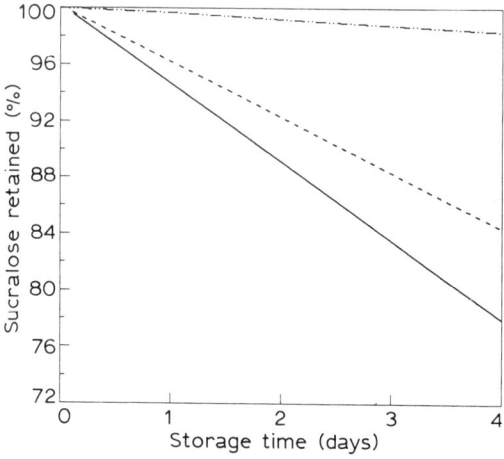

FIG. 7.    Aqucous stability of sucralose (0·1%); effect of pH, 75°C. ——, pH 3·0; –··–, pH 5·0; – – –, pH 7·0.

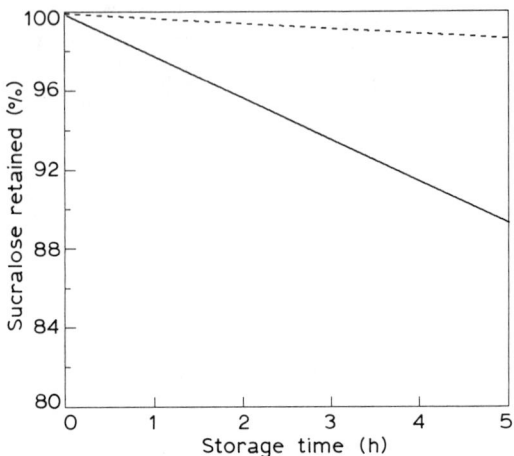

FIG. 8.    Aqueous stability of sucralose (0·1%); effect of pH, 100°C. ——, pH 3·0, 7·0;
––– , pH 5·0.

at 20°C, decomposition following simple first-order kinetics. At higher
temperatures, sucralose degrades more rapidly. Figures 7 and 8 show the
situation at 75 and 100°C, respectively. The maximum stability at both
temperatures is achieved at approximately pH 5. At pH 3 the rate of loss of
sucralose is approximately 0·2%/h at 75°C and approximately 2%/h at
100°C.

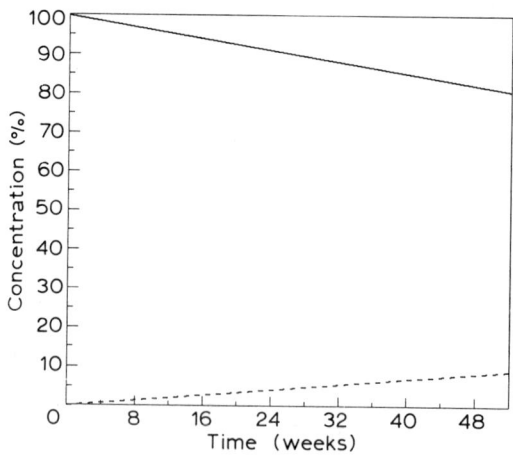

FIG. 9.    Aqueous stability of [$^{36}$Cl]sucralose; buffered pH 3·0 solution, 40°C. ——,
Sucralose; ––– , 1,6-DCF and 4-CG.

The pattern of hydrolysis of sucralose in acidic products has been confirmed using radiolabelled sucralose to facilitate greater analytical accuracy. Solutions were prepared using $^{36}Cl$-labelled sucralose at pH 2·5, 3·0 and 3·5 with storage at 40°C for 1 year. Samples were removed after 8, 16, 26 and 52 weeks, and analysed by thin-layer chromatography (TLC). Radioactive spots were removed from the TLC plates and the activity determined using liquid scintillation counting. The results are presented in Fig. 9, which shows the two products of hydrolysis increasing in total at the same rate as the sucralose is decreasing. Table 2 shows the results in the form of a mass balance table. This was prepared by comparing the amount of sucralose found at each sampling point with the amount of sucralose which would have given rise to the quantity of acid-catalysed hydrolysis products (1,6-DCF and 4-CG) determined by liquid scintillation counting. The results in Table 2 show that at every sampling point essentially quantitative recovery was achieved by determination of these three components. This confirms that hydrolysis to the component mono-saccharide derivatives is the only mechanism operating under these conditions.

This places sucralose as the most stable of all the available high-intensity sweeteners at the pH of soft drinks. The benefit to the manufacturer and the consumer results from a consistent product which can be marketed with a long shelf-life and no reduction of sweetness resulting from degradation. In addition, there are fewer constraints on the food technologist when formulating soft drinks. High-quality diet drinks can be formulated at lower pH values than are possible with other sweeteners, resulting in certain advantages in flavour as well as taste. Furthermore, low-pH concentrates, such as the fountain syrups used to produce carbonated soft drinks in places of public refreshment, can be successfully formulated with sucralose, an application that has not been possible with other high-intensity sweeteners.

TABLE 2
AQUEOUS STABILITY OF SUCRALOSE AT 40°C

| Solution | Mass balance (%) | | | |
|---|---|---|---|---|
| | 8 weeks | 16 weeks | 26 weeks | 52 weeks |
| pH 2·5 | 99·8 | 99·8 | 100·0 | 100·0 |
| pH 3·0 | 99·9 | 99·9 | 99·6 | 100·0 |
| pH 3·5 | 100·0 | 100·0 | 100·0 | 100·1 |

## 2.4 Stability of Dry Sucralose

The stability of dry crystalline sucralose is adequate for its anticipated storage and distribution cycles and use in dry-mix products. Sucralose offers no handling problems provided it is stored under cool, dry conditions. For example, it is stable for 4 years at 20°C in the dry state. This property is a function of temperature and time, and if the raw material is subjected to elevated temperatures the shelf-life will be reduced. For example, at 35°C sucralose has a shelf-life of 12 weeks while at 70°C its shelf-life is 12 h. The first detectable indication that sucralose has been exposed to too high a temperature is the development of a tan colour, providing a quick and simple check. Careful control of temperature is only important for storage of the raw material.

## 2.5 Potential Interactions between Sucralose and Other Food Ingredients

Sucralose is a remarkably unreactive food ingredient since it lacks chemically reactive functional groups. In a study set up to examine the potential for interaction between sucralose and typical food ingredients covering a range of conditions (Table 3), representative examples of bases, oxidising and reducing agents, aldehydes, ketones and metal salts were evaluated. They were added at a concentration of 0·1% to 1·0% sucralose solutions buffered to pH 3, 4, 5 and 7. The resulting solutions were then stored at 40°C for 7 days, after which time the sucralose level was

TABLE 3

INTERACTIONS OF SUCRALOSE WITH FOOD INGREDIENTS:
STUDY DESIGN

| | |
|---|---|
| Sucralose solution | 1·0% w/v |
| Additive level | 0·1% w/v |
| pH of solutions | 3·0, 4·0, 5·0, 7·0 |
| Additives | Bases |
| | Niacinamide |
| | Monosodium glutamate |
| | Oxidising and reducing agents |
| | Hydrogen peroxide |
| | Sodium metabisulphite |
| | Aldehydes and ketones |
| | Acetaldehyde |
| | Ethyl acetoacetate |
| | Metal salts |
| | Ferric chloride |
| Duration of study | 7 days |
| Temperature | 40°C |

TABLE 4
SUCRALOSE RETENTION (%) AFTER 7 DAYS AT 40°C

| Sample | pH 3·0 | pH 4·0 | pH 5·0 | pH 7·0 |
|---|---|---|---|---|
| Control | 100 | 99·8 | 98·9 | 99·5 |
| Sucralose + hydrogen peroxide | 100 | | 97·7 | |
| Sucralose + sodium metabisulphite | 99·9 | | 99·9 | |
| Sucralose + acetaldehyde | 100 | | 100 | |
| Sucralose + ethyl acetoacetate | 98·8 | | 100 | |
| Sucralose + ferric chloride | 95·9 | | 98·0 | |
| Sucralose + niacinamide | | 100 | | 100 |
| Sucralose + monosodium glutamate | | 99·8 | | 100 |

determined by HPLC analysis, a technique which typically has a coefficient of variation of around 2%. The results, summarised in Table 4, show that there was no significant interaction between sucralose and any of the components tested, apart from a slight increase in the rate of hydrolysis in the presence of ferric chloride. This appeared to be greater at pH 3 than at pH 5, and is not expected to have any impact on the performance of sucralose in any food application.

## 3 ORGANOLEPTIC PROPERTIES OF SUCRALOSE

Sucralose is 400–800 times sweeter than sugar, the sweetness factor depending on the application and the pH as well as on the level of sweetness chosen. Sucralose has a sweetness time–intensity profile very similar to that of sugar, and its flavour profile combines the characteristics of sugar without negative features such as the bitter aftertaste associated with saccharin.

### 3.1 Sweetness Threshold
The taste threshold of sucralose in water was determined using a group of 41 taste panellists. The methodology was derived from the 'ascending concentration series method of limits',[3] which allows an estimation to be made of each panellist's individual threshold. The results revealed that the most sensitive panellist had an estimated taste threshold of 0·00014% w/v of sucralose whilst the least sensitive was just able to detect 0·00113% w/v sucralose. The group mean threshold was found to be 0·00038% w/v sucralose. A spread of sensitivity of this magnitude is often observed in

sensitivity measurements. A similar study with sugar revealed corresponding values of 0·07% and 1·13%, with a group mean threshold of 0·31% w/v sugar. Calculating from the group mean threshold values, the taste panel found sucralose to be approximately 815 times sweeter than sugar.

## 3.2 Sweetness Values of Sucralose in Solution

In common with other high-intensity sweeteners,[4,5] the relative sweetness of sucralose against sugar varies as a function of the concentration. Thus at a concentration equivalent in sweetness to 4% sugar, sucralose is approximately 800 times sweeter, while at a concentration equivalent to 10% sugar it is 450 times sweeter. The picture is further complicated by the fact that the apparent sweetness varies as a function of pH, and other food ingredients may also affect the sweetness perception.

Figure 10 shows the relative sweetness of sucralose against concentrations of 4% to 12% sugar at pH 2·75, 3·1 and 7·6 in water at 20°C. It can be seen that the relative sweetness of sucralose increases inversely with pH and that this effect is most apparent at lower sweetness levels. At higher sweetness levels corresponding to the typical use range, the effect of pH becomes less important whilst the effects of excipients and other food ingredients have a greater effect, as demonstrated in Section 4.

## 3.3 Sweetness Time–Intensity Profile

The sweet sensation created by different sweeteners has been found to develop in the mouth at different rates.[6,7] Some sweeteners have a delayed

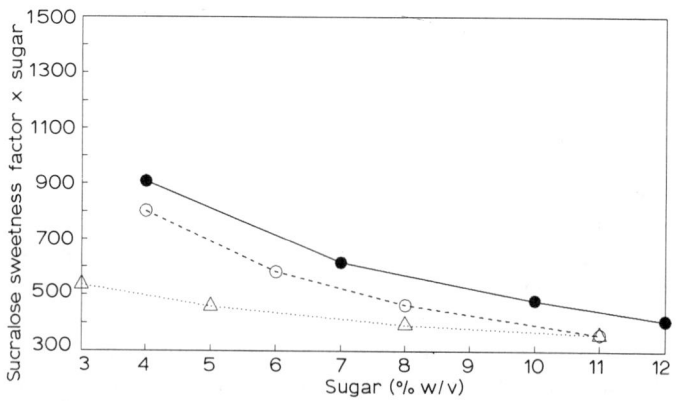

FIG. 10.    Relative sweetness of sucralose against sucrose at pH 2·75, 3·1 and 7·6.
——, pH 2·75; – – –, pH 3·1; ···, pH 7·6.

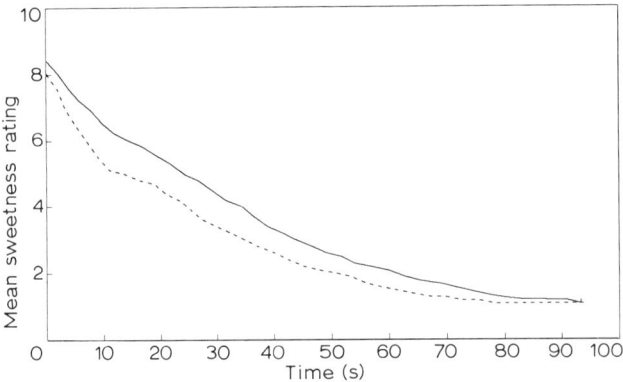

FIG. 11.    Sweetness intensity versus time for sucralose and sucrose in Malvern water; pH 7·6, 20°C. ——, Sucralose; – – –, sugar.

onset of sweetness; others give an immediate impression of sweetness. Similarly, the time taken for the sweetness to subside varies. These effects can be quantified by sweetness time–intensity measurements, revealing the development and decay of sweetness over a period of time. Sucralose is characterised by a sweetness time–intensity profile which is very similar to that of sugar. The onset of sweetness perception is very rapid, and the sweet taste persists in a similar way to that of sugar, as shown in Fig. 11.

### 3.4 Flavour Profile
An assessment of the flavour profile of a sweetener in aqueous solution without the complication of other food components provides a means of

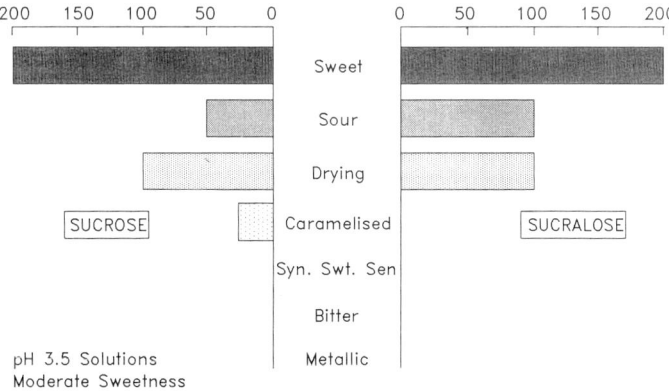

FIG. 12.    Flavour profile for sucrose and sucralose.

judging the quality of the sweetener and, by analogy, the sweetness quality of products using that sweetener. A number of techniques are available, and one that has been applied to sucralose is the 'flavour profile' method developed by Arthur D. Little Inc., Massachusetts, USA. This is an empirical approach relying on perceptual judgements by carefully selected and extensively trained panellists working as a team to reach a consensus evaluation. The flavour profile of sucralose was determined at a moderate sweetness level corresponding approximately to 8% sugar in a solution acidulated to pH 3·5 by the addition of citric acid. The results are presented in Fig. 12 alongside a similar evaluation of sugar. The conclusion was that sucralose is the high-intensity sweetener whose flavour profile is most like that of sugar. It is of particular interest that sucralose does not possess the bitter, metallic or synthetic sweetener sensations often associated with other high-intensity sweeteners.[8]

## 4 APPLICATIONS

In February 1987 a food additive petition was submitted to the Food and Drug Administration (FDA) in the USA for the use of sucralose in the 14 food categories listed in Table 5. The list demonstrates the wide range of products in which sucralose has been proved useful. Examples of foods in each category were produced, and analytical methods were developed for stability trials over an appropriate shelf-life. In addition, sensory studies were conducted to establish a technologically desirable level (TDL) for the addition of sucralose, providing a sweetness equivalence value (SEV) against a sugar-sweetened control. A self-limiting value (SLV) was also determined when an excessively high use level would result in unacceptably sweet products. The SLV was determined by dividing the TDL by a self-limiting level which was defined as the sucralose addition level at which

TABLE 5
FOOD CATEGORIES IN WHICH SUCRALOSE IS USED

| | |
|---|---|
| Baked goods and baking mixes | Fruit and water ices |
| Beverages and beverage bases | Gelatins and puddings |
| Chewing gum | Jams and jellies |
| Coffee and tea | Milk products |
| Dairy product analogues | Processed fruits and fruit juices |
| Fats and oils | Sugar substitutes |
| Frozen dairy desserts and mixes | Sweet sauces, toppings and syrups |

TABLE 6
SUMMARY TABLE OF SENSORY DATA

| Product | $TDL^a$ (%) | $SEV^b$ | $SLV^c$ |
|---|---|---|---|
| Cake | 0·048 | 590 | 3·5 |
| Cookie | 0·033 | 967 | 5·5 |
| Brownie | 0·068 | 320 | — |
| Cola (carbonated) | 0·0285 | 449 | 2·4 |
| Lemon/lime (carbonated) | 0·0246 | 438 | 1·8 |
| Powdered cherry drink | 0·0216 | 440 | 2·0 |
| Peppermint gum | 0·0843 | 747 | 3·0 |
| Grape bubblegum | 0·0626 | 1 007 | 2·4 |
| Iced tea mix | 0·0142 | 528 | 1·4 |
| Canned iced tea | 0·0147 | 510 | 2·9 |
| Whipped topping | 0·024 | 875 | 2·3 |
| French salad dressing | 0·0125 | 531 | 2·2 |
| Vanilla ice milk | 0·027 | 498 | 1·2 |
| Cheesecake | 0·03 | 873 | — |
| Grape water ice | 0·022 | 909 | 1·5 |
| Chocolate pudding mix | 0·0203 | 419 | 2·0 |
| Orange gelatin mix | 0·0314 | 446 | 2·2 |
| Grape jelly | 0·035 | $320^d$ | 2·9 |
| Strawberry jam | 0·045 | $541^d$ | 2·2 |
| Strawberry flavoured milk | 0·0059 | 677 | 2·1 |
| Strawberry yoghurt | 0·02 | 450 | 2·3 |
| Hot chocolate mix | 0·0076 | 658 | 2·0 |
| Canned peaches | 0·057 | 526 | 2·6 |
| Aseptically processed orange drink | 0·0183 | 492 | 2·1 |
| Spoon for spoon (table-top) | 0·0074–0·0178 | 534–680 | 1·7 |
| Sachet (table-top) | 0·0069–0·0074 | 676–719 | 1·8 |
| Tablets (table-top) | 0·0076–0·0077 | 649–658 | 1·3 |
| Chocolate syrup | 0·107 | 689 | 1·8–1·9 |

$^a$ Technologically desirable level.
$^b$ Sweetness equivalence value determined against sugar-sweetened product.
$^c$ Self-limiting value (number of times TDL).
$^d$ Sweetness equivalence value determined against a reduced-sugar product.

50% of the taste panellists said that the sample was too sweet or much too sweet. The TDL, SEV and SLV values for the products developed for the FDA petition are presented in Table 6. Other than the obvious fact that sweeter products require more sweetener, there is no clear pattern which would enable prediction of an exact sucralose level for a new formulation. The most appropriate sucralose level for any formulation can only be determined by the use of an expert taste panel or a consumer trial. Not only do the other components in a product influence the perception of sweetness as part of the whole organoleptic sensation, but cultural and regional differences also influence the most desirable level. For example, it is generally understood that the USA market requires products at higher sweetening levels than those for the UK or Canadian markets.

## 5 FORMULATIONS

Many hundreds of high-quality products have been formulated with sucralose. Their range and complexity is limited only by the imagination of

### TABLE 7
#### LEMON/LIME CARBONATED DRINK

Sugar can be readily substituted by sucralose in carbonated soft drinks to produce low-calorie beverages with good sweetness and flavour characteristics.

*Syrup formulation*
| | | |
|---|---|---|
| 1. | Sodium benzoate (10% w/v) | 10·00 litres |
| 2. | Citric acid (anhydrous) | 22·85 kg |
| 3. | Trisodium citrate | 3·58 kg |
| 4. | Flavour: lemon/lime 17.40.3524 | 1·00 litre |
| 5. | Flavour: lemon/lime 60132-56 | 0·15 litre |
| 6. | Flavour: lemon/lime 74940-74 | 0·10 litre |
| 7. | Sucralose | 1·24 kg |
| 8. | Water to | 1 000·00 litres |

*Raw materials suppliers*
| | |
|---|---|
| 1–3. | Hays Chemicals Ltd, London Division |
| 4. | International Flavours & Fragrances (GB) Ltd |
| 5, 6. | Givaudan & Co. Ltd |

*Product make-up procedure*
1. Dissolve all the ingredients individually in some water.
2. Mix the ingredients in the order stated and make up to volume.
3. Dose 1 part syrup to 5·5 parts carbonated water, to 4 volumes carbonation.
4. Cap, invert to mix, and store.

*Comparative product data*

| Sweetener | Conc. (%) in syrup | Conc. (%) in finished beverage | Sweetness equivalence value | Kcal/100 ml finished beverage | Caloric reduction (%) |
|---|---|---|---|---|---|
| Sugar | 70·000 | 10·77 | 1 | 43 | — |
| Sucralose | 0·124 | 0·019 | 567 | 1·0 | 97·6 |

## TABLE 8
### ORANGE DRINK

Sugar can be readily substituted by sucralose in soft drinks to produce low-calorie beverages with good sweetness and flavour characteristics.

*Syrup formulation*
| | | |
|---|---|---|
| 1. | Sodium benzoate (10% w/v) | 0·7875 kg |
| 2. | Concentrated orange base ($SO_2$ preserved) | 150·0000 NEL[a] |
| 3. | Anhydrous citric acid | 11·6250 kg |
| 4. | Ascorbic acid | 0·2250 kg |
| 5. | Colour: $\beta$-carotene (5%) | 0·4375 kg |
| 6. | Flavour: natural jaffa 730082E | 0·6250 kg |
| 7. | Flavour: natural orange 511107E | 0·8750 kg |
| 8. | Sucralose | 1 000·0000 litres |
| 9. | Water to | |

*Raw materials suppliers*
1, 3, 4. Hays Chemicals Ltd, London Division
2. Gerald McDonald & Co. Ltd
5. Roche Products Ltd
6, 7. PFW Ltd

*Product make-up procedure*
1. Dissolve all the ingredients individually in some water.
2. Mix the ingredients in the order stated and make up to volume.
3. Mix the syrup well prior to bottling.
4. Dilute 1 part syrup with 4 parts of water to taste.

*Comparative product data*

| Sweetener | Conc. (%) in syrup | Conc. (%) in finished beverage | Sweetness equivalence value | Kcal/100 ml finished beverage | Caloric reduction (%) |
|---|---|---|---|---|---|
| Sugar | 45·0000 | 9·0000 | 1 | 37·3 | — |
| Sucralose | 0·0875 | 0·0175 | 514 | 1·0 | 97·3 |

[a] Natural equivalent litres.

the food technologist, as the stability and lack of chemical reactivity of sucralose enable a broad range of processing conditions and ingredients to be used.

Formulations for 5 typical products are presented in Tables 7–11. These cover a lemon/lime carbonated drink, an orange drink, peppermint chewing gum, chocolate dessert and tomato ketchup, which have been designed as high-quality products particularly appropriate for the UK palate.

## TABLE 9
### PEPPERMINT CHEWING GUM

Sucralose can be used to replace sugar when formulating a chewing gum. The use of alternative carbohydrates and glycerine is necessary to maintain the physical properties of the gum.

| Formulation | | % w/w |
|---|---|---|
| 1. | Sorbitol: Neosorb (60w powder) | 54·473 |
| 2. | Sugar-free gum base | 24·480 |
| 3. | Sorbitol solution (70%) | 14·905 |
| 4. | Glycerine | 4·800 |
| 5. | Flavour: peppermint 17.92.2110 | 1·000 |
| 6. | Flavour: peppermint 17.41.0105 | 0·300 |
| 7. | Sucralose | 0·042 |

*Raw materials suppliers*
1, 3.  Roquette UK Ltd
2, 5, 6.  International Flavours & Fragrances (GB) Ltd
4.  British Fermentation Products Ltd

*Product make-up procedure*
1. Heat the Z-blade mixer to 60–80°C.
2. Heat gum base, glycerine, sorbitol solution and sorbitol powder individually at 90–95°C for 20 min to aid softening.
3. Mix softened gum base in pre-heated Z-blade mixer.
4. Add pre-heated sorbitol solution and sorbitol powder and mix.
5. Add remaining ingredients and mix.
6. Roll out and dust with unperfumed talcum powder.

*Comparative product data*

| Sweetener | Conc. (%) in finished gum | Sweetness equivalence value[a] |
|---|---|---|
| Sugar | 59·200 | 1 |
| Sucralose | 0·042 | 448 |

[a] The SEV of sucralose does not correlate directly to the sugar level due to sweetness contribution from other ingredients.

## TABLE 10
### CHOCOLATE DESSERT

Sucralose can be used to replace sugar when formulating a reduced-calorie instant dessert mix. A slight increase in the starch level is required to maintain a comparable mouthfeel.

| *Formulation* | | *% w/w* |
|---|---|---|
| 1. | Spray dried fat: Silkido W60 | 50·0 |
| 2. | Starch: Instant Pureflo | 19·0 |
| 3. | Carrageenan: Genuvisco CSM-1 | 6·0 |
| 4. | Tetrasodium pyrophosphate | 3·0 |
| 5. | Disodium hydrogen phosphate | 1·5 |
| 6. | Cocoa powder: AA10 alkalised | 16·2 |
| 7. | Flavour: chocolate 17.42.4055 | 4·0 |
| 8. | Sucralose | 0·3 |

*Raw materials suppliers*
1. L. E. Pritchett & Co. Ltd
2. National Starch & Chemical
3. Hercules Ltd
6. British Cocoa Mills
7. International Flavours & Fragrances (GB) Ltd

*Dry mix preparation*
1. Pre-blend the sucralose with the Instant Pureflo to aid the uniform dispersion of sucralose in the final mix.
2. Add the pre-mix to the remaining ingredients and dry-blend the ingredients until a uniform mix is obtained.
3. Fill the final mix into laminated foil sachets (20·1 g per sachet) and heat-seal.

*Product make-up procedure*
1. Empty the contents of a sachet into 200 ml of whole milk and whisk until thick.
2. Leave to set in the refrigerator for 5 min prior to serving.

*Comparative product data*

| Sweetener | Conc. (%) in dry mix | Conc. (%) in finished product | Sweetness equivalence value | Kcal/100 ml finished product | Caloric reduction (%) |
|---|---|---|---|---|---|
| Sugar | 56·15 | 10·41 | 1 | 143 | — |
| Sucralose | 0·30 | 0·03 | 347 | 102 | 28 |

## TABLE 11
### TOMATO KETCHUP

The replacement of sugar with sucralose in a tomato ketchup produces a low-calorie product with good sweetness characteristics. A reduction in the acid level is required to maintain the flavour balance of the product together with a marginal increase in the stabiliser level to maintain viscosity.

| Formulation | | % w/w |
|---|---|---|
| 1. | Tomato purée (28/30 solids) | 44·3400 |
| 2. | Water | 33·4968 |
| 3. | Distilled malt vinegar (5·2% acetic) | 13·3100 |
| 4. | Distilled malt vinegar (5·2% acetic) | 4·4360 |
| 5. | Salt | 3·4700 |
| 6. | Stabiliser: Frimulsion E | 0·8729 |
| 7. | Spices: mustard (50%) | 0·01485 |
| 8. | Spices: onion powder (50%) | 0·01485 |
| 9. | Sucralose | 0·0446 |

Raw materials suppliers
1. Cirio & Co. Ltd
3. Hammonds & Co. Ltd
4. Ellsey & Co. Ltd
6. Hercules Ltd
7. J. Dalton (Seasoning & Spices Ltd)
8. G. Fiske & Co. Ltd

Product make-up procedure
1. Pre-blend the dry ingredients and add to the warmed water under continuous agitation.
2. Add the tomato purée and vinegar and heat the sauce to 80–85°C, whilst stirring, in a covered vessel.
3. Hot-fill immediately. Do not let temperature fall below 85°C prior to filling to avoid the need for pasteurisation.

Comparative product data

| Sweetener | Conc. (%) in finished product | Sweetness equivalence value | Kcal/100 g | Caloric reduction (%) |
|---|---|---|---|---|
| Sugar | 25·5000 | 1 | 126·0 | — |
| Sucralose | 0·0446 | 571 | 30·9 | 75·47 |

# 6 CONCLUSION

Sucralose offers a unique combination of features including a clean, sweet taste and no undesirable off-flavours. It is a versatile ingredient, very stable under the processing and storage conditions used for food products, and it does not present any problems of interaction with other food components. These features, combined with ease of handling due to its high water solubility, suggest that, upon approval, sucralose will soon become the sweetener of choice in industry.

## REFERENCES

1. Jenner, M. R. & Waite, D. (Tate & Lyle). UK patent 2065646 (1980).
2. Christofides, J. C., Davies, D. B., Martin, J. A. & Rathbone, E. B. *J. Amer. Chem. Soc.*, **108** (1986) 5738–43.
3. Standard practice for determination of odour and taste thresholds by a forced-choice ascending concentration series method of limits. In *Manual on Sensory Testing Methods*. ASTM E679, American Society for Testing and Materials, 1979, pp. 51–7.
4. Ripper, A., Homler, B. E. & Miller, G. A., Aspartame. In *Alternative Sweeteners*, ed. L. O'B. Nabors & R. C. Gelardi. Marcel Dekker, New York, 1986, pp. 43–70.
5. von Rymon Lipinski, G.-W., Acesulfame-K. In *Alternative Sweeteners*, ed. L. O'B. Nabors & R. C. Gelardi. Marcel Dekker, New York, 1986, pp. 89–102.
6. Larson-Powers, N. & Pangborn, R. M., Paired comparison and time–intensity measurements of the sensory properties of beverages and gelatins containing sucrose or synthetic sweeteners. *J. Food Sci.*, **43** (1978) 41–6.
7. Swartz, M., Sensory screening of synthetic sweeteners using time–intensity evaluations. *J. Food Sci.*, **45** (1980) 577–81.
8. Redlinger, P. A. & Setser, C. S., Sensory quality of selected sweeteners: aqueous and lipid model systems. *J. Food Sci.*, **52** (1987) 451–4.

Chapter 6

# PALATINOSE—AN ISOMERIC ALTERNATIVE TO SUCROSE

Ichiro Takazoe

Department of Microbiology, Tokyo Dental College, Japan

## SUMMARY

*The chemical nature, production method, utilization, cariogenicity and nutritional characteristics of palatinose (6-O-α-D-glucopyranosyl-D-fructose) are described. Compared with sucrose, palatinose is less soluble, less sweet and less thermostable. Principally palatinose is produced by enzymic conversion from sucrose. The conversion process with immobilized α-glucosyltransferase from* Protaminobacter rubrum *is used for large-scale production in Japan. The low cariogenicity of palatinose was demonstrated by experiments in vitro. Various serotypes of* Streptococcus mutans *do not utilize palatinose. The acid production activity of* Streptococcus mutans *from palatinose was quite low. Furthermore, dental plaque samples produced little or no lactate from palatinose. No water-insoluble glucan was synthesized from palatinose by* Streptococcus mutans *glucosyltransferase. Also palatinose was found to inhibit insoluble polyglucan synthesis from sucrose, functioning as a receptor of glucosyl base. In experimental caries studies with rats the low cariogenicity and inhibitory action of insoluble polyglucan synthesis from sucrose were also demonstrated. A human study with frequent oral rinses with palatinose solution indicated that palatinose is considerably less acidogenic than glucose. Metabolic studies show that hydrolysis of palatinose is slower than that of sucrose. As a potential parenteral nutrient, palatinose is indicated to be valuable clinically for both diabetic and non-diabetic patients. Palatinose is considered as a promising alternative to sucrose.*

## 1 INTRODUCTION

In 1957 Weidenhagen and Lorenz[1] found a disaccharide other than sucrose in the beet sugar manufacturing process and called it palatinose in view of the location of the factory, Palatine. As a chemical nomenclature isomaltulose was proposed. The same authors discovered that *Protaminobacter rubrum* formed palatinose from sucrose, and later it was found in honey[2] and cane-juice.[3] Thus the history of palatinose is rather new, but much attention has already been paid to this naturally occurring disaccharide because of its special biological characteristics and low cariogenicity. This chapter deals with the chemical nature, production method, utilization, cariogenicity and nutritional character of palatinose.

## 2 GENERAL CHARACTERISTICS OF PALATINOSE

Palatinose (6-*O*-α-D-glucopyranosyl-D-fructose) is a crystalline-reducing disaccharide. It crystallizes with $1\,H_2O$ per molecule, but the dehydrated form is not crystalline. The crystals are of rhombic form, like fructose, of molecular weight 360 including water of crystallization. The melting point is 122–123°C, which is much lower than that of sucrose (182°C). Specific rotation $[\alpha]_D^{20}$ is 97·2°. Reduction activity is 52% of that of glucose. The chemical formula of palatinose is shown in Fig. 1.

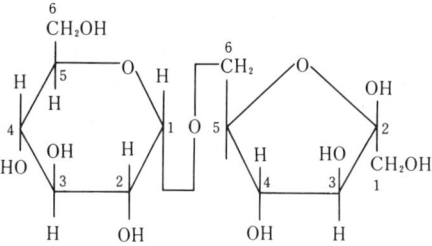

FIG. 1.   Chemical structure of palatinose.

Palatinose possesses a good sweetness profile similar to that of sucrose. The initial rise of sweetness is perceived more rapidly than that from sucrose; a strong sweetness similar to that of sucrose then persists, and at the final stage a taste weaker than that of sucrose remains. It has no foreign taste.[4] The sweetness is 42% of that of sucrose, and no change in sweetness with temperature has been noticed. When used in foods like candy and chocolate no major difference in sweetness between palatinose and sucrose has been noticed.

The solubility of palatinose at room temperature is about half that of sucrose.[4] As the temperature rises the solubility increases steeply and at 80°C it reaches 85% of that of sucrose (Fig. 2). There is a possibility

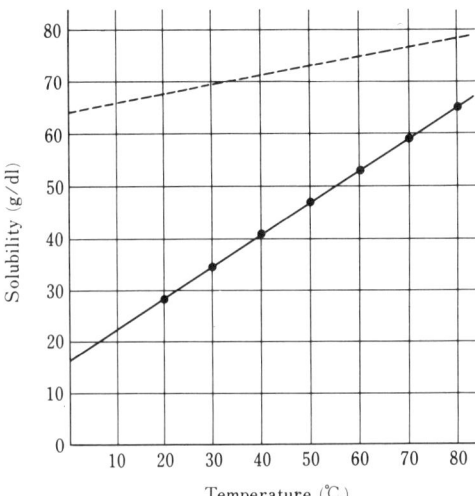

FIG. 2.   Solubility of palatinose.

therefore that, in palatinose-containing foods prepared at relatively high temperatures, crystallization of palatinose may sometimes occur when they are kept at room temperature. The viscosity of palatinose is slightly lower than that of sucrose at the same concentration.

Palatinose alone does not possess hygroscopic properties like those of granular sucrose or lactose.[4] Even on adding citric acid at 1·5–15% to palatinose, the hygroscopicity did not increase as it does with granular sucrose. In fact, in the mixture of palatinose and citric acid no invert sugar was formed in 22 days of incubation. These observations imply that palatinose-containing foods acidified with organic acids or vitamin C are more stable than those containing sucrose.

Palatinose is quite resistant to acid. Acidified (pH 2) 20% solutions of palatinose and sucrose were boiled and their hydrolysis rates compared. While sucrose was completely hydrolysed after 60 min, palatinose was not hydrolysed. Tests in confectionery indicated that up to 120°C there was no change in the taste but some slight colouration. Above 140°C colouration, decomposition and polymerization were observed, and these were acutely intensified beyond 160°C. Thus the thermostability[4] of palatinose is slightly weaker than that of sucrose.

Palatinose is not fermented by most bacteria and yeasts. In acidic beverages and bread containing palatinose and sucrose, the palatinose was found to remain intact. This property of palatinose certainly facilitates the maintenance of sweetness and taste in fermented foods and beverages. The fact that palatinose is not fermented by oral bacteria, including cariogenic strains, predicts its low cariogenicity, which will be discussed later.

## 3 PRODUCTION OF PALATINOSE

Enzymic conversion from sucrose is the principle of palatinose production. The enzyme is α-glucosyltransferase. The first record of this enzyme was in *Protaminobacter rubrum*, from which it was isolated by Weidenhagen and Lorenz, and later identified by Windish[5] in 1958. Later Schmidt-Berg-Lorenz and Mauch[6] (1964) found a similar enzyme of high activity in a strain of *Serratia plymuthica*. For large-scale production of palatinose, simplification of the conversion process was attempted by Nakajima[7] using a column reactor containing immobilized α-glucosyltransferase. This production method is now established and used on an industrial scale.

To accumulate induced cellular glucosyltransferase, *Protaminobacter rubrum* or *Serratia plymuthica* were cultivated under aerobic conditions in a sucrose-containing medium. The cells were harvested from the culture broth by centrifugation and entrapped in calcium alginate gel in a granular form. The granulated enzyme was soaked in a slightly acidified 2% polyethyleneimine solution for 5 min, and then stirred in a 0·5% glutaraldehyde solution below 5°C for 30 min to convert it into an immobilized enzyme form. The treatment with polyethyleneimine followed by glutaraldehyde seems to stabilize the granulated enzyme with calcium alginate gel.

The α-glucosyltransferase activity of *Protaminobacter rubrum* was found to be more stable than that of *Serratia plymuthica*. The half-lives of immobilized enzymes from these two organisms were estimated at 73 and 23 days respectively in successive batch reactions at 25°C (Fig. 3). *Protaminobacter rubrum* has consequently been selected for practical use.

The optimum pH for the immobilized enzyme is about 5·5, while that for the original cellular enzyme was 6·8 (Fig. 4). Though a considerable improvement in the temperature stability was expected by the immobiliz-ation, it is not realistic to use the enzyme above 30°C for more than a few days. The Michaelis constants ($K_m$) of the immobilized and original cellular enzymes were determined to be 0·14 and 0·12M respectively. Thus the immobilization does not affect $K_m$ of the enzyme. The product inhibition constant was estimated at 0·31M by observing the course of batch reactions with immobilized enzyme. The product inhibition constant is apparently

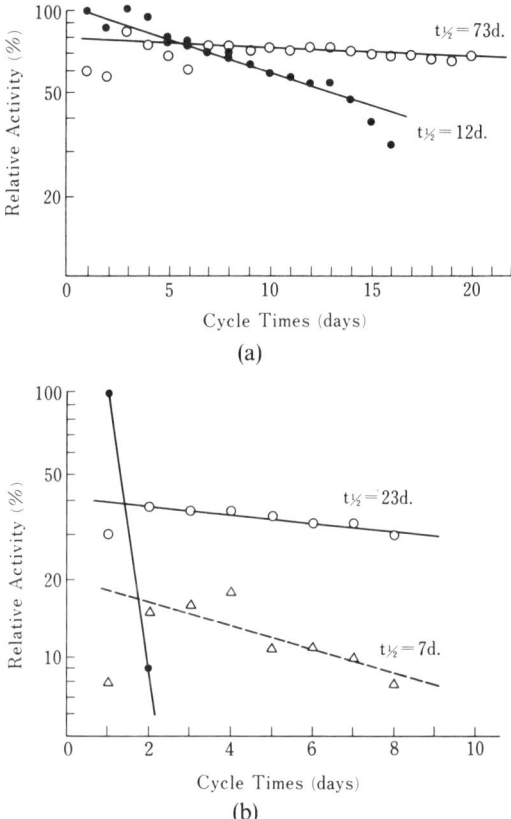

FIG. 3.    (a) Activity of granulated and immobilized α-glucosyltransferase preparations from *P. rubrum* cells: ●, granulated; ○, immobilized. (b) Activity of various granular α-glucosyltransferase preparations from *S. plymuthica* cells: ●, Ca alginate entrapped; △, GA treated without PEI; ○, GA treated with PEI.

larger than the $K_m$, indicating that the affinity of the enzyme to sucrose is somewhat stronger than that to the product, so that sucrose concentration in the reaction mixture was lowered.

The efficiency of the enzyme activity decreased as the Bx value of the reaction mixture increased, particularly in the range over 30, but a Bx value of 40 has been selected so as to save energy in the later evaporation process.

A Bx 40 sucrose solution, adjusted to pH 5·5, sterilized at 120°C for 15 min and cooled to 25°C, was passed through a column containing the immobilized enzyme. The flow rate was controlled to keep the residual

ICHIRO TAKAZOE

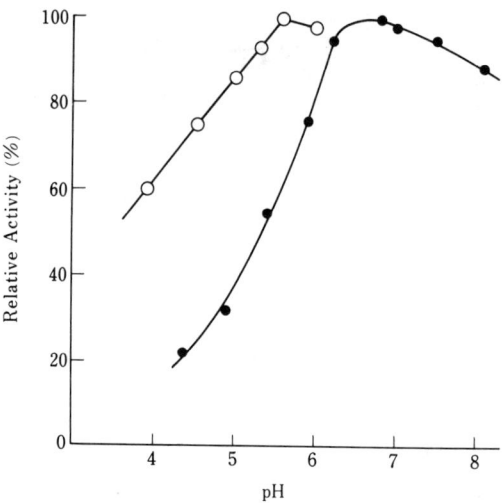

Fig. 4. pH–activity profiles of immobilized and free cellular α-glucosyltransferase: ○, immobilized; ●, free cellular.

sucrose concentration in the effluent below 0·8% (w/w). This way more than
85% of applied sucrose was converted into palatinose, though small
amounts of 1-O-α-D-glucosyl-D-fructose, isomaltose, isomelezitose,
fructose and glucose were also found in the effluent. The effluent was refined
with ion-exchange resins and evaporated under reduced pressure.
Palatinose was crystallized from the concentrated solution by seeding and

TABLE 1

ANALYTICAL VALUES OF PALATINOSE

|  | Crystal | Syrup (molasses) |
|---|---|---|
| Water content | 0·1% | — |
| Solid content (including crystal water) | 99·9% | — |
| Sugar composition (%) |  |  |
|     Fructose | 0 | 14·8 |
|     Glucose | 0 | 13·0 |
|     Sucrose | 0 | 3·5 |
|     Palatinose | 99·5 | 19·4 |
|     Trehalulose | trace | 39·4 |
|     Isomaltose | 0 | 5·2 |
|     Isomelezitose | 0 | 2·2 |

cooling. The crystals were separated from the molasses by centrifugation. The final molasses contained 1-$O$-α-D-glucosyl-D-fructose as the largest component (Table 1) and tasted sweet.

The molasses is called palatinose syrup and is also used in foods. Its solubility is fairly high, and the sweetness is 70% of that of sucrose. As an alternative to palatinose syrup production the refined solution can be concentrated, totally solidified and pulverized into a crude solid material.

In Japan approximately 3000 tons of palatinose was produced in 1987 and utilized for foods such as candies, toffee, chewing gum, chocolate, jam, spreads, soft drinks and cakes.

## 4 CARIOGENICITY OF PALATINOSE

The low cariogenicity of palatinose was predicted from a pilot study by Gehring,[8] but it has recently been studied more systematically by several authors, examining its acid production, polyglucan synthesis in an *in vitro* system, and its cariogenicity in animals and man.

### 4.1 Acid Production from Palatinose

*4.1.1 Acid Production by Oral Microorganisms*[1]
Representative strains of each serotype of *Streptococcus mutans* were used for fermentation tests. In addition, *Streptococcus sanguis* ATCC 10556, 10557, 10558, *Streptococcus mitis* 9811, *Streptococcus salivarius* 9759 and *Streptococcus* sp. MG9895 were also tested. Fermentation in inoculated culture media was studied first. Palatinose was filter-sterilized separately and added to the medium at a final concentration of 1%. The medium alone, and with 1% glucose, served as controls. Approximately 0·04 ml of culture grown in Todd Hewitt broth for 24 h was inoculated into 4·0 ml of the test medium. After 7 days of anaerobic incubation the pH of the culture medium was measured and the growth of the culture was assayed.

The pH of all the strains in the glucose media dropped sharply to between 4·1 and 4·8. All the strains of *S. mutans* belonging to serotypes *a*, *d* and *g*, and one of serotype *e*, failed to grow in the palatinose medium, and the pH remained unchanged. However, the other strains of *S. mutans* did lower the pH to some extent. Strains such as JC-2, PS-14, LM-7 and OMZ 115 lowered the pH to less than 5·5, with some growth[9] (Table 2). These findings agreed with those reported by Gehring.[8]

Acid production from palatinose by cell suspensions of various serotypes

## TABLE 2
FINAL pH AND GROWTH OF *STREPTOCOCCUS* STRAINS IN CULTURE AFTER 7 DAYS OF INCUBATION

| Strain number and serotype (in parentheses) | Glucose | Palatinose | No sugar |
|---|---|---|---|
| *S. mutans* | | | |
| E49 (a) | $4·32^a(0·65)^b$ | $6·77 (0·0)^c$ | 6·77 (0·0) |
| OMZ61 (a) | 4·36 (0·66) | 6·78 (0·0) | 6·78 (0·0) |
| AHT (a) | 4·35 (0·80) | 6·77 (0·0) | 6·78 (0·0) |
| BHT (b) | 4·50 (0·86) | 6·65 (0·09) | 6·87 (0·02) |
| 107P (b) | 4·31 (0·76) | 6·66 (0·10) | 6·85 (0·02) |
| Ingbritt (c) | 4·14 (0·78) | 6·42 (0·08) | 6·77 (0·0) |
| JC-2 (c) | 4·11 (0·69) | 5·15 (0·23) | 6·78 (0·0) |
| PS-14 (c) | 4·13 (0·73) | 5·03 (0·31) | 6·77 (0·0) |
| OMZ176 (d) | 4·23 (0·84) | 6·76 (0·0) | 6·77 (0·0) |
| B-13 (d) | 4·41 (0·56) | 6·76 (0·0) | 6·78 (0·0) |
| LM-7 (e) | 4·47 (0·47) | 5·05 (0·27) | 6·78 (0·0) |
| AT-10 (e) | 4·65 (0·38) | 6·75 (0·0) | 6·78 (0·0) |
| OMZ175 (f) | 4·42 (0·76) | 5·27 (0·21) | 6·80 (0·0) |
| QP50-1 (f) | 4·44 (0·55) | 5·95 (0·13) | 6·78 (0·0) |
| 6715 (g) | 4·30 (0·54) | 6·76 (0·0) | 6·77 (0·0) |
| K1R (g) | 4·20 (0·74) | 6·78 (0·0) | 6·77 (0·0) |
| *S. sanguis* | | | |
| ATCC 10556 | 4·56 (0·49) | 6·85 (0·04) | 6·89 (0·04) |
| ATCC 10557 | 4·47 (0·57) | 6·66 (0·02) | 6·76 (0·0) |
| ATCC 10558 | 4·55 (0·37) | 6·84 (0·02) | 6·90 (0·02) |
| *S. mitis* | | | |
| ATCC 9811 | 4·86 (0·40) | 6·54 (0·07) | 6·76 (0·0) |
| *S. salivarius* | | | |
| ATCC 9759 | 4·84 (0·55) | 6·76 (0·0) | 6·79 (0·0) |
| *S.* sp. *MG* | | | |
| ATCC 9895 | 4·49 (0·62) | 6·90 (0·02) | 6·92 (0·02) |
| Control | 6·77 | 6·77 | 6·78 |

All data expressed as average of duplicate tubes.
[a] Final pH of the culture.
[b] Optical density at 550 nm.
[c] Indicates less than 0·01.

## TABLE 3
ACID PRODUCTION ACTIVITIES WITH SUCROSE, GLUCOSE AND PALATINOSE IN CELL
SUSPENSIONS OF S. MUTANS STRAINS

| Strain number and serotype (in parentheses) | Acid production activities[a] | | | |
|---|---|---|---|---|
| | Sucrose | Glucose | Palatinose | No sugar (endogenous activity) |
| E49 (a) | 253 | 240 | 1·1 | 0·9 |
| OMZ 61 (a) | 200 | 187 | 0·7 | 0·7 |
| AHT (a) | 187 | 201 | 0·5 | 0·5 |
| BHT (b) | 315 | 322 | 1·0 | 0·4 |
| 107P (b) | 238 | 283 | 0·9 | 0·7 |
| Ingbritt (c) | 238 | 248 | 0·3 | 0 |
| JC-2 (c) | 281 | 315 | 1·0 | 0·4 |
| PS-14 (c) | 274 | 286 | 1·3 | 0 |
| ONZ 176 (d) | 307 | 299 | 0·7 | 0·6 |
| B-13 (d) | 272 | 310 | 0·5 | 0·4 |
| LM-7 (e) | 210 | 190 | 1·3 | 0·7 |
| AT-10 (e) | 443 | 326 | 0·8 | 0·3 |
| OMZ 175 (f) | 414 | 372 | 1·5 | 0·5 |
| QP50-1 (f) | 321 | 359 | 2·9 | 1·8 |
| 6715 (g) | 364 | 378 | 0·7 | 0·7 |
| K1R (g) | 253 | 240 | 1·6 | 1·5 |

[a] Expressed as equivalent weight $\times 10^{-9}$ per mg dry weight of cells per min.

of S. mutans, in total 16 strains, was also determined by the automatic titration method described by Larje and Frostell[10] and Birkhed.[11] The strains of S. mutans were cultured in Todd Hewitt broth to the early stationary phase.

The acid production activity of S. mutans strains grown in Todd Hewitt broth from sucrose or glucose varied from 187 to 378. However, acid production activity of S. mutans strains with palatinose was quite low, almost the same as the endogenous activity or only slightly higher (Table 3).

The acid production activity of 5 S. mutans strains grown in partially defined media containing either 0·5% glucose or 0·5% palatinose + 0·05% glucose was also examined. S. mutans cells grown in the medium containing only glucose as a carbon source did not produce acid from palatinose beyond the level of endogenous activity. On the other hand, cells grown in the medium containing 0·5% palatinose together with 0·05% glucose produced acid from palatinose to a minor extent.

The acid production from palatinose by dental plaque suspensions was compared with that from sucrose and sugar alcohols. One series of experiments was carried out in Tokyo, Japan, and one series in Stockholm, Sweden. Acid production from palatinose was noticeably lower than that from sucrose.[12]

### 4.1.2 *Lactate Production by Dental Plaque Suspensions*

Dental plaque samples were collected from 5 subjects who had refrained from daily oral hygiene measures for 3 days and who had been instructed not to eat or drink for 4 h prior to sampling. Collected samples were washed once with 0·05M potassium phosphate buffer (pH 6·8) and homogenized by agitation in the same buffer with glass beads for 1 min using a vortex mixer. A 2-ml sample of reaction mixture prepared to contain 2 mg wet weight of dental plaque and 1% sugar in 5 mM $MgCl_2$ containing 0·05M potassium phosphate buffer (pH 6·8) was incubated at 37°C for 1 h. The reaction was stopped by the addition of 0·2 ml of 25% (w/v) metaphosphoric acid solution. After centrifugation the supernatant was analysed, and it was found that the plaque suspensions produced little or no lactate from palatinose but considerably more from sucrose and glucose.[9]

### 4.2 Effect of Palatinose on Insoluble Glucan Synthesis by Glucosyltransferase

Crude glucosyltransferase was obtained from the supernatant of *S. mutans* strain 6715 culture grown in TTY broth for 18 h. Reaction mixtures were composed of 0·2 ml of the enzyme preparation, 0·01% merthiolate and various concentrations of sugars in 0·05M potassium phosphate buffer (pH 6·8) in a total volume of 2·0 ml. After incubation for 17 h at 37°C, the water-insoluble glucan formed was collected by centrifugation, and was washed 3 times with distilled water. The amount of insoluble glucan was determined by the phenol sulphuric acid method, using glucose as the standard.

No water-insoluble glucan was synthesized from palatinose by the crude glucosyltransferase preparation obtained from *S. mutans* strain 6715. To investigate whether palatinose affected the synthesis of insoluble glucan from sucrose by this enzyme, palatinose to give 0·25%, 0·5% and 1·0% concentrations was added to reaction mixtures containing various concentrations of sucrose. As Fig. 5 indicates, the addition of palatinose significantly reduced the yield of insoluble glucan.[9] The effect of palatinose addition at various concentrations on insoluble glucan synthesis from 1% sucrose is shown in Fig. 6. Addition of 4% palatinose resulted in a reduction of synthesis to 4%. Even the addition of 0·125% palatinose

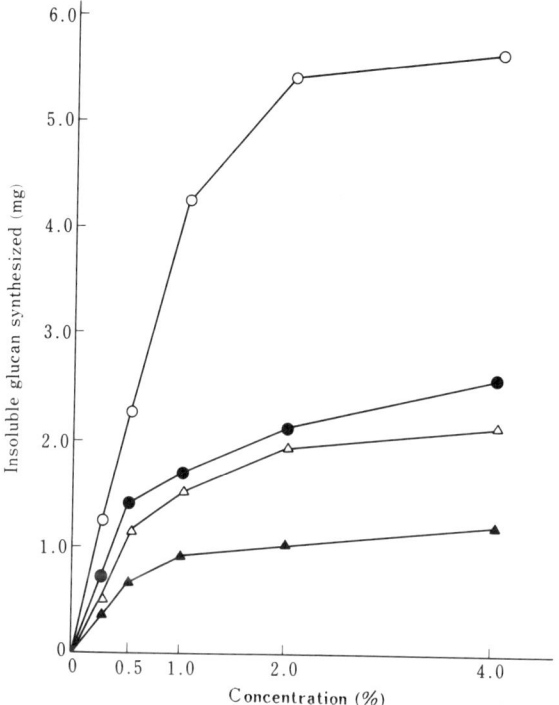

FIG. 5.   Insoluble glucan synthesis from various concentrations of sucrose (○),
and the effect of 0·25% (●), 0·5% (△) and 1% (▲) palatinose addition.

reduced it to approximately 50%. The inhibition rate correlated directly
with the concentration of palatinose.

The time sequence of glucan synthesis from 20 mM of sucrose with 20 mM
of palatinose is shown in Fig. 7. The inhibition rates calculated for each step
were almost identical, and were between 40% and 45%.[13]

Sugars in the supernatant of the reaction mixture were analysed by thin-
layer chromatography and high-speed liquid chromatography. When only
sucrose was the substrate, spots identical with sucrose, glucose and
fructose, plus a vague spot for palatinose, were recognized. In contrast,
when palatinose was added together with the sucrose, spots identical to
sucrose, palatinose, glucose and fructose were observed. Furthermore,
three additional spots of unidentified oligosaccharides appeared in the
thin-layer chromatogram. The synthesis of oligosaccharides was confirmed
by high-speed liquid chromatography. Unlike the results with sucrose only,
unidentified peaks following the sucrose peak were recognized when

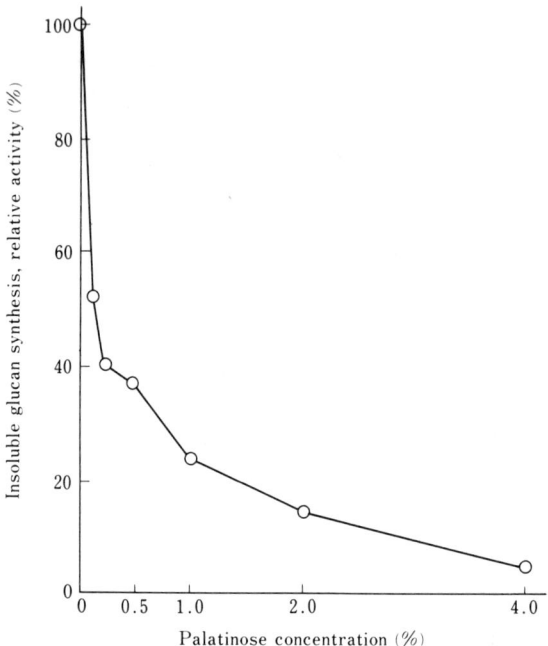

FIG. 6.   Effect of palatinose concentration on insoluble glucan synthesis from 1%
sucrose.

TABLE 4

RELATIVE CONCENTRATION RATIOS OF FRACTIONS IN TLC AUTORADIOGRAM OF
SUPERNATANT OF GTF-LABELLED SUCROSE REACTION MIXTURE IN THE PRESENCE OF
PALATINOSE

| Fraction 9 | + | — | 1·33 | — | 1·18 | — |
|---|---|---|---|---|---|---|
| 8 | — | 26·93 | — | 54·87 | — | 70·86 |
| 7 | 5·57 | — | 11·96 | — | 14·55 | — |
| 6 | 60·81 | 61·15 | 31·29 | 32·90 | 17·13 | 19·96 |
| 5 | — | 3·20 | 6·38 | 12·23 | 8·86 | 8·96 |
| 4 | 7·05 | 6·52 | 8·51 | + | 4·83 | + |
| 3 | 4·69 | — | 8·51 | — | 11·81 | — |
| 2 | 9·38 | — | 9·56 | — | 13·29 | — |
| 1 | 12·50 | 2·21 | 22·46 | — | 28·35 | 1·18 |
| Sample number | 1(G)[a] | 2(F)[a] | 3(G) | 4(F) | 5(G) | 6(F) |
| Incubation time | 1 h | | 2 h | | 3 h | |

[a] G, glucose; F, fructose.
+ indicates slight spot below quantifiable amount.

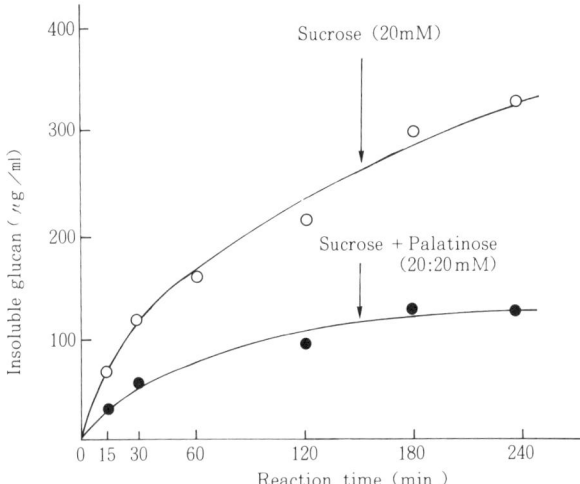

FIG. 7.    Insoluble glucan synthesis during 4 h by GTF from sucrose with and without palatinose.

palatinose was added to the reaction mixture. When the sugar content in the supernatants of the reaction mixtures from sucrose alone and from sucrose together with palatinose were compared, the consumption of sucrose was found to be much faster when palatinose was added than with sucrose alone. Among the resultant sugars, an increase in fructose was promoted whereas the increase of glucose was hindered.

The fate of sucrose in the supernatant was also studied by thin-layer chromatography–autoradiography using ([14]C-glucose)-sucrose and ([14]C-fructose)-sucrose as substrates.[14] Both forms of labelled sucrose were incubated with a GTF sample together with palatinose for 1–3 h at 37°C. Then the supernatants were subjected to thin-layer chromatography and the plates were exposed to X-ray film for 2 weeks. Nine spots were recognized and numbered 1 to 9 from the origin to the front on the TLC autoradiogram. The relative concentration ratios of each spot were expressed as products of their denseness and areas. They are summarized in Table 4. Fraction 6, which was identified as sucrose, was consumed at a similar level when either labelled form of sucrose was used. The glucose, fraction 7, showed a tendency to increase slightly over the 3-h period. In contrast, fructose, fraction 8, increased to a great extent in 3 h, and it finally comprised more than 70% of the total sugar in the supernatant. Accordingly, the released fructose is not likely to be utilized for oligosaccharide or polysaccharide synthesis as much as glucose is. This was supported by the fact that the density scores of fractions 1 to 5 were quite low when ($^{14}$C-fructose)-sucrose was used as tracer. Fraction 5 was identical to palatinose, and was recognized in the autoradiogram in both cases when either ($^{14}$C-glucose)-sucrose or ($^{14}$C-fructose)-sucrose was used as a tracer. This indicates that the transposition of $^{14}$C-glucose and $^{14}$C-fructose in the molecule of palatinose occurs to a slight extent. In fraction 4, which is identical to unidentified spot 1 on the TLC plate, the transposition of glucose and fructose was recognized. The transposition of glucose to fractions 2 and 3, which were identical to unidentified spots 2 and 3 respectively on the TLC plate, is also clear. Fraction 9 in the autoradiogram, which was not detected on the TLC chromatogram, appeared only when ethanol was present in the reaction mixture and when ($^{14}$C-glucose)-sucrose was used as a substrate. Therefore this fraction is assumed to be ethyl α-glucoside.

These findings indicate that palatinose functions as a receptor for the glucosyl base originating from sucrose, resulting in the formation of certain oligosaccharides. These have been identified. The supernatant of a reaction mixture of *S. mutans* glucosyltransferase plus palatinose was fractionated on Sephadex G-10 gel; each eluate was then subjected to TLC. Oligosaccharides were recovered from the chromatogram, partially hydrolysed, and analysed by TLC. The components of the oligosaccharides were also analysed by the periodate oxidation method. Of the 5 oligosaccharides, 3 were isolated and identified as 3G-*O*-α-glucosyl-isomaltulose, isomaltotriulose and isomaltotetraulose respectively. It is therefore possible that the supply of glucosyl base to insoluble glucan synthesis was inhibited. The addition of palatinose resulted in the promotion of sucrose consumption. An increase in the palatinose concentration of the reaction mixture resulted in an increase in the relatively low-molecular-weight soluble polysaccharide fraction. Accordingly, another possible mechanism is that the palatinose promotes water-soluble glucan synthesis, which results in the limitation of glucosyl base available for insoluble glucan synthesis.

### 4.3 Experimental Caries Studies

The development of caries in groups of rats infected with *S. mutans* and fed on Keyes diet 2000 containing palatinose (56%) instead of sucrose was examined.[15]

Three groups of 19–20 female weanling Wistar rats (18–22 days old) were used. They were housed in pairs in macrolon cages with stainless steel bottoms. The diet 2000 (56% sucrose) described by Keyes and Jordan[16] was fed to the control animals. In one group the sucrose was completely replaced by palatinose and in another by a mixture of sucrose (38·5%) and palatinose (17·5%). All diets were finely powdered and then offered *ad libitum*. Once a week the animals were given a portion of mixed, minced vegetable. Deionized drinking water was provided in glass flasks. Half of the animals were kept on the diet for 8 weeks and half for 14 weeks.

A 0·2-ml aliquot of a 6–8 h culture in Jordan's broth of a streptomycin-resistant mutant of *S. mutans* E-49 was inoculated into the oral cavity of each animal, and 5 ml was added to 100 ml of drinking water at the beginning of the experiment and once every week. Samples of dental plaque flora were obtained after 4, 8 and 14 weeks. After swabbing, the sticks were immediately put into tubes containing 2·0 ml of yeast extract medium, vigorously rotated and shaken. The suspension was then homogenized. The sample was diluted serially in the same medium, and 20-ml portions were dropped on Mitis-Salivarius agar with or without 200 IU streptomycin/ml of medium. All the plates were incubated anaerobically at 37°C for 1 day and then aerobically for 1 day. All colony-forming units (CFU) were carefully counted. The number of CFUs were determined for both media; then the percentage of *S. mutans* CFUs versus the total number of CFUs on streptomycin-free MS-agar was calculated.

The animals were weighed 3 times during the experiment. On termination of the experimental periods they were decapitated and the heads were defleshed by cooling in water and rinsing in 10% hydrogen peroxide. The teeth were stained and scored for caries.

The scores for carious enamel and dentine lesions on the bucco-lingual, approximal and sulcal surfaces are summarized in Table 5. In the group fed palatinose (group I), only sulcal lesions were found after 8 and 14 weeks. On the other hand, in the group fed sucrose (group S), high numbers of carious enamel lesions on the bucco-lingual, approximal and sulcal surfaces were found. In the group fed the mixture (group M, 38·5% sucrose plus 17·5% palatinose), lower numbers of enamel lesions were found than in group I in both experiments. Although rats fed the mixture exhibited carious enamel lesions on bucco-lingual surfaces after the 8-week experiment and in bucco-lingual and approximal surface lesions after the 14-week experiment, their scores were consistently lower than those of animals fed the sucrose diet (group S).

The average weight gains for all groups were high. No statistically significant differences were found between the average initial and final weights in the three groups.

The bacteriological study showed that *S. mutans* E-49 had colonized in

## TABLE 5

CARIES SCORES OF RATS FED DIETS CONTAINING EITHER PALATINOSE (GROUP I), SUCROSE (GROUP S) OR A MIXTURE OF PALATINOSE AND SUCROSE (GROUP M)

*Mean caries scores and total caries scores*

| Group diet | Bucco-lingual | | | | Approximal | | | | Sulcal | | | | Total |
|---|---|---|---|---|---|---|---|---|---|---|---|---|---|
| | $E$ | $D_s$ | $D_m$ | $D_x$ | $E$ | $D_s$ | $D_m$ | $D_x$ | $E$ | $D_s$ | $D_m$ | $D_x$ | |
| *8-week experiment* | | | | | | | | | | | | | |
| S Sucrose (56%) | 7·5 | 3·6 | 2·9 | 1·5 | 0·4 | 0·2 | 0 | 0 | 17·2 | 14·3 | 0·4 | 0·2 | 48·2 |
| | (4·2) | (3·1) | (2·9) | (1·9) | (1·0) | (0·6) | | | (5·3) | (5·6) | (0·6) | (0·4) | |
| M Mixture of sucrose (38·5%) | 2·2 | 1·3 | 1·0 | 0·3 | 0 | 0 | 0 | 0 | 14·6 | 11·7 | 1·4 | 0 | 32·5 |
| and isomaltulose (17·5%) | (2·5) | (1·4) | (1·2) | (0·6) | | | | | (6·2) | (6·1) | (3·2) | | |
| I Palatinose (56%) | 0 | 0 | 0 | 0 | 0 | 0 | 0 | 0 | 5·9 | 4·5 | 0·1 | 0 | 10·5 |
| | | | | | | | | | (5·9) | (1·7) | (0·3) | | |
| *14-week experiment* | | | | | | | | | | | | | |
| S Sucrose (56%) | 14·9 | 10·5 | 10·5 | 8·5 | 2·9 | 2·1 | 1·4 | 1·3 | 17·0 | 14·1 | 5·0 | 3·2 | 91·4 |
| | (8·6) | (9·0) | (9·0) | (9·0) | (3·1) | (2·6) | (2·0) | (2·0) | (7·6) | (7·6) | (5·7) | (4·0) | |
| M Mixture of sucrose (38·5%) | 4·7 | 2·1 | 1·8 | 1·6 | 0·4 | 0·2 | 0·1 | 0·1 | 14·4 | 11·1 | 4·2 | 1·5 | 42·2 |
| and isomaltulose (17·5%) | (5·8) | (3·8) | (3·9) | (4·0) | (1·0) | (0·4) | (0·3) | (0·3) | (8·4) | (8·7) | (4·9) | (3·4) | |
| I Palatinose (56%) | 0 | 0 | 0 | 0 | 0 | 0 | 0 | 0 | 7·5 | 6·1 | 1·0 | 0 | 14·6 |
| | | | | | | | | | (5·7) | (4·9) | (1·6) | | |

Figures in parentheses are standard deviations.
E, enamel lesions; $D_s$, slight dentin lesions; $D_m$, moderate dentin lesions; $D_x$, extensive dentin lesions.

TABLE 6

PERCENTAGES OF *S. MUTANS* CFU OF TOTAL STREPTOCOCCI CFU ISOLATED FROM THE TEETH OF RATS FED A SUCROSE, A PALATINOSE AND A MIXED SUCROSE–PALATINOSE DIET

| Diet | Weeks after infection | | |
|------|-------|-------|-------|
| | 4 | 8 | 14 |
| Sucrose (56%) | 54·4  (20) | 44·5 (20) | 30·9  (10) |
| Mixture of sucrose and palatinose (38·5/17·5%) | 28·9  (20) | 9·9 (19) | 23·2   (9) |
| Palatinose (56%) | 0·34 (20) | 0·0014 (20) | 0·49 (10) |

Figures within parentheses denote the number of rats.

high numbers in all animals in the sucrose group (Table 6). In the palatinose diet group, the percentage of *S. mutans* E-49 had decreased significantly by 8 weeks, compared with the sucrose group and the sucrose/palatinose group. The percentages of *S. mutans* E-49 in animals fed the sucrose/palatinose mixture were lower than those in animals fed sucrose, especially after 8 weeks. The recovery rates of *S. mutans* were higher in the sucrose and mixture groups than in the palatinose group at the end of the 14-week experiment. The study confirmed that palatinose is significantly less cariogenic than sucrose and does not support the growth of *S. mutans* to the same extent as sucrose. Similar results were reported by Ooshima *et al.*[17]

When palatinose (17·5%) was included in a diet containing sucrose (38·5%), palatinose reduced the incidence of caries. This phenomenon is in agreement with previous experiments on acid production rates and the inhibition of insoluble glucan synthesis from sucrose by *S. mutans* glucosyltransferase *in vitro*.

The weight gains of the animals were similar in all three groups. None had diarrhoea or loose stools. These findings are in agreement with the findings in an earlier pilot study[18] in rats which showed no clinical symptoms of intestinal disorder, in contrast to experience with sugar alcohols, which produced diarrhoea.

The effect of combining palatinose and fructose on experimental caries in animals was also tested. Addition of fructose to a palatinose diet (1:1 or 1:2 fructose:palatinose) did not significantly change the caries incidence in comparison with a palatinose diet alone. The palatinose and palatinose–fructose diets did not induce smooth-surface caries. Addition of fructose to the palatinose diet changed the percentage of *S. mutans* E-49 on the teeth to only a small extent (Takazoe, I., unpublished).

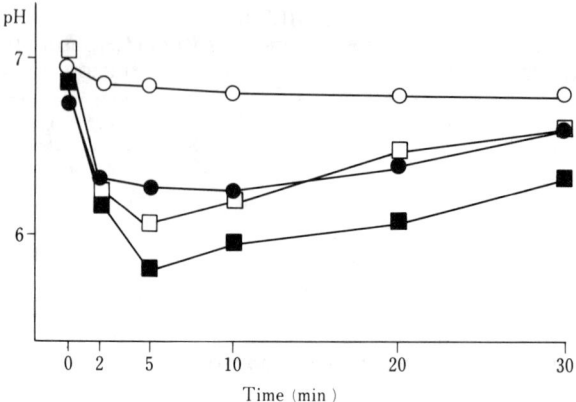

FIG. 8.  Changes in the pH of plaque after mouth rinsing for 30 s with 15%
solutions of palatinose (○) and glucose (□) before and with palatinose (●) and
glucose (■) after the rinsing period; means from 18 subjects.

## 4.4 Palatinose in Human Studies

The effect of frequent rinses with palatinose solution on acid production
from sugars by human dental plaque has been examined.[19] The adaptation
period lasted 6 weeks, during and after which experiments *in vivo* and *in
vitro* were carried out. The 18 subjects between 14 and 33 years old who
participated in the study were instructed to rinse their mouths 6 times per
day, each time for approximately 2 min, with 10 ml of a 15% (w/v)
palatinose solution. Special instructions were given to maintain reprodu-
cible conditions.

In the fourth and fifth weeks, pH changes in plaque samples of the
subjects were measured *in vivo*, after a mouth rinse with glucose or
palatinose. In the sixth week of the rinsing period a plaque sample was
taken and subjected to an acid production determination *in vitro* using
glucose and palatinose as substrates.

The pH changes in dental plaque exposed to aqueous mouth rinses
containing 15% (w/v) glucose or palatinose are presented in Fig. 8. All
mean pH values in the experiment with glucose were lower after the rinsing
period than before. The greatest before/after pH decreases with glucose
were found at 5 and 20 min ($p < 0.001$ and $0.01$). Palatinose caused lower
pH values after the rinsing period at all time intervals, with significantly
larger pH changes at 2, 5 and 10 min ($p < 0.01$, $0.05$ and $0.05$).

The results from the study *in vitro* are given in Fig. 9. The mean acid
production activity from glucose after the rinsing period was found to be

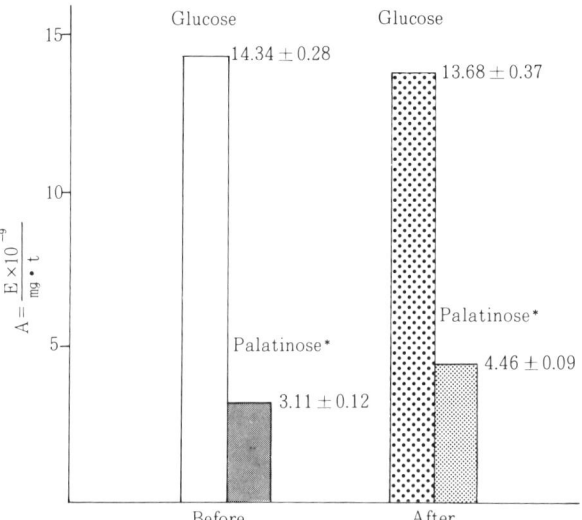

FIG. 9.    Acid production activity from glucose and palatinose before and after the rinsing period; $A$ = equivalent weight of acid per mg of plaque per min. Means ($\pm$ SE) from 18 subjects. Differences significant: $*p < 0.05$.

about 4·6% lower than before, but this difference was not statistically significant. All the subjects showed higher acid production activity from palatinose after the rinsing period than before ($p < 0.05$). When the acid production was expressed as a percentage of that from glucose, an increase was also noticed. However, irrespective of the time of measurement before or after the rinsing period, palatinose showed significantly lower acid production activity than glucose ($p < 0.001$).

The initial plaque pH values in the experiments *in vivo* were lower after the rinsing period than before, both in the experiments with glucose and in those with palatinose. Moreover, the calculated pH differences in the experiments with palatinose, especially at 5 and 10 min, showed statistically significant differences, in contrast to the results with glucose. Frequent exposure of the oral flora *in situ* to palatinose produced a somewhat greater fall in pH, but the values never fell below pH 6·2. These results suggest that adaptation does occur. Perhaps the mechanism involved is an ecological change in the mouth or an induction phenomenon. However, the increase in fermentation was moderate, and even after the rinsing period the difference between glucose and palatinose was still highly significant. Thus it is not likely that palatinose loses its low cariogenic potential through this adaptive mechanism.

The cariogenicity of palatinose in the form of candy has also been examined. Acid production from palatinose candy by various serotypes of *S. mutans* was not observed (unpublished data).

The cariogenicity of palatinose syrup was studied by Ooshima.[20] The development of caries in groups of rats infected with *S. mutans* and fed on diets containing 20% and 40% of sucrose or palatinose syrup was examined. While severe caries occurred in the sucrose groups, only sulcal caries occurred in the group fed palatinose syrup. Caries scores on palatinose syrup were one-third to one-fifth of those for sucrose. Between the group fed a mixture of 20% sucrose and 20% palatinose syrup and the 20% sucrose group, significant differences in caries scores were not found. The findings indicate that the cariogenicity of palatinose syrup is also low but not as low as that of palatinose.

## 5 DIGESTION AND ABSORPTION

It is known that palatinose is hydrolysed to glucose and fructose by a disaccharidase in the small intestine of humans, and these sugars are then absorbed.[21-23] Isomaltase was suspected of being responsible for this.[22] The characteristics of palatinose-hydrolysing activity have recently been studied in the sucrase–isomaltase complex purified from adult rat intestine.[24] Palatinose-hydrolysing specific activity was increased by purification of the sucrase–isomaltase complex in parallel with sucrase and isomaltase activity. Palatinose was hydrolysed by the complex in competition with isomaltose. Palatinose-hydrolysing activity, like iso-maltase activity, was not inhibited by $\alpha$-glucosidase inhibitor, acarbose, which inhibits sucrase and maltase activity. These findings collectively indicate that palatinose is hydrolysed at the active site of isomaltase of the sucrase–isomaltase complex.

The velocity of palatinose hydrolysis is slower than that of sucrose, and absorption could be slow.[25] This was demonstrated by Kawai *et al.*[26] Changes in plasma glucose and insulin concentration in response to palatinose ingestion were compared with those for sucrose in 8 normal volunteers. When 50 g palatinose was administered the plasma glucose gradually increased to its peak of $110.9 \pm 4.9$ mg/dl at 60 min after administration and maintained a plateau during 120 min of the experiment. The peak value of plasma glucose in response to 50 g sucrose in the same group was $143.3 \pm 8.8$ mg/dl at 30 min after administration, and then the value sharply decreased to the fasting level (Fig. 10).

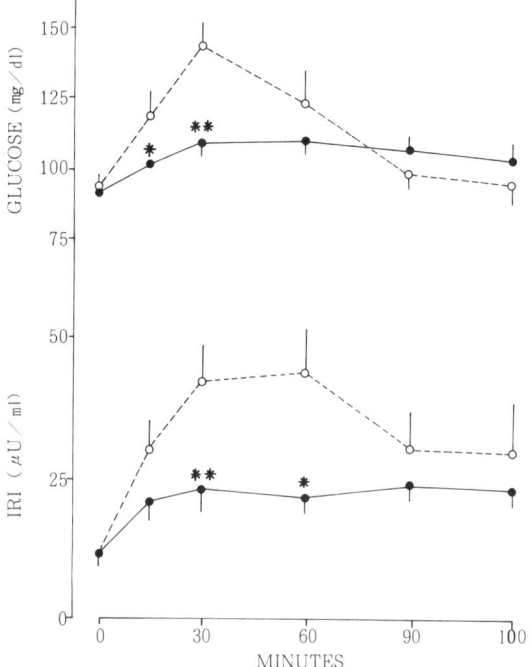

Fɪɢ. 10.   Changes in plasma glucose and IRI (immunoreactive insulin) in response to oral administration of palatinose (●) or sucrose (○).[26] $N = 8$, mean $\pm$ SEM; *$p < 0.05$, **$p < 0.01$; palatinose versus sucrose.

The cumulative increase in plasma glucose in response to palatinose was significantly smaller than that from sucrose. The changes in the plasma insulin level almost paralleled those in the plasma glucose level. These results indicate that palatinose is more slowly absorbed than sucrose and therefore useful as a sweetener for diabetic patients. The metabolic effects and fate of palatinose after intravenous injection were also studied by the same authors.[27] Using dogs as the experimental model, it was found that the behaviour of parenterally administered palatinose is in many respects similar to that of maltose, and is likely to be clinically valuable for both diabetic and non-diabetic patients. The advantage of palatinose over maltose is that it can be converted into both glucose and fructose, whereas maltose is only converted into glucose. Fructose has a potent antiketogenic activity[28] and is more efficiently utilized by the liver than glucose.[29] A recent report describes a sharp reduction of postprandial serum glucose and insulin response after fructose ingestion as compared with glucose or sucrose ingestion.[30] However, infusion of fructose alone causes hyper-uricaemia, hypertriglyceridaemia and a decrease in the ATP level.[31-33] Palatinose infusion causes mild increases in glucose and plasma insulin concentrations without any increase in the plasma triglyceride concentration. Further studies on the effects of long-term infusion of palatinose are necessary before any clinical application.

## 6 SAFETY OF PALATINOSE

As described above, palatinose is decomposed by sucrase–isomaltase complex in the intestinal mucosa into glucose and fructose, which are then absorbed, and should therefore be regarded as safe. Its safety has been extensively studied. First, palatinose was shown by the Ames test to possess no mutagenic effect.[34] A chronic toxicity study was carried out using Sprague–Dawley rats of both sexes.[35] The substance was administered orally to the rats at dose levels of 1500, 3000 and 4500 mg/kg daily for 26 weeks. It caused no deaths or any noticeable clinical signs of toxicity. Nor did it affect the body weight, food consumption or urine volume. Urine analysis showed no significant abnormality; the results for ketone bodies and bilirubin were negative. Ophthalmology and haematology showed no significant abnormalities. Blood analysis revealed lowered concentrations of uric acid and creatinine. Pathological examination revealed no toxic changes in any organs.[35] An acute toxicity test in rats by oral administration of 4, 8, 16 and 32 g/kg caused no deaths or any observable clinical signs of toxicity or pathological changes.[35] A chronic toxicity test of palatinose syrup was carried out in a similar way, and no degree of toxicity was found in any of the features tested.[36]

## 7 CONCLUSION

Palatinose, a naturally occurring disaccharide, is produced by enzymic conversion from sucrose, and is sweet. A crystalline form and a syrup (molasses) are available at the industrial level. Both palatinose and palatinose syrup have been shown to be low cariogenic. Furthermore, palatinose was found to inhibit insoluble polyglucan synthesis from sucrose. Addition of palatinose to sucrose therefore reduces the cariogenicity of sucrose. Palatinose is hydrolysed by sucrase–isomaltase complex in the small intestine into glucose and fructose. Consequently it is a caloric sweetener which can be an important requirement for a sugar substitute.[37] Hydrolysis of palatinose is slower than that of sucrose, so as a potential parenteral nutrient it may be clinically valuable for both diabetic and non-diabetic patients. The safety of palatinose in foods has also been confirmed. It is utilized in many kinds of food in Japan, and is considered a promising alternative to sucrose.

## REFERENCES

1. Weidenhagen, R. & Lorenz, S., Palatinose [6-(α-Glucopyranoside)-fructo-furanose], ein neues bakterielles Umwandlungsprodukt der Saccharose. *Z. Zuckerindust.*, **7** (1957) 533–4.
2. Siddique, I. R. & Furgala, B., Isolation and characterization of oligosaccharides from honey. I. Disaccharides. *J. Apicult. Res.*, **6** (1967) 139–45.
3. Schiweck, H., Palatinit-Herstellung, technologische Eigenschaften und Analitik palatinithaltiger Lebensmittel. *Alimenta*, **19** (1980) 5–16.
4. Koga, T. & Mizutani, T., Applications of palatinose for foods. *Proc. Res. Soc. Jap. Sugar Refineries Tech.*, **34** (1985) 45–57 (in Japanese).
5. Windish, S., Über einen Farbstoffbildner von Zuckerruben, *Protaminobacter rubrum* (den Dooren de Jong). *Z. Zuckerindust.*, **9** (1958) 446.
6. Schmidt-Berg-Lorenz, S. & Mauch, W., Ein weiterer Isomaltulose bildender Bakterienstamm. *Z. Zuckerindust.*, **14** (1964) 625–7.
7. Nakajima, Y., Palatinose production by immobilized α-glucosyltransferase. *Proc. Res. Soc. Jap. Sugar Refineries Tech.*, **33** (1983) 55–63.
8. Gehring, F., Über die Säurenbildung kariesätiologisch wichtiger Strepto-kokken aus Zuckern und Zuckeralkoholen unter besonderer Berücksichtigung von Isomaltit und Isomaltulose. *Z. Ernährungswissenschaft*, Suppl. 15 (1973) 16–27.
9. Ohta, K. & Takazoe, I., Effect of isomaltulose on acid production and insoluble glucan synthesis by *Streptococcus mutans. Bull. Tokyo Dent. Coll.*, **24** (1983) 1–11.
10. Larje, O. & Frostell, G., Acid production activities of caries-inducing Streptococci. *Acta Pathol. Microbiol. Scand.*, **72** (1968) 453.
11. Birkhed, D., Automatic titration method for determination of acid production from sugars and sugar alcohols in small samples of dental plaque material. *Caries Res.*, **12** (1978) 128–36.
12. Maki, Y., Ohta, K., Takazoe, I., Matsukubo, Y., Takaesu, Y., Topitsoglou, V. & Frostell, G., Acid production from isomaltulose, sucrose, sorbitol and xylitol in suspensions of human dental plaque. *Caries Res.*, **17** (1983) 335–9.
13. Takazoe, I., Ohta, K. & Kadomura, Y., Inhibitory mechanism of isomaltulose for insoluble glucan synthesis by *Streptococcus mutans. J. Dent. Res.*, **61** (1982) 340.
14. Takazoe, I., Ohta, K. & Takaoka, M., Oligosaccharide produced by *Str. mutans* GTF from sucrose with isomaltulose. *J. Dent. Res.*, **62** (1983) 684.
15. Sasaki, N., Topitsoglou, V., Takazoe, I. & Frostell, G., Cariogenicity of

isomaltulose (palatinose), sucrose and mixtures of these sugars in rats infected with *Streptococcus mutans* E-49. *Swed. Dent. J.,* **9** (1985) 149–55.

16. Keyes, P. H. & Jordan, H. V., Periodontal lesions in the Syrian hamster. III. Findings related to an infectious transmissible component. *Arch. Oral Biol.,* **9** (1964) 377–400.

17. Ooshima, T., Izumitani, A., Sobue, S., Ikahashi, N. & Hamada, S., Non-cariogenicity of the disaccharide palatinose in experimental caries of rats. *Infect. Immun.,* **39** (1983) 43–9.

18. Takazoe, I., Frostell, G., Ohta, K., Topitsoglou, V. & Sasaki, N., Palatinose—a sugar substitute; pilot studies. *Swed. Dent. J.,* **9** (1985) 81–7.

19. Topitsoglou, V., Sasaki, N., Takazoe, I. & Frostell, G., Effect of frequent rinses with isomaltulose (palatinose) solution on acid production in human dental plaque. *Caries Res.,* **18** (1984) 47–51.

20. Ooshima, T., Cariogenicity of palatinose syrup. *Proc. First Conf. on Palatinose,* ed. T. Kawachi. Mitsui Sugar Co., Tokyo, Japan, 1985, p. 15 (in Japanese).

21. Auricchio, S., Rubino, A., Tosi, R., Semenza, G., Laudholt, M., Kistler, H. & Prader, A., Disaccharidase activities in human intestinal mucosa. *Enzymol. Bio. Clin.,* **3** (1963) 193–208.

22. Dahlqvist, A., Auricchio, S., Semenza, G. & Prader, A., Human intestinal disaccharidase and hereditary disaccharidase intolerance. The hydrolysis of sucrose, isomaltose, palatinose (isomaltulose), and a 1,6-oligosaccharide (isomalt-oligosaccharide) preparation. *J. Clin. Invest.,* **42** (1963) 556–62.

23. Welsh, J. D., Rohrer, G. V. & Walker, A., Human intestinal disaccharidase activity. *Arch. Intern. Med.,* **117** (1966) 488–94.

24. Goda, T. & Hosoya, N., Hydrolysis of palatinose by rat intestinal sucrase-isomaltase complex. *J. Jap. Soc. Nutr. Food Sci.,* **36** (1983) 169–73.

25. Yamada, K., Shinohara, H. & Hosoya, N., Hydrolysis of 1-*O*-α-D-gluco-pyranosyl-D-fructofuranose (trehalulose) by rat intestinal sucrase-isomaltase complex. *Nutr. Rep. Int.,* **32** (1985) 1211–20.

26. Kawai, K., Okuda, Y. & Yamashita, K., Changes in blood glucose and insulin after an oral palatinose administration in normal subjects. *Endocrinol. Japon.,* **32** (1985) 933–6.

27. Kawai, K., Okuda, Y., Chiba, Y. & Yamashita, K., Palatinose as a potential parenteral nutrient; its metabolic effects and fate after oral and intravenous administration to dogs. *J. Nutr. Sc. Vitaminol.,* **32** (1986) 297–306.

28. McGarry, J. D. & Foster, D. W., The regulation of ketogenesis from oleic acid and the influence of antiketogenic agents. *J. Biol. Chem.,* **246** (1971) 6247–53.

29. Mäenpää, P. H., Raivio, K. O. & Kekomäki, M. P., Liver adenine nucleotides, fructose-induced depletion and its effect on protein synthesis. *Science,* **161** (1968) 1253–4.

30. Crapo, P. A., Kolterman, O. G. & Olefsky, J. M., Effects of oral fructose in normal, diabetic, and impaired glucose subjects. *Diabetic Care,* **3** (1980) 375–82.

31. Woods, H. F. & Alberti, K. G. M. M., Dangers of intravenous fructose. *Lancet,* **ii** (1972) 1354–7.

32. Sestoft, L. & Fleron, P., Determination of the kinetic constants of fructose transport and phosphorylation in perfused rat liver. *Biochim. Biophys. Acta,* **345** (1974) 27–38.

33. Sestoft, L., Fructose and the dietary therapy of diabetes mellitus. *Diabetologia*, **17** (1979) 1–3.
34. Suzuki, K., A new sweetener, palatinose, and its utilization. *Jap. Food Sci.*, **24** (1985) 1–9.
35. Yamaguchi, K., Yoshimura, S., Inada, H., Matsui, E., Ohtaki, T. & Ono, H., A 26-week oral toxicity study of palatinose in rats. *Pharmacometrics*, **31** (1986) 1015–31 (in Japanese).
36. Yamaguchi, K., Yoshimura, S., Inada, H., Ozawa, K., Kato, H. & Ono, H., A 26-week oral toxicity study of palatinose syrup in rats. *Pharmacometrics*, **34** (1987) 1–16.
37. Takazoe, I., New trends on sweeteners in Japan. *Int. Dent. J.*, **35** (1985) 58–65.

Chapter 7

# THE USE OF INTENSE SWEETENERS IN SOFT DRINKS

A. G. WELLS

*Beecham Products Research Department,*
*Coleford, Gloucestershire, UK*

## SUMMARY

*Soft drinks are a major market for intense sweeteners, and the potent sweetener saccharin has a long history of use in these products dating from the late 1880s. There are technical reasons, economic objectives and health considerations supporting the need for intense sweeteners in soft drinks. The properties of acesulfame-K, aspartame, saccharin, thaumatin, cyclamate, neohesperidin dihydrochalcone, stevioside, alitame and sucralose are reviewed in relation to the requirements of the ideal intense sweetener for soft drinks. It is concluded that no single sweetener is ideally suited to meet all the requirements, but by using them in combination the limitations of one sweetener can be offset by the strengths of another. Examples are given of the synergistic effects of mixtures of intense sweeteners, and the ways in which malic acid and fructose can be used to advantage in the formulation of low-calorie soft drinks are reviewed. The legislation controlling the use of intense sweeteners in soft drinks is outlined, and the implications of the proposed changes to the soft drinks regulations are discussed. Finally the results are presented of a survey of the use of intense sweeteners in national brand soft drinks on the market in the UK.*

## 1 INTRODUCTION

Soft drinks make a valuable contribution to liquid intake and have become established as an important part of the diet. They are consumed by all age groups and social and economic classes, and are readily available for both at-home and on-premises consumption.

TABLE 1
SOFT DRINK SALES BY UK MANUFACTURERS AND THE USE OF
INTENSE SWEETENERS

| Year | Volume$^a$ (Ml) | Intense sweeteners$^b$ (kg) |
|---|---|---|
| 1975 | 4 163 | 275 874 |
| 1976 | 4 475 | 302 636 |
| 1977 | 4 145 | 277 263 |
| 1978 | 4 283 | 274 598 |
| 1979 | 4 807 | 301 928 |
| 1980 | 4 408 | 290 678 |
| 1981 | 4 280 | 301 681 |
| 1982 | 4 372 | 318 961 |
| 1983 | 4 694 | 318 058 |
| 1984 | 4 814 | 327 216 |
| 1985 | 4 703 | 329 350 |
| 1986 | 5 471 | 306 708 |
| 1987 | 5 952 | 335 794 |

Based on Business Monitor PQ4283 (Soft Drinks) published by HMSO.
$^a$ Total of ready-to-drink and concentrates, expressed on a ready-to-drink basis.
$^b$ Saccharin only until 1983.

A satisfactory soft drink formula calls for a balanced harmony of contributing ingredients, namely water, fruit juice, acid, sweetener, flavour and carbon dioxide. The balance between sweetness and acid is one of the most important characteristics of a soft drink, and has to be varied depending on the flavour. Sweetness can be contributed by carbohydrate sweeteners such as sucrose, or by the use of intense sweeteners such as saccharin, or by a blend of the two.

Soft drinks are a major market for intense sweeteners,[1] particularly in the UK where, unlike many other countries, they can be used in conjunction with nutritive sweeteners and are not therefore limited to use in dietetic beverages. The statistics on usage and the legal aspects which follow therefore mainly relate to the UK. Sales of soft drinks have gradually increased over the years, and there has been a concomitant rise in the use of intense sweeteners (Table 1). Intense sweeteners make the most contribution to sweetness in low-calorie soft drinks, where they often replace the added carbohydrate sweetener completely.

The market for low-calorie soft drinks, particularly carbonates

TABLE 2
LOW-CALORIE CARBONATED DRINKS MARKET

| Year | Value (£m) | Growth (%) | Volume (Ml) | Growth (%) |
|------|-----------|-----------|------------|-----------|
| 1983 | 96  | —  | 136 | —  |
| 1984 | 110 | 15 | 157 | 15 |
| 1985 | 150 | 36 | 220 | 40 |
| 1986 | 235 | 53 | 350 | 60 |

Source: Leatherhead Food Research Association (from Ref. 2).

(carbonated drinks), has been increasing as the general public has become more diet-conscious; it has grown from a £28 million market in 1976 to £235 million in 1986.[2] The most dramatic growth has taken place over the past 3 years, triggered by the introduction of the range of new intense sweeteners permitted by the 1983 Sweeteners in Food Regulations[3] (Table 2). Low-calorie carbonates now represent about 12% of the market value and volume of the total carbonates market, compared with 6% in 1977. Diet cola drinks now represent 20% of the total cola sales, and diet mixers constitute 28% of the total mixer sales.[2]

The availability of intense sweeteners for the formulation of diet drinks has also increased consumption in other countries. Sales of diet soft drinks in Canada rose from virtually nil in 1981 to 19% of the soft drink market 20 months later, following the approval of aspartame.[4] The approval of aspartame in the USA in 1983 has also seen a rapid growth in the diet drink market to about 25% of the total soft drinks market in 1986.[5]

As the diet soft drink sector continues to expand worldwide, the demand for intense sweeteners will continue to grow.

## 2 HISTORY OF USE OF INTENSE SWEETENERS IN SOFT DRINKS

### 2.1 UK Background

The intense sweetener saccharin has a long history of use in soft drinks dating from the late 1880s.[6] In the early part of the 20th century a few soft drink manufacturers made a range of sugarless carbonated beverages, although production was quite small. The real impetus for the use of intense sweeteners in soft drinks came with the two world wars when sugar was in short supply and was rationed. During the 1939–45 war and for some little time afterwards, intense sweeteners other than saccharin could be used for the manufacture of diabetic drinks in Britain. Substances used for this purpose included Dulcin or p-phenetyl urea, n-hexylchlorimalonamide and the alkoxy-amino-nitrobenzenes, particularly n-propoxy-2-amino-4-nitrobenzene, popularly known as P4000, rated at 4000 times as sweet as sucrose. However, with the exception of saccharin, these sweeteners were eventually banned in 1953.

In Britain, the 1943 Soft Drinks Order[7] stipulated that unconcentrated, sweetened soft drinks should have a fixed sugar content of 18 oz per 10 gallons, and a fixed saccharin content of 82 grains per 10 gallons (equivalent to 1·12% w/v sugar and 0·0117% w/v saccharin respectively). Concentrated citrus squashes were required to contain 7·5 lb of sugar per 10 gallons and 1·00 oz of saccharin per 10 gallons (equivalent to 7·5% w/v sugar and 0·062% w/v saccharin). In these systems approximately 80% of the total sweetness is contributed by the intense sweetener. In the following 4 years the sugar content was gradually increased and the saccharin content decreased, but in the 1947 Soft Drinks Order[8] they returned to the values of 1943.

The 1959 Food Standards Committee Report on Soft Drinks[9] proposed the prohibition of the use of saccharin in soft drinks other than those manufactured for diabetics. Trade representations persuaded the Ministry of Agriculture, Fisheries and Food not to embody this proposal in legislation, and the 1963 Soft Drinks Regulations[10] permitted ready-to-drink soft drinks to contain the equivalent of 0·008% w/v of saccharin (0·011% w/v in brewed ginger beer and herbal or botanical beverages) with a minimum quantity of added sugar of 4·5% w/v (3% w/v for dry ginger ale, 2% w/v for brewed ginger beer, etc.). Concentrated soft drinks were permitted a maximum content of 0·040% w/v of saccharin with a minimum quantity of added sugar of 22·5% w/v. Diabetic soft drinks, containing no added sugar, were exempt from the saccharin limitations.

A revolution in the use of intense sweeteners in soft drinks came in 1964 in the provisions of the 1964 Soft Drinks Regulations[11] which were brought into operation in 1965. These permitted, for the first time in Britain, the use of the intense sweetener cyclamate, either alone or in combination with saccharin, subject to specified limits. Unconcentrated soft drinks were allowed a maximum of 0·133% w/v of cyclamic acid (0·19% w/v in brewed ginger beer, etc.), whilst concentrated soft drinks were permitted a maximum of 0·665% w/v cyclamic acid. The permitted maximum levels of addition of saccharin remained as specified in 1963. In addition, these regulations provided for a category of semi-sweet soft drinks with maximum limits for sugar and intense sweeteners to provide products with approximately half the sweetness of the fully sweetened drinks. A low-calorie category of soft drinks was also introduced, with maximum calorie limits, which were exempt from the limitations imposed on the use of saccharin and cyclamate, as also were drinks suitable for diabetics.

Mixtures of saccharin and cyclamate provided very acceptable sweetness characteristics, and this combination was extensively used not only in low-calorie soft drinks but also as a partial substitute for sugar in full-calorie soft drinks.[12]

Following adverse toxicity reports emanating from the USA,[13] the use of cyclamate was prohibited in the UK in 1970.[14] From 1970 until 1983 saccharin remained the only intense sweetener permitted, but this situation was dramatically changed by the provisions of the 1983 Sweeteners in Food Regulations.[3] In addition to saccharin, the use of acesulfame potassium, aspartame and thaumatin was permitted. Whilst the amount of saccharin that can be added to soft drinks remains limited to the 1964 levels, the addition of the other intense sweeteners is limited only by practical applications, there being no statutory limits imposed.

## 2.2 History in the USA

The use of intense sweeteners in soft drinks in the USA shows a similar pattern. Saccharin was introduced commercially into foods and beverages in 1900 but found little use in soft drinks until World War I, when 50% or more of the sugar in soft drinks was replaced by saccharin.[15] At the end of the war some manufacturers continued using saccharin and sugar together for economic reasons but many States passed laws banning saccharin in soft drinks.

In 1938 Federal Food and Drug laws provided for saccharin in foods and beverages accompanied by appropriate labelling, and between 1950 and 1956 it became permitted in most States. In 1950 cyclamate was authorised for use in foods and beverages; by 1960 it was used extensively in low-calorie soft drinks, mainly in conjunction with saccharin. Following adverse toxicity reports,[13] it was banned in 1970,[16] leaving saccharin as the only permitted intense sweetener. Saccharin itself was removed from the 'generally recognised as safe' (GRAS) list in 1971 but the sweetener continues to be widely used under repeated Congressional moratoria accompanied by mandatory warning labelling.

Aspartame was first approved by the Food and Drug Administration in

1974,[17] but approval was subsequently stayed in 1975.[18] Following much debate, it was authorised for dry uses in July 1981,[19] and finally approved for use in carbonated beverages and their syrups in July 1983.[20] Acesulfame-K has been approved for use in a number of products including powdered beverages in the USA since 28 July 1988. Further petitions for extended approvals are to be filed.

# 3 THE NEED FOR INTENSE SWEETENERS IN SOFT DRINKS

There are technical reasons, economic purposes and health considerations which support the use of intense sweeteners in soft drinks, as either a complete or a partial replacement for carbohydrate sweeteners.

## 3.1 Technical Reasons

A major application of intense sweeteners is in soft drinks where sweetness is an essential characteristic but changes in food structure consequent upon the partial or complete removal of sugar may be relatively unimportant or desirable.[12]

(a) For certain concentrated citrus soft drinks it is difficult to provide an adequate degree of sweetness and maintain a stable appearance using carbohydrate as the only sweetener. The higher the sugar solids, the higher the density of the product and the greater will be the propensity of immiscible flavour components to separate out. Because of the low density of citrus oils and other flavouring materials, they readily separate in a high-density soft drink to form an unsightly 'ring' at the neck of the container. The partial substitution of carbohydrate sweetener with an intense sweetener reduces the density of the product, thereby improving the stability of dispersions of citrus oils.

(b) For reasons of palatability soft drinks are often formulated to contain in excess of 10% w/v sugar in ready-to-consume form. With certain flavours, such drinks can be somewhat cloying to the palate,[12] and they benefit from an alteration in 'mouthfeel' by a reduction in sugar, providing sweetness is maintained by replacement of the carbohydrate by an intense sweetener.

(c) In mixer drinks, such as tonic water and bitter lemon, which are formulated for mixing with spirits, an excess of sugar can spoil the flavour. In this situation the use of a mixture of sugar and intense sweetener gives a product much more acceptable to the consumer.

## 3.2 Economic Reasons

With the exception of soft drinks formulated with beer, e.g. shandies, carbohydrate sweeteners contribute more to the cost of a product than any other single ingredient. In terms of sweetening power, i.e. cost per unit of sweetness, sugar is usually more expensive than intense sweeteners. Partial replacement of sugar by an intense sweetener can therefore effect a cost saving.

In 1983 the UK sales of soft drinks amounted to an estimated 4694 million litres of ready-to-consume product in which was used 318 058 kg of saccharin, the only intense sweetener permitted at that time. To replace this amount of saccharin by sugar to achieve an equivalent sweetness would have cost the soft drink manufacturers an additional £55 million approximately at current prices, which would have to be passed on to the consumer.

## 3.3 Health Considerations

### 3.3.1 Body Weight

A report on obesity from the Royal College of Physicians[21] indicated that a substantial proportion of the UK adult population are overweight: 39% of men and 32% of women, with 6% of men and 8% of women being classified as obese. Excess body weight is a contributory factor to both high blood pressure and heart disease, the primary cause of mortality in the UK today. The report recommends that, in order to achieve a loss in weight, foods rich in fats and/or sugar need to be restricted on a long-term basis. There is evidence[22] to suggest that, at least in the short term, a reduction in the energy density of the diet can lead to a fall in energy consumption.

Soft drinks sweetened with mixtures of sugar and intense sweeteners are less caloric than drinks sweetened entirely with sugar, and low calorie soft drinks contribute a maximum of 5·3 kcal/100 ml. The use of 318 058 kg of saccharin in soft drinks in the UK in 1983 represents a reduction of approximately $5 \times 10^{11}$ kcal in the amount contributed to the diet by soft drinks.

The recent trend towards healthier eating has led to an increase in demand for low-calorie foods. The availability of intense sweeteners allows the formulation of low-calorie soft drinks with good consumer acceptability as part of a calorie-controlled diet.

The need for dietetic products of this nature is recognised by the provisions of the UK 1964 Soft Drinks Regulations[11] which exempt low-calorie soft drinks from the limitations imposed on sugar and saccharin for soft drinks in general. The relatively low caloric limits imposed on soft drinks compared with other foods[23] make it virtually impossible to include carbohydrate sweeteners in low-calorie soft drinks, and simultaneously to satisfy the minimum standards for fruit/juice content required by the Soft Drinks Regulations. Low-calorie soft drinks cannot therefore be formulated without the use of intense sweeteners.

### 3.3.2 Dental Caries

Dental caries is dependent on bacterial growth on the tooth surface, metabolism of sugars in the mouth by these bacteria and the formation of acid which attacks the teeth. It is well known that one of the main aetiological factors is the length of time sugar is in the mouth. The use of intense sweeteners in soft drinks reduces their viscosity and may assist in shortening the exposure time.

In addition, the complete replacement of carbohydrates by intense sweeteners will give a less cariogenic product. In the last decade, increasing interest has been shown in the dental properties of the intense sweeteners, and this subject has recently been reviewed.[24] Studies have indicated that acesulfame-K,[25] saccharin[26,27] and aspartame[27,28] are non-cariogenic and that they may be anti-cariogenic as well. Mixtures of intense sweeteners have been shown to be even more effective in caries reduction.[29]

### 3.3.3 Diabetes

Intake of glucose and lack of insulin in diabetics result in diminished activity of the normal energy provision pathways. Although soft drinks may be formulated initially with sucrose, the formation of glucose by hydrolysis in storage and in metabolism makes the use of soft drinks containing sucrose unsuitable for diabetics. The 1964 Soft Drinks Regulations[11] consequently prohibit the addition of any carbohydrate sweetener to soft drinks suitable for diabetics.

The use of sweeteners such as sorbitol is permitted by the 1983 Sweeteners in Food Regulations.[3] However, sorbitol is only about half as sweet as sucrose on a weight-for-weight basis,[30] and another drawback is its laxative effect when ingested in large amounts. For this reason it has been recommended that no more than 30 g per day of sorbitol is consumed,[31] and in the UK foods which claim to be suitable, or have been made specially for diabetics, must carry the warning that it is best to eat less than 25 g a day.[23]

In order to make up the deficiency in sweetness resulting from the use of sorbitol, and to reduce the daily intake, the availability of intense sweeteners to produce palatable soft drinks for diabetics is essential. Sweet-tasting products have always been in great demand by persons with this particular metabolic disorder. The replacement of sugar by an intense sweetener is therefore essential. This is recognised by the provisions of the 1964 Soft Drinks Regulations[11] which exempt such products from the limits imposed by the addition of saccharin to soft drinks in general.

## 4 REQUIREMENTS OF THE IDEAL INTENSE SWEETENER IN SOFT DRINKS

### 4.1 Relative Sweetness

The material must have a good intensity of sweetness so that a minimal amount is required to achieve the required sweetness in the product. The relative sweetness is a major factor in determining the cost per unit of sweetness, and the caloric value per unit of sweetness.

### 4.2 Sweetness Characteristics

Since the material is to be used as a substitute for sugar, the closer its taste to sugar the better. The material should therefore match the sweetness quality and aftertaste of sucrose and the sweetness/time profile should be similar to that of sucrose. Sucrose has a relatively quick sweetness impact followed by a fairly sharp, clean cut-off (perception in less than 1 s and sweet taste remaining for about 30 s). The sweetness intensity of the sugar substitute should therefore build and then recede in the mouth in a similar way.

### 4.3 Stability

The material must be stable as a bulk chemical and in liquids of a low pH at moderate temperatures over long storage periods. The pH range of most soft drinks lies in the range 2·0–3·5 and suitable sweeteners must therefore be stable within this range. Concentrated soft drinks are normally expected to have a shelf-life of 12–18 months under fluctuating temperature conditions, and the intense sweetener should therefore show no loss of sweetness or any off-flavour development due to degradation in this time.

It must also be capable of withstanding the pasteurisation conditions used in the manufacture of soft drinks, typically 86–90°C for 30–45 s for products which are flash-pasteurised, or 60–70°C for 15–20 min for those which are in-pack pasteurised.

Ascorbic acid and sulphur dioxide used in soft drinks produce a strong reducing environment. The intense sweetener must be stable under such conditions.

#### 4.4 Solubility
The material must be readily soluble in cold water and in acid solution.

#### 4.5 Compatibility
It must be compatible with the other components of the drink such as aldehydes in flavours.

#### 4.6 Caloric Value
The intense sweetener should have zero caloric value or a low caloric value per unit of sweetness.

#### 4.7 Cariogenicity
It should be non-cariogenic or anti-cariogenic if possible.

#### 4.8 International Regulatory Status
The material should be non-toxic, with wide regulatory approval, and an acceptable daily intake that does not limit its use.

#### 4.9 Availability
It should be readily available on a commercial scale.

#### 4.10 Synergism
Ideally it should be synergistic with carbohydrate sweeteners and other intense sweeteners in enhancing taste.

#### 4.11 Relative Cost
It should be economic in terms of sweetening power compared with sucrose, i.e. the relative cost per unit of sweetness should be less than that of sucrose.

### 5 ORGANOLEPTIC AND FUNCTIONAL PROPERTIES OF INTENSE SWEETENERS IN RELATION TO THE REQUIREMENTS OF SOFT DRINKS

Since the prohibition of cyclamate in the UK and the USA, the food industry has been searching for a satisfactory alternative. The controversy which continues to surround the toxicological status of saccharin has added impetus to this quest. As a result, several new materials are available commercially and have been approved in various countries, whilst others are still awaiting approval or are at the development stage.

## 5.1 Approved Intense Sweeteners in the UK

### 5.1.1 Acesulfame-K

*5.1.1.1 Relative sweetness.* Acesulfame-K has a sweetening power about 200 times that of a 3% sucrose solution.[32] At higher concentrations of sucrose the relative sweetness decreases, so that at 6% sucrose it is 100. Normally acesulfame-K can be considered to be about half as sweet as sodium saccharin and 4 times sweeter than cyclamate. In hot drinks the sweetness of intense sweeteners is usually lower than at ambient temperature, but in the case of acesulfame-K this effect is less pronounced.[33] The sweetening power of acesulfame-K is also slightly higher in acidic drinks than it is in neutral aqueous solutions.[34]

*5.1.1.2 Sweetness characteristics.* The sweet taste of acesulfame-K is perceived rapidly, without any noticeable delay. It diminishes slowly, without lingering unacceptably, although the sweet taste may persist slightly longer than that of sucrose.[32] At high concentrations a lingering bitter and chemical synthetic taste can be perceived by some individuals, but at moderate concentrations it has little aftertaste and the quality of sweetness is higher than that of saccharin.[35]

*5.1.1.3 Stability.* The stability of acesulfame-K in the dry form appears to be virtually unlimited at room temperature under dry conditions, and is independent of exposure to, or protection from, light.[32] It also shows good stability in aqueous media at pH 3 and above, even at temperatures as high as 40°C. Stability falls at pH values below 3·0, but at pH 2·5 only a few per cent of the compound decomposes within 6 months. Hydrolytic decomposition is generally faster at elevated temperatures, and some loss in sweetness may be noticed on prolonged storage at pH 2·5 at 40°C.[32] This is not likely to be experienced for long periods, and in practice no real stability problem exists. Beverages containing acesulfame-K can be pasteurised under normal conditions without loss of sweetness. No decomposition of acesulfame-K was observed after pasteurising at 90°C for several seconds or after pasteurisation at 72–74°C for longer periods.[32]

*5.1.1.4 Solubility.* Acesulfame-K is a white, crystalline solid, readily soluble in water to produce clear, colourless solutions. The solubility at 20°C is about 270 g/litre.[36] With increasing temperatures the solubility rises sharply to more than 1000 g/litre at 100°C. Stable, concentrated stock solutions can easily be prepared as long as precautions are taken to prevent microbiological contamination. Aqueous solutions are almost neutral.

*5.1.1.5 Compatibility.* Compatibility differs with the different flavours used for soft drinks production.[37] In our own laboratories satisfactory results have been obtained with shandy, lemonade, cola, iron brew, blackcurrant, American ginger ale, tonic water, bitter lemon, lemon and orange. However, partial replacement of sucrose in a strawberry flavour cordial resulted in the development of an unacceptable off-flavour, after storage for only 1 month at room temperature, which was traced to interaction between acesulfame-K and the strawberry flavour.

*5.1.1.6 Caloric value.* Acesulfame-K is excreted unmetabolised and therefore has no caloric value.

*5.1.1.7 Cariogenicity.* Acesulfame-K is not affected by the bacteria of the mouth and does not therefore contribute to the development of caries, and it may be anti-cariogenic as well.[25]

*5.1.1.8 International regulatory status.* The Joint Expert Committee on Food Additives of the FAO/WHO has evaluated the available toxicological data and allocated acesulfame-K an ADI of 0–9 mg/kg of body weight.[38] It has been approved for use in beverages in a total of more than 15 countries. Petitions for the approval of acesulfame-K have been filed in other countries.

*5.1.1.9 Availability.* Acesulfame-K is readily available on a commercial scale, and is marketed in the UK by Hoechst AG under the trade-mark Sunett.

*5.1.1.10 Synergism.* Acesulfame-K shows a pronounced synergistic action with aspartame and cyclamate but little with saccharin. It also has a synergistic effect with nutritive sweeteners such as sorbitol, isomalt and fructose.[32]

*5.1.1.11 Relative cost.* 0·4–0·9 (sucrose = 1).

*5.1.2 Aspartame*

*5.1.2.1 Relative sweetness.* The relative sweetness of aspartame varies with the concentration and the food system in which it is used, but it has generally of the order of 160–220 times the sweetness of sucrose.[39,40] In soft drinks a good average value to use is 180 but this can vary according to the flavour system, the pH and the presence of sucrose and other intense sweeteners.

*5.1.2.2 Sweetness characteristics.* Aspartame is reputed to have a clean sugar-like taste, without the bitter aftertaste often associated with intense sweeteners.[40-44] In a consumer evaluation of the sweetness of a 4% sucrose solution and a 0·02% aspartame solution, no significant difference in sweetness ratings was found, but both aspartame and sucrose were found to have an aftertaste.[45] The aspartame aftertaste was described as having a lingering sweetness, a bitter-sweetness and a slightly powdery sensation, whereas the aftertaste of sucrose was described as sweet and drying with only a slight bitterness. The sweetness/time profile of aspartame is quite similar to that of sucrose, but its sweetness develops more gradually and persists slightly longer.[46]

Aspartame was evaluated in a range of soft drinks in our own laboratories, and the sweetness was found to be clean and quite similar to that of sucrose, but with a slight delay in the perceived sweetness, which persisted slightly longer than that of sucrose. No aftertaste was apparent in any of the drinks, and in the case of fruit drinks the fruit flavour was intensified. The enhancement of flavour, particularly acid fruit flavours such as orange, lemon and grapefruit, has been reported in the literature.[47]

*5.1.2.3 Stability.* Stability in the dry form is excellent. Temperatures well over 150°C are necessary for breakdown to become substantial. Regular storage conditions of 20–25°C and 50% relative humidity do not cause any stability problems provided that the moisture content is kept below 8%.[48]

In liquids, under certain conditions of temperature and pH, the ester bond is hydrolysed, forming aspartylphenylalanine and methanol. Alternatively, methanol may be eliminated by the cyclisation of aspartame to form the diketopiperazine. The diketopiperazine ring could, in turn, open to form aspartylphenylalanine, which can be hydrolysed to its individual amino acids—aspartic acid and phenylalanine.[49-52]

The stability of aspartame in solution is affected by time, temperature and pH, with pH being particularly critical. Aspartame is most stable in the acid range of pH 3–5, with optimum stability at pH 4·3. Most carbonated soft drinks have a pH within this range, but with high-acid drinks (pH in the region of 2·0–2·5) some hydrolysis can be expected during storage, at a rate which is temperature-dependent. This will result in a gradual loss of perceived sweetness, but without the development of off-flavours because the conversion products are tasteless.[39,40]

In carbonated beverages stored for 8 weeks, 11–16% and 28% of added aspartame was lost under storage conditions of 20 and 30°C respectively.[53] However, extensive testing has shown that a loss of up to 40% aspartame will be tolerated before beverages become unacceptable,[54] and storage at a constant temperature of 30°C is unlikely to be experienced for long periods.

For concentrated soft drinks which have a pH in the region of 2–3 and for which a shelf-life of at least 12 months is required, the loss of sweetness may be a limiting factor. Improvement in the stability of aspartame in solution by means of aspartame–cyclodextrin–sucrose–fatty acid compounds has been reported.[55] Alternatively, aspartame may be combined with other intense sweeteners to extend the sweetness stability.[54]

Beverages containing fruit juice can be pasteurised at 93–96°C for 15–30 s with a minimal loss, no greater than 4%, or at 136°C for 6–8 s with a maximum loss of 12%. These results show adequate stability to allow the addition of aspartame before pasteurisation of the product.[56]

*5.1.2.4 Solubility.* Aspartame is only slightly soluble in water (about 1·0% at 25°C), solubility improving as the pH is reduced or the temperature raised. Maximum solubility is at pH 2·2. For optimum rate of dissolution, a temperature of 40°C is recommended at pH 4, at which it is most stable.[48] Alternatively, aspartame can be pre-dissolved in a citric acid solution without the need for raising the temperature, provided that the solution is used without delay.

*5.1.2.5 Compatibility.* No problems of compatibility have been observed in a wide range of soft drinks. Aspartame has the property of enhancing flavours, and this is best displayed with flavours that are naturally derived rather than artificial flavours.[47]

*5.1.2.6 Caloric value.* Aspartame is metabolised to its component amino acids via normal metabolic processes. It has an energy value of 4 kcal/g. Assuming a relative sweetness of 180, the caloric value per unit of sweetness is approximately 0·02 kcal/g, which is a negligible contribution on a sweetness basis.

*5.1.2.7 Cariogenicity.* Studies have shown that aspartame is non-cariogenic and may be anti-cariogenic.[24,27,28]

*5.1.2.8 International regulatory status.* Aspartame is permitted in soft drinks in more than 18 countries, including Australia, Belgium, Canada, Hong Kong, Japan, South Africa, Switzerland, the UK and the USA.[57] It has been reviewed and found to be safe by the EEC Scientific Committee for Food and the Joint Expert Committee on Food Additives of the FAO/ WHO, and has been allocated an ADI of 0–40 mg per kg body weight.[58]

Aspartame contains phenylalanine which, like all proteins, cannot be tolerated by phenylketonurics. In the USA the label of any food or beverage containing aspartame must alert those who have to restrict their intake of phenylalanine by bearing the warning statement. 'Phenyl-ketonurics: contains phenylalanine'.

*5.1.2.9 Availability.* Aspartame is commercially available to food manufacturers worldwide, and is marketed in Europe by NutraSweet AG under the trade-mark NutraSweet.

*5.1.2.10 Synergism.* Aspartame is synergistic with a wide range of carbohydrate sweeteners and intense sweeteners.

*5.1.2.11 Relative cost.* 0·9–1·2 (sucrose = 1).

### 5.1.3 Saccharin

*5.1.3.1 Relative sweetness.* Saccharin is 300–600 times as sweet as sucrose, the relative sweetness declining with increasing concentration. It is also affected by acidity, temperature and flavour.[59]

*5.1.3.2 Sweetness characteristics.* Unlike sucrose, saccharin has a slow sweetness impact which builds up to a maximum intensity and then persists. Many people find that it leaves bitter, metallic and astringent aftertastes, which tend to increase with rising concentrations of the sweetener.[60] This aftertaste is an intrinsic characteristic of the saccharin molecule and is not caused by the presence of impurities.[61] It severely limits the use of saccharin on its own as a sweetener in low-calorie soft drinks. Removing or masking this aftertaste has been the subject of numerous 'recipes' over the years. Dulcin was originally used for this purpose;[62] when cyclamate was a permitted sweetener, combinations of saccharin and cyclamate were used to produce taste qualities very similar to sugar. Since the ban on cyclamate, and before the approval of other intense sweeteners, various additives have been used in combination with saccharin to mask the aftertaste; these include sucrose,[63] lactose,[64] tartrates,[65] cornstarch hydrolysate[66] and fructose with gluconate salts.[67] More recently, saccharin has been combined with other approved intense sweeteners, such as aspartame, to overcome the problem in low-calorie soft drinks.

*5.1.3.3 Stability.* In its bulk form, saccharin and its salts have been shown to be stable over several years.[68] Saccharin exhibits satisfactory stability in most soft drink applications, but in low-pH fruit juices it hydrolyses at a slow, temperature-dependent rate to produce 2-sulphobenzoic acid and 2-sulphamoylbenzoic acid. In tests in our own laboratories, we have observed a 20% loss in lemon juice stored at ambient temperatures after 12 months. Storage at 30°C led to a 16% loss in 1 month and 34% loss in 6 months. Saccharin can withstand the pasteurisation conditions normally used for soft drink production.

*5.1.3.4 Solubility.* The acid form of saccharin is only sparingly water-soluble, but its sodium and calcium salts are readily soluble in water and acidic aqueous solutions. The easily dissolved sodium salt is most commonly used, as a white, granular dihydrate, 82 g dissolving in 100 g of water.[68]

*5.1.3.5 Compatibility.* Saccharin is compatible with other soft drink components.

*5.1.3.6 Caloric value.* Saccharin is not metabolised and does not contribute energy to the diet.

*5.1.3.7 Cariogenicity.* Studies have indicated that saccharin is non-cariogenic and may be anti-cariogenic.[24,26,27]

*5.1.3.8 International regulatory status.* Saccharin has gained wide regulatory approval and is used in more than 80 countries. It is approved by the Joint Expert Committee on Food Additives of the FAO/WHO and the EEC Scientific Committee for Food and has been allocated a temporary ADI of 0–2·5 mg/kg body weight.[69] The use of saccharin in the USA must be accompanied by the mandatory warning statement 'Use of this product may be hazardous to your health. This product contains saccharin which has been determined to cause cancer in laboratory animals'.

*5.1.3.9 Availability.* Saccharin is readily available on a commercial scale.

*5.1.3.10 Synergism.* Saccharin is synergistic with a number of intense sweeteners, but shows little synergism with acesulfame-K or stevioside.

*5.1.3.11 Relative cost.* 0·02–0·03 (sucrose = 1).

### 5.1.4 Thaumatin

*5.1.4.1 Relative sweetness.* Thaumatins are a group of intensely sweet proteins extracted from the fruit of the West African plant *Thaumatococcus danielli* (Katemfe). Thaumatin is claimed to have a sweetness 2000–2500 times that of an 8–10% sucrose solution.[70]

*5.1.4.2 Sweetness characteristics.* The sweetness time/intensity profile of thaumatin differs considerably from that of sucrose. There is a slow onset of sweetness which builds up to a maximum intensity, followed by a long, lingering sweet liquorice-like aftertaste.[35,71] It does not, however, have the bitter, metallic aftertaste associated with some of the other intense sweeteners.[71] Attempts have been made to modify the sweetness profile by chemical manipulation, but without commercial success. The use of additives has also been tried, and a slight reduction in the aftertaste has been achieved using arabinogalactan,[72] or aldohexuronic acids and their salts, amides or lactones.[73] In Japan a proprietary mixture is available marketed under the name San Sweet T-100 in which thaumatin is mixed with alanine, food acids and fillers. This mixture not only slightly reduces the initial delay in sweetness and the aftertaste, but almost doubles the relative sweetness of the thaumatin. Elimination of the lingering aftertaste is also the subject of a Japanese patent[74] in which thaumatin is mixed with phosphates, the sodium salts of food acids and amino acids.

*5.1.4.3 Stability.* Thaumatin is extremely stable in aqueous solutions in the pH range 2·7–6·0 with an optimum around pH 2·8–3·0, which is ideal for soft drinks application.[71] At low pH solutions can be heated at 100°C for several hours without loss of sweetness, so soft drinks and their syrups can be pasteurised without loss of sweetness.

*5.1.4.4 Solubility.* Thaumatin is a cream-coloured powder which is extremely soluble in water, enabling 60% solutions to be prepared.[71]

*5.1.4.5 Compatibility.* The high overall positive ionic charge on the thaumatin molecule causes it to associate with suitably shaped, negatively charged ingredients such as synthetic colours and food gums.[71] Synthetic colours having suitably spaced sulphonate groups can, under certain conditions of temperature, pH and concentration, link thaumatin molecules together, eventually causing precipitation. Even if precipitation does not occur there may be a loss of colour. The interaction of thaumatin with synthetic colours can be prevented by the use of gum arabic.[71]

Loss of sweetness can also occur with gums such as xanthan, pectin, carboxymethylcellulose, carrageenan, locust bean gum and alginates, when they are present in great excess. Thaumatin can enhance certain flavours, particularly peppermint, spearmint, coffee and ginger flavour; it can also enhance aromas. Flavour enhancement can be achieved at levels below the sweet taste threshold.[71]

*5.1.4.6 Caloric value.* Since thaumatin is a protein it does have an energy value, 4 kcal/g. However, because of its high sweetness potency, the caloric value per unit of sweetness is less than 0·002 kcal, so its contribution is negligible.

*5.1.4.7 Cariogenicity.* Thaumatin is non-cariogenic but it is not cariostatic or anti-cariogenic.[71]

*5.1.4.8 International regulatory status.* Thaumatin is permitted as an intense sweetener for use in soft drinks in the UK, Australia and Japan. The safety of thaumatin was reviewed by the Joint Expert Committee on Food Additives of the FAO/WHO. In June 1985 they agreed it was safe for food use without specifying an acceptable daily intake.[71]

*5.1.4.9 Availability.* Thaumatin is available commercially and is marketed by Tate & Lyle, Reading, England, under the trade name Talin.

*5.1.4.10 Synergism.* Thaumatin is synergistic with saccharin, acesulfame-K and stevioside, but not with cyclamate or aspartame. It can be used to mask the bitterness of sweeteners such as saccharin, as its sweetness is still perceived when the bitterness of saccharin would normally be at a maximum.[71]

## 5.2 Intense Sweeteners Permitted in Other Countries
### 5.2.1 Cyclamate
*5.2.1.1 Relative sweetness.* Cyclamate is generally considered to have a sweetness 30–40 times that of sugar, based on a weight-for-weight equivalence.[75] Used in food applications, it exhibits a greater sweetness factor than 30, and in fruit drinks a factor of 40 can usually be applied.

*5.2.1.2 Sweetness characteristics.* Cyclamate has a slow onset of sweetness, and the duration of the lingering sweet-sour aftertaste makes it difficult to determine exact sweetness equivalents at high concentrations.[35]

At high concentrations a distinct off-taste is noticeable, but a higher sweetness level can be achieved by cyclamate than by saccharin before the off-taste becomes detectable.[76]

*5.2.1.3 Stability.* Sodium cyclamate shows good stability in the dry form. Aqueous solutions are stable to heat, light and air throughout the pH range 2–10,[77] and soft drinks and their syrups can be pasteurised without loss of sweetness.

*5.2.1.4 Solubility.* Sodium cyclamate is readily soluble in water, about 19·5 g dissolving in 100 ml of water at 20°C to produce a clear, colourless solution.[75]

*5.2.1.5 Compatibility.* Cyclamate is compatible with a broad range of other ingredients, including natural and artificial flavouring agents, and it enhances the natural flavour of a fruit, making some re-balancing necessary.[78] When using calcium cyclamate it is necessary to use a high-acid-proof caramel to avoid precipitation. The calcium salt can also cause gelling problems with pectin, and in concentrates with high acidity precipitates of calcium citrate can form.[79]

*5.2.1.6 Caloric value.* Cyclamate can be metabolised to cyclohexylamine by a small fraction of the population,[80] but is considered to be non-caloric.

*5.2.1.7 Cariogenicity.* Cyclamate does not promote tooth decay, but it does not have the anti-cariogenic effect demonstrated by acesulfame-K,[25] saccharin[26,27] and aspartame.[27,28,81]

*5.2.1.8 International regulatory status.* Cyclamate is approved by the EEC Scientific Committee for Food and the Joint Expert Committee on Food Additives of FAO/WHO and has been allocated a temporary ADI of 0–11 mg/kg body weight.[82] It is permitted in Australia and New Zealand and several countries in Europe, including Switzerland, Spain and Germany. However, it may not be used in any drugs, other than those with approved new drug applications, or in any foods (including beverages) intended for use in the United States.

*5.2.1.9 Availability.* Cyclamate is available in the United States, as well as from several international suppliers.

*5.2.1.10 Synergism.* Cyclamate is synergistic with saccharin, aspartame, acesulfame-K, sucralose, alitame and stevioside. The use of mixtures of cyclamate and saccharin eliminates both the bitter, metallic aftertaste of saccharin and the off-taste associated with cyclamate. Mixtures of cyclamate and saccharin produce excellent sweetness characteristics. The blend gives an intense bright sweetness similar to that of sucrose, and also potentiates fruit flavours, imparting the sensation of more body to low-calorie soft drinks.

*5.2.1.11 Relative cost.* 0·4–0·5 (sucrose = 1).

### 5.2.2 Neohesperidin Dihydrochalcone (NHDC)

*5.2.2.1 Relative sweetness.* At the 5% sucrose level NHDC is about 250 times sweeter, but this reduces to about 170 at a sweetness equivalent to 10% sucrose.[83]

*5.2.2.2 Sweetness characteristics.* NHDC has a pleasant sweet taste, but it has a delayed onset and a lingering liquorice/menthol aftertaste.[84] It has no bitter aftertaste and it has the ability to reduce the perception of bitterness in grapefruit juice when added at 6–12 ppm.[84] The slow onset of sweetness is reputed to be overcome by the use of $\beta$-gluconolactone,[85] but the lingering aftertaste is a limiting factor in soft drink applications.

*5.2.2.3 Stability.* Aqueous solutions of NHDC are stable in the pH range 2·5–3·5 under normal storage conditions. Only below pH 2 does the material hydrolyse to free sugars and the aglycone.[86,87] Excessively high temperatures are required to break the glycosidic bonds of NHDC, so beverages and syrups with NHDC can be safely pasteurised without loss of sweetness.

*5.2.2.4 Solubility.* At room temperature NHDC is only sparingly soluble in water (0·05 g/100 ml),[86] but it is soluble in ethanol (2·04 g/100 ml at 25°C), and this vehicle can be used to add NHDC to soft drinks.

*5.2.2.5 Compatibility.* No problems of incompatibility with soft drinks ingredients have been reported.

*5.2.2.6 Caloric value.* NHDC has a calculated caloric value of approximately 2 kcal/g.[84] Because of its high potency, NHDC would contribute less than 1/1000 as many calories as an equivalent amount of sucrose in terms of sweetness.

*5.2.2.7 Cariogenicity.* NHDC is relatively inert to the action of cariogenic bacteria.[88]

*5.2.2.8 International regulatory status.* It is permitted in Belgium, where it can be used for the total replacement of sugar in lemonade (max. 50 mg/litre), but not allowed in Canada, the USA or the UK.

*5.2.2.9 Availability.* Commercially available.

*5.2.2.10 Synergism.* Combinations of saccharin and NHDC are reputed to be synergistic.[89]

### 5.2.3 Stevioside

This is a natural intense sweetener extracted from the leaves of the plant *Stevia rebaudiana* Bertoni, indigenous to Paraguay. The leaves contain a complex mixture of sweet diterpene glycosides, including stevioside, steviolbioside, rebaudiosides A, B, C, D, E and dulcoside A. Stevioside is the main sweet glycoside and the others are present in varying concentrations depending on the extraction and modifications used. The following details relate to the commercially available material sold under the brand name Sato Stevia,[90] which contains many glycosides all extracted from the plant without modification. Many other variations are commercially available in Japan.

*5.2.3.1 Relative sweetness.* Sato Stevia has a sweetness 160 times that of sucrose at a concentration equivalent to 4% sucrose. The relative sweetness decreases as the equivalent concentration of sucrose increases.

*5.2.3.2 Sweetness characteristics.* Sato Stevia resembles sucrose in its sweetness profile, which can be varied by blending two different base products. Some stevioside extracts exhibit a menthol-like, persistent bitter aftertaste, which diminishes with increasing purity.[91] The Sato stevioside material tested in our own laboratories did not suffer from this disadvantage, although some tasters detected a liquorice aftertaste. Laboratory tests established that a 25% replacement of the added sugar in a blackcurrant juice drink could be achieved with no significant difference in sensory evaluation from the all-sugar product.

*5.2.3.3 Stability.* The bulk material is very stable even under extremes of temperature, and for practical industrial purposes has an unlimited shelf-life and good stability in acid solutions. The long-term stability of stevioside in a citrus beverage acidified with citric acid, and a cola beverage acidified with phosphoric acid, was studied by Chang and Cook.[92] No significant changes were detected after 5 months storage at room temperature (22°C). A 36% loss was observed after 4 months at 37°C in the cola beverage, and a 17% loss after 4 months in the citrus beverage. Stevioside solutions in the pH range 3–9 can be maintained at 100°C for 1 hour with no significant loss.[93] Soft drinks and their syrups can therefore be safely pasteurised without loss of sweetness.

*5.2.3.4 Solubility.* This material is freely soluble in water; 800 g will dissolve in 800 ml of water.

*5.2.3.5 Compatibility.* Sato Stevia is compatible with the ingredients normally found in soft drinks and will enhance certain fruit flavours.

*5.2.3.6 Caloric value.* Sato Stevia is not absorbed into the human system and has no energy value.

*5.2.3.7 Cariogenicity.* It will not ferment and does not cause or contribute to tooth decay.

*5.2.3.8 International regulatory status.* Extensive toxicological testing has been carried out on the material in Japan, where it has been approved for use since 1972. It is also approved in Paraguay, Brazil, South Korea and the People's Republic of China. In the UK it is currently being reviewed by the Committee on Toxicity of Chemicals in Food, Consumer Products and the Environment. An application has also been submitted to the Scientific Committee for Food of the EEC.

*5.2.3.9 Availability.* Readily available on a commercial scale in Japan.

*5.2.3.10 Synergism.* Stevioside has been found to be synergistic with aspartame, acesulfame-K and cyclamate, but not with saccharin.[94]

*5.2.3.11 Relative cost.* 0·9–1·5 (sucrose = 1).

## 5.3 New Intense Sweeteners Under Regulatory Review

### 5.3.1 Alitame

Alitame is a new dipeptide-based, high-intensity sweetener developed by Pfizer.

*5.3.1.1 Relative sweetness.* Alitame is intensely sweet, about 2000 times as sweet as sucrose, 12 times sweeter than aspartame and 6 times sweeter than saccharin at a level comparable to 10% sucrose.[95]

*5.3.1.2 Sweetness characteristics.* Alitame has a clean, sweet taste with no unpleasant aftertaste. The onset of sweetness is very similar to that of aspartame, but it is delayed in comparison with sucrose, with a tendency to linger.

*5.3.1.3 Stability.* In the neutral pH range (6–8), at room temperature, alitame is stable in solution for more than a year. In acid solution (pH 2–4) it can hydrolyse at the aspartylalanine dipeptide bond to give aspartic acid and alanine amide, but the solution is less labile than that of aspartame. The decomposition products are tasteless, so there is a gradual loss of sweetness but no off-flavour development. In tests in our own laboratories we have observed a slight loss of sweetness in soft drinks containing alitame after 3 months shelf storage at ambient temperature. It has been found to be stable to pasteurisation at 90°C for 42 s and 63°C for 10 min.

*5.3.1.4 Solubility.* Alitame is a crystalline, odourless, non-hygroscopic powder which dissolves readily in water to produce clear, colourless solutions. Minimum solubility in water at isoelectric pH 5·7 is 13% at 25°C. The solubility rises as the pH increases or decreases, and at pH 3·0 and 20°C the solubility is 39% w/v.

*5.3.1.5 Compatibility.* Prolonged storage of alitame in a few standard acidic beverage recipes may result in an incompatibility as measured organoleptically. This is not reflected in storage stability as measured by chemical assay for alitame and its degradation products. Levels of off-flavourant(s) are below the limits of modern analytical detection. Substances which may produce off-flavours on storage with alitame in liquid products are hydrogen peroxide and sodium bisulphite. Adjustment of formulations to eliminate or modify incompatible ingredients may resolve such problems.

*5.3.1.6 Caloric value.* Alitame is partially caloric, with a maximum contribution of 1·4 kcal/g. Because of its high intensity of sweetness, its contribution to energy intake is negligible in the amounts used (about 0·02% of the replaced sucrose in the product).

*5.3.1.7 Cariogenicity.* Alitame is non-cariogenic.

*5.3.1.8 International regulatory status.* Alitame has been subjected to extensive animal and human studies which document its safety. It has not yet been approved for use anywhere, but a Food Additive Petition requesting broad clearance was submitted to the US Food and Drug Administration in August 1986. Submissions have also been accepted for review in Canada, the UK, Switzerland and the EEC/SCF, and regulatory activities are underway in a number of other countries throughout the world (Pfizer, personal communication, June 1988).

*5.3.1.9 Availability.* Not yet available on a commercial scale.

*5.3.1.10 Synergism.* Alitame is synergistic with acesulfame-K and cyclamate and combines well with these and other sweeteners, including saccharin, to produce a good quality of sweetness.

*5.3.1.11 Cost.* It is expected to be competitive with aspartame on a cost per unit of sweetness basis.

### 5.3.2 Sucralose

Sucralose is the generic name for 1,6-dichloro-1,6-dideoxy-$\beta$-D-fructofuranosyl-4-chloro-4-deoxy-$\alpha$-D-galactopyranoside, a chlorinated derivative of sucrose possessing a high degree of sweetness (see Chapter 5). Its intense sweetness was discovered in 1976 at the Tate & Lyle research laboratories. It is being jointly developed for commercialisation, in the UK by Tate & Lyle Speciality Sweeteners and in the US by Johnson & Johnson's McNeil Speciality Products Company.

*5.3.2.1 Relative sweetness.* The relative sweetness of sucralose varies with pH, temperature, concentration and ingredient mix. In the pH range of soft drinks it varies from 450 at a concentration equivalent of 10% sucrose to 650 at the equivalent of 5% sucrose. The lower the pH the greater the relative sweetness.[96]

*5.3.2.2 Sweetness characteristics.* The sweetness time/intensity profile of sucralose is very similar to that of sucrose. There is an almost instant onset of clean sweetness, which lingers slightly longer than that of sucrose. There is no bitter aftertaste. In tests in our own laboratories, evaluation of sucralose in lemonade, a sparkling orange drink and a concentrated orange drink demonstrated good sweetness quality when it was used as either a full or partial replacement of sucrose, and also when used in conjunction with saccharin as a complete replacement for sucrose.

*5.3.2.3 Stability.* The stability of sucralose in the solid form is very good. Bulk sucralose can be stored at 20°C for several years without degradation. In acid solution sucralose hydrolyses slowly to its constituent mono-saccharides at a rate dependent on pH and temperature. At normal storage temperatures sucralose is extremely stable in the pH range of soft drinks. At pH 3·0 and 20°C there is a loss of less than 1% over a storage period of 1 year.

In our own laboratories no loss of sweetness was perceived in low-calorie carbonated beverages sweetened with sucralose after storage for 6 months at 20°C. A slight loss of sweetness was observed in a low-calorie concentrated orange drink (pH 2·3) stored at 20°C for 6 months. In a cola formulation at pH 3·0 and 40°C, sucralose was found to be more stable than cyclamate, saccharin, acesulfame-K and aspartame (Tate & Lyle, personal communication, January 1987). It can readily withstand the pasteurisation conditions for soft drinks and soft drink syrups.

*5.3.2.4 Solubility.* Sucralose is readily soluble in water to give a 26% w/w solution at 20°C.

*5.3.2.5 Compatibility.* The interaction of sucralose with a wide variety of soft drink components, including minerals, vitamins, preservatives and flavourings, has been studied, and no adverse interactions have been observed.[96]

*5.3.2.6 Caloric value.* Sucralose is quite poorly absorbed from the intestine and is not broken down during its passage through the body. It has no caloric value.

*5.3.2.7 Cariogenicity.* Sucralose is not metabolised by microorganisms in the mouth and does not therefore promote tooth decay.

*5.3.2.8 International regulatory status.* Sucralose has been subjected to a comprehensive safety evaluation programme to meet the requirements of both national and international regulatory authorities. The Joint Expert Committee on Food Additives of the FAO/WHO has evaluated the available toxicological data and allocated sucralose a temporary ADI of 0–3·5 mg/kg of body weight.[97] At the time of writing, the approval of sucralose is being sought in a number of countries, including Canada, the USA and the UK. The EEC Scientific Committee for Food is currently assessing sucralose as part of their further review of sweeteners.

*5.3.2.9 Synergism.* Synergistic action has been found between sucralose and cyclamate and acesulfame-K, but little between sucralose and saccharin and aspartame.

*5.3.2.10 Relative cost.* No specific information is available, but it is expected to be competitive with aspartame.

## 6 SENSORY EVALUATION IN THE PRODUCT

Among the most important attributes of a successful intense sweetener for use in soft drinks are the quality and characteristics of the sweetness it produces. Several studies have been devoted to these considerations.

Using similarity and descriptive scales, Schiffman *et al.*[98] compared the sensory properties of a lemon/lime and a cola flavour beverage, sweetened with either sucrose, sodium saccharin, aspartame, acesulfame-K or two calcium cyclamate/sodium saccharin blends (10:1 and 3·5:1). Using sweetness intensity factors of 250 for saccharin, 200 for aspartame, 150 for acesulfame-K in lemon/lime, 50 for acesulfame-K in cola, 100 for the 10:1 calcium cyclamate/saccharin blend and 125 for the 3·5:1 calcium cyclamate/saccharin blend, they were able to formulate drinks iso-sweet with a 10·8° Brix lemon/lime product and a 10·7° Brix cola drink. The results indicated that the drinks containing sucrose and aspartame could not be distinguished from each other in either beverage, and were statistically equivalent on every descriptive scale. The calcium cyclamate/saccharin blends also rated highly. On both similarity judgements and descriptive scales, acesulfame-K and saccharin differed most from sucrose.

Askar *et al.*[99] examined the sweetness quality of a carbonated cola beverage sweetened with sucrose (10·8° Bx), aspartame, acesulfame-K, aspartame/sorbitol blend, acesulfame-K/sorbitol blend or a saccharin/

sorbitol blend. The aspartame drink achieved a taste quality equivalent to that of the sucrose product, and had a clean, sweet taste. The beverage with acesulfame-K showed a slight decrease in taste quality which could be improved by the addition of maltol or ethyl maltol.

Larson-Powers and Pangborn[43] developed time–intensity (T–I) curves for flavour, sweetness, sourness and bitterness in orange, strawberry and lemon drinks, each of which was sweetened with sucrose, saccharin, cyclamate or aspartame. In most sensory attributes aspartame closely resembled sucrose. The saccharin sample was more sour, considerably more bitter and significantly less sweet than the other three. Cyclamate imparted moderate bitterness, little sourness, but a persistent sweetness. The bitterness of saccharin and cyclamate persisted significantly longer than the taste of the other sweeteners. Similar results were obtained in a subsequent study using anchored and unanchored descriptive analyses.[44]

Cloninger and Baldwin[42] also concluded that a carbonated orange-flavoured beverage sweetened with 0·075% aspartame did not differ significantly in sweetness, sourness or acceptability from a 10% sucrose sweetened beverage.

## 7 USE OF MIXTURES OF INTENSE SWEETENERS: MULTIPLE SWEETENER CONCEPT

A study of the organoleptic and functional properties of the intense sweeteners clearly shows that none alone can match sucrose (Table 3). Using intense sweeteners as the single sweetener in soft drinks often results in poor or objectionable sensory properties. Off- or after-tastes can sometimes be detected; loss of sweetness can occur on storage if the intense sweetener is unstable; lack of 'body' or mouthfeel is also a major objection to the incorporation of intense sweeteners in low-calorie soft drinks. Cost restraints, legal restrictions and low acceptable daily intakes can also be limiting factors.

By using intense sweeteners in combination, the limitations of one sweetener can be offset by the strengths of others. This is known as the multiple sweetener concept; it is supported by the International Sweetener Association and the US Calorie Control Council. Mixtures of intense sweeteners can be used to

(a)  produce a sweetness profile similar to that of sucrose,
(b)  mask aftertaste,
(c)  improve stability,

## TABLE 3
TECHNICAL ADVANTAGES AND DISADVANTAGES OF THE CHARACTERISTICS OF INTENSE
SWEETENERS FOR SOFT DRINK APPLICATION

| Sweetener | Advantages | Disadvantages |
|---|---|---|
| Acesulfame-K | Good stability<br>Readily soluble<br>Synergistic with wide range of intense sweeteners | Lingering bitter aftertaste at high concentrations<br>Variable compatibility with flavourings |
| Aspartame | Good quality sweetness<br>Enhances fruit flavours<br>Good compatibility | Lack of stability<br>Low solubility |
| Saccharin | Good stability | Slow onset of sweetness<br>Lingering bitter/metallic aftertaste |
| Thaumatin | Good stability<br>Readily soluble<br>Very high sweetness potency, so low use levels<br>Natural | Slow onset of sweetness<br>Lingering liquorice aftertaste<br>Incompatible with certain beverage ingredients |
| Cyclamate | Good stability, solubility and compatibility<br>Enhances fruit flavours | Slow sweetness onset<br>Lingering aftertaste at high concentrations<br>Low sweetness intensity, so relatively high use levels |
| Neohesperidin dihydrochalcone | Good stability and compatibility<br>Masks bitterness | Delayed sweetness onset<br>Lingering liquorice/menthol aftertaste<br>Poor solubility |
| Sato Stevia | Good stability, solubility and compatibility<br>Natural | Liquorice aftertaste to some tasters |
| Alitame | Good quality sweetness<br>Very high sweetness potency, so low use levels<br>Readily soluble | Some instability<br>Incompatible with certain beverage ingredients |
| Sucralose | Good quality sweetness<br>Sugar-like sweetness profile<br>Good stability, solubility and compatibility | |

(d)  meet cost restraints,
(e)  comply with legal limits,
(f)  reduce the daily intake of any one particular sweetener.

## 7.1  Masking the Aftertaste of Saccharin
### 7.1.1  Saccharin/Cyclamate Mixtures
High levels of saccharin and cyclamate each exhibit their own off-tastes in addition to sweetness. A higher level of sweetness can be achieved by cyclamate before this off-taste can be detected. Thus Vincent et al.[76] found that 20% of their taste panel could detect an off-taste at a saccharin concentration which was iso-sweet with 3% sucrose, whereas a 6% sweetness level could be achieved with cyclamate. By combining cyclamate and saccharin in the ratio 10:1 a sweetness equivalent to 26% sucrose could be achieved before the off-taste could be detected to the same degree. This concept was eventually patented by Helgren.[100] Thus saccharin boosts the sweetening power of the less sweet cyclamate, while cyclamate helps to mask the bitter aftertaste experienced with saccharin alone.

Mixtures of saccharin and cyclamate were extensively used in soft drinks in the UK and the USA in the period when both sweeteners were permitted in these countries. In a survey of dietetic beverages on the US market in 1965 it was found that more than 70% used this blend.[101]

### 7.1.2  Saccharin/Thaumatin Mixtures
The delayed perception of sweetness of thaumatin, the slow build-up to maximum intensity and the lingering sweet aftertaste can also be used to advantage in masking the bitter aftertaste of saccharin. The masking is probably a function of both the area of the tongue with maximum sensitivity to thaumatin, and also the characteristic that its sweetness is still perceived while the bitterness of saccharin would normally be at a maximum.[71] Thaumatin/saccharin combinations can be further improved by the addition of carbohydrate taste-modifiers to minimise the lingering thaumatin aftertaste.[73]

### 7.1.3  Saccharin/Aspartame Mixtures
In order to mask the bitter aftertaste of saccharin the addition of very small amounts of aspartame can be effective.[102] The addition of 7 mg/litre of aspartame to a cola beverage sweetened with 330 mg/litre of saccharin produces a drink with improved consumer acceptability.[103] The use of aspartame at this level falls below the sweetness threshold of aspartame (0·0007–0·001%).

### 7.1.4 Saccharin/Neohesperidin Dihydrochalcone Mixtures

The taste qualities of NHDC limit its potential as a single sweetener in soft drinks, but it can be effective in improving the taste perception of saccharin when used in mixtures.[103]

## 7.2 Improving the Sweetness Stability of Aspartame

### 7.2.1 Aspartame/Saccharin Mixtures

Aspartame-based beverages lose sweetness as a function of storage time, temperature and pH. To improve stability, mixtures of aspartame and saccharin can be used.[104] Cola beverages sweetened with aspartame and saccharin at a ratio of 1:1 show significantly better sweetness stability than with aspartame alone.[103] A blend of 18 mg of saccharin and 8 mg of aspartame per 100 ml is also in commercial use in a cola beverage in the USA.[103] Mixtures of aspartame and saccharin are currently being extensively used in low-calorie soft drinks on the UK market in both carbonates and concentrates.

### 7.2.2 Aspartame/Cyclamate Mixtures

Aspartame/cyclamate combinations also exhibit improved sweetness stability and superior taste profiles,[103] and the cost per unit of sweetness is also reduced. Cyclamate extends the shelf-life of aspartame and produces a better taste profile than aspartame/saccharin mixtures.[105]

### 7.2.3 Aspartame/Acesulfame-K Mixtures

The use of acesulfame-K as a single sweetener in soft drinks appears to have some limitations, although it can be used successfully with many flavours. By combining acesulfame-K with aspartame, the stability properties of acesulfame-K can be used to advantage in extending the sweetness stability of aspartame, and the taste quality of aspartame broadens the taste profile of acesulfame-K, bringing it closer to the sweetness profile of sucrose.[103] In tests in our own laboratories we have found that this mixture enhances blackcurrant flavour better than an aspartame/saccharin mixture when used as a partial replacement for glucose syrup, producing a better-tasting quality product. Aspartame/acesulfame-K mixtures are also currently being used in low-calorie orange and lemon concentrated fruit drinks on the UK market.

## 7.3 Other Intense Sweetener Combinations

Acesulfame-K/cyclamate combinations yield excellent taste qualities and exceptional storage stability.[103] In cola beverages mixtures of sucralose and aspartame provide a sweetness which is very like that of sugar.[106]

In addition to binary combinations of intense sweeteners, saccharin/ aspartame/cyclamate mixtures in the ratio 1:5:8 result in an improved sweetness profile over those of each single sweetener, and the sweetness quality is also improved.[103]

Combinations of NHDC with saccharin and cyclamate, in a mixture in which NHDC contributes 25% of the total sweetness, saccharin 64% and cyclamate 11%, resulted in a taste sensation that was considered to be of a good quality and gave an apparent increase in flavour.[87]

### 7.4 Synergism

Another advantage of using certain intense sweeteners in combination is that they are synergistic, i.e. the sweetness of the combination is greater than the predicted sweetness of the sum of the individual components, and this can result in a significant cost saving.

Combinations of intense sweeteners which have been shown to exhibit synergy are

| | | |
|---|---|---|
| (a) | Acesulfame-K | + alitame, aspartame, cyclamate, thaumatin, stevioside, sucralose |
| (b) | Alitame | + acesulfame-K, cyclamate |
| (c) | Aspartame | + acesulfame-K, saccharin, cyclamate, stevioside |
| (d) | Cyclamate | + alitame, aspartame, acesulfame-K, saccharin, stevioside, sucralose |
| (e) | NHDC | + saccharin |
| (f) | Saccharin | + aspartame, cyclamate, NHDC, thaumatin |
| (g) | Stevioside | + aspartame, acesulfame-K, cyclamate, thaumatin |
| (h) | Sucralose | + acesulfame-K, cyclamate |
| (i) | Thaumatin | + acesulfame-K, saccharin, stevioside |

Optimal efficiency for the different components is reached when each contributes about the same amount of sweetness.[107] Mixtures of cyclamate and saccharin in the ratio 10:1 exhibit a synergism amounting to 50–100%.[76] A 1:1 blend of acesulfame-K and aspartame has a sweetness factor of more than 300 times that of sucrose, a 2:1 blend of acesulfame-K and aspartame has a sweetness factor of 270–280, and a 3:1 blend a factor of approximately 250 (G. W. Von Rymon Lipinski, personal communication, 1988).

Some examples of synergism between intense sweeteners in low-calorie soft drinks are provided by Houghton[108] and Lanton.[109] The effectiveness of multiple sweeteners in food formulation and the quantitative prediction of synergism are discussed by Porter.[110]

## 7.5 Use of Acids and Sugars in Combination with Intense Sweeteners

### 7.5.1 Malic Acid

The balance between sweetness and acid is one of the most important characteristics of a soft drink. By selecting an acid with a time/intensity profile similar to that of the sweetener with which it is being used, a better-balanced flavour profile can be achieved.

The time/intensity profiles of citric acid and malic acid are not similar. Citric acid has a clean, sharp taste that disappears quickly and therefore combines well with sucrose, the sweetness of which also disappears rapidly. Malic acid is also a clean-tasting acid, but its flavour persists much longer on the palate than that of citric acid.[111] This feature of malic acid can be used to advantage with some of the intense sweeteners, particularly saccharin and acesulfame-K, as it complements their lingering sweetness to produce a better sweetness/acid balance. It can also be used in combination with aspartame/acesulfame-K mixtures to produce a better-balanced flavour, and this combination is currently being used by some low-calorie soft drinks on the UK market.

### 7.5.2 Sugars

7.5.2.1 Fructose. In order to mask aftertastes and give 'mouthfeel' or 'body' to low-calorie soft drinks, mixtures of sugars and intense sweetener combinations have been developed. Among the sugars, fructose is the most interesting because it is the sweetest sugar found in nature, and particularly sweet at low concentrations and low temperatures.

Van Tornout et al.[112] established non-linear sweetness/concentration relations against sucrose in acidified, non-carbonated mineral water (pH 3·0) for fructose, saccharin, aspartame and acesulfame-K. They found that equi-sweet (ES) sucrose concentrations (% w/w) could be related to concentration (C, % w/w) for each sweetener by the following equations:

$$\text{Fructose} \qquad ES = 1\cdot767 C_f^{0\cdot848}$$

$$\text{Aspartame} \qquad ES = 0\cdot118 C_a^{0\cdot684}$$

$$\text{Saccharin} \qquad ES = 0\cdot451 C_s^{0\cdot509}$$

$$\text{Acesulfame-K} \qquad ES = 0\cdot237 C_k^{0\cdot541}$$

The relative sweetness (ES/C) increased with decreasing sweetener concentration, particularly in the case of the intense sweeteners. In the case of saccharin, for example, the relative sweetness doubles from 330 to 660 for a decrease in concentration from 0·02% to 0·005%.

Using these equations, the authors were able to predict, within an accuracy of $\pm 10\%$, the sweetness values of 2·5%, 4% and 5% fructose mixtures with saccharin, aspartame and acesulfame-K respectively, by the algebraic summation of the ES values of the individual components. They concluded that the synergisms reported by other authors were 'false synergisms'. Because of the non-linear relationship between relative sweetness and concentration, each component of a mixture can be used with greater efficiency than alone at higher concentrations.

In addition to a higher sweetness efficiency, the use of fructose in combination with the intense sweeteners resulted in a masking of off-tastes. These were only detected at a relatively high intense sweetener concentration, equivalent to a sucrose concentration greater than about 5%. Reducing the contribution of sweetness from the intense sweetener and increasing the concentration of fructose drastically reduced the detection of off-tastes.

The concept of predicting the sweetness values of mixtures fructose and intense sweeteners was also evaluated in cola drinks and orange drinks, when similar results were obtained. Except for the cola drink based on a fructose/saccharin mixture, drinks could be formulated with 3% fructose in combination with each of the intense sweeteners, which were indistinguishable from all sugar standards. In the case of the cola drink, a higher level of fructose and reduction in saccharin was necessary to reduce the bitter aftertaste.

Hyvonen et al.[113] also evaluated the use of mixtures of fructose and saccharin in citrus and cola-type drinks and found that the aftertaste of saccharin could be minimised by limiting the amount of saccharin in the mixture to 0·3–0·4%. A drink containing 3·5% fructose and 0·0136% saccharin was indistinguishable from a citrus drink sweetened with 10% sucrose. A cola drink sweetened with a mixture of 3·2% fructose and 0·0097% saccharin was indistinguishable from the cola drink sweetened with 10% sucrose. They also established that maximum synergism was achieved when each of the components of the mixture contributed 50% to the total sweetness.

In separate studies, Hyvonen[114] and Johnson[115] found that combinations of fructose and aspartame were synergistic, and the sweetening effect minimised any lingering non-sweet flavours due to the aspartame.

Fructose can also be used to mask the liquorice-like taste of stevioside. This combination is successfully used in Japan to produce reduced-calorie soft drinks which enjoy good consumer acceptability.[93] High-fructose syrup can also be used in combination with stevioside to produce the same effect.[116]

*7.5.2.2 Other sugars.* In addition to fructose, Hyvonen *et al.*[113] have shown that glucose and sucrose can also be used in combination with saccharin to mask its bitter aftertaste successfully. In a citrus drink a mixture of 4·9% glucose and 0·0195% saccharin was indistinguishable from a 10% sucrose product. In a cola-type product a mixture of 4·3% sucrose and 0·013% saccharin was found to match the 10% sucrose-sweetened product. Mixtures of sucrose and saccharin are extensively used in regular soft drinks on the UK market.

## 8 FORMULATION OF SOFT DRINKS USING INTENSE SWEETENERS

In order to achieve a maximum synergistic effect, intense sweeteners should be used in a combination such that each contributes an equal amount of sweetness to the mixture. Other factors often have to be taken into consideration, however, and in practice the mixture used will be a compromise, taking into account such limitations as cost, aftertaste, stability, legal limits, caloric limits and contractual agreements with suppliers.

The economics of sweetness mixtures in which pairs of sweeteners are used in combination to increase the overall sweetness of a product has been studied by Moskowitz *et al.*[117] Equations were developed from which it was possible to minimise overall cost whilst maintaining sweetness, or maximising sweetness while maintaining cost.

The soft drinks formulator is also concerned with the establishment of an appropriate level of sweetness for a particular flavour. When liking is evaluated on a scale ranging from 'dislike extremely' to 'like extremely', the respondent will give low ratings to samples that are not sweet enough and also to those that are too sweet. Therefore the liking curve will have a maximum at the most liked sweetness level.[118] Having established the appropriate sweetness level for a particular flavour, it then becomes necessary to decide on the most appropriate sweetener system for the product. Moskowitz *et al.*[118] developed an equation relating total mixture sweetness to concentration for mixtures of aspartame, saccharin and cyclamate in cola-flavoured beverages. This concept can also be used for other mixtures of intense sweeteners.

In addition to providing sweetness, nutritive sweeteners provide 'body', and this must be taken into account in the formulation of low-calorie or reduced-calorie soft drinks. A carbohydrate sweetener cannot therefore simply be replaced by an intense sweetener. Natural or synthetic gums can be used to increase viscosity and 'mouthfeel', with carboxymethylcellulose often used for this purpose. Increase in flavour level and acid will often help overcome the difference in mouthfeel.[79] Flavours ideally should be of a high quality and free from any undesirable notes that would be masked in a sucrose-sweetened product. Flavour blends have been developed especially for use with intense sweeteners.

Carbonation level also influences flavour and mouthfeel, and this effect is even more pronounced in low-calorie beverages. A higher degree of carbonation can help to increase the mouthfeel in low-calorie carbonates.

To improve the sweetness stability of aspartame, the pH of the product can be raised by the addition of sodium citrate to bring it closer to the optimum pH for stability.[108] Consequently an increase in the addition of acid will be required to maintain the original sweetness/acid balance. The increase in pH will result in the product being more prone to microbiological spoilage, so it should subsequently be pasteurised and/or preserved by the addition of permitted preservatives.


## 9 LEGISLATION

### 9.1 UK Legislation at the Time of Writing

The sweeteners and sugars that can be used in the formulation of soft drinks are controlled in the UK by the Sweeteners in Food Regulations[3] and the Soft Drinks Regulations.[11] The permitted list of intense sweeteners includes acesulfame-K, aspartame, saccharin (including sodium saccharin and calcium saccharin) and thaumatin.

Intense sweeteners are permitted in all categories of soft drinks except those specially prepared for babies and young children, unless they are to meet special dietary requirements.

The intense sweeteners acesulfame-K, aspartame and thaumatin can be used in soft drinks without limitation, but the amount of saccharin is controlled by the Soft Drinks Regulations[11] as follows (calculated as metric equivalents):

| | |
|---|---|
| Soft drinks for consumption after dilution | max. 400 mg/litre |
| Soft drinks for consumption without dilution | max. 80 mg/litre |
| Brewed ginger beer and herbal and botanical beverages | max. 114 mg/litre |

| Semi-sweet soft drinks for consumption after dilution | max. 200 mg/litre |
|---|---|
| Semi-sweet soft drinks for consumption without dilution | max. 40 mg/litre |

These specified limits for saccharin do not apply to low-calorie soft drinks or to soft drinks for diabetics. Neither do they apply to super-concentrated soft drinks (requiring a dilution greater than $1+4$ with water), provided that when these drinks are diluted according to the labelling instructions the ready-to-drink product complies with the levels for the corresponding soft drink for consumption without dilution.

In addition to setting legal limits for saccharin, the Soft Drinks Regulations[11] stipulate the amount of sugar which must be added to each category of soft drink as follows:

| Soft drinks for consumption after dilution | min. $22\frac{1}{2}$% w/v |
|---|---|
| Soft drinks for consumption without dilution | min. $4\frac{1}{2}$% w/v |
| Dry ginger ale | min. 3% w/v |
| Brewed ginger beer and herbal and botanical beverages | min. 2% w/v |
| Semi-sweet soft drinks for consumption after dilution | min. $11\frac{1}{4}$% w/v |
| | max. 15% w/v |
| Semi-sweet soft drinks for consumption without dilution | min. $2\frac{1}{4}$% w/v |
| | max. 3% w/v |

These requirements do not apply to soft drinks for diabetics, which must contain no added sugar, or to low-calorie soft drinks.

All categories of soft drinks must comply with the minimum fruit/juice requirements, unless they are designated as 'flavour' drinks or '-ades', and do not show pictures of fruit on their labels. The minimum fruit/juice content requirements are as follows:

(a) Soft drinks for consumption without dilution

| Citrus juice and barley water | min. 3% v/v citrus juice |
|---|---|
| Lime crushes, lime juice and soda | min. 3% v/v lime juice |
| Citrus crushes (including those containing a bitter principle) | min. 5% v/v citrus juice |
| Citrus and non-citrus fruit juice mixtures | min. 5% v/v fruit juice |
| Comminuted citrus fruit and barley drinks | min. $1\frac{1}{2}$% w/v potable citrus |
| Comminuted citrus and citrus or non-citrus juice (including those containing a bitter principle) | min. 2% w/v potable citrus |

| | |
|---|---|
| Fermented apple or fermented pear juice drinks | min. 5% v/v fermented juice |
| Any other soft drink containing fruit juice | min. 5% v/v fruit juice |

(b) Soft drinks for consumption after dilution

| | |
|---|---|
| Citrus juice and barley water | min. 15% v/v citrus juice |
| Citrus squashes (juice) including those containing a bitter principle | min. 25% v/v citrus juice |
| Comminuted citrus fruit and barley drinks | min. 7% w/v potable citrus |
| Comminuted citrus and citrus or non-citrus fruit juice (including those containing a bitter principle) | min. 10% w/v potable citrus |
| Non-citrus fruit squashes (juice) | min. 10% v/v juice |
| Citrus and non-citrus juice mixtures | min. 25% v/v fruit juice |

The Soft Drinks Regulations[11] limit the caloric value of low-calorie soft drinks to 1·5 kcal per fluid ounce (5·3 kcal/100 ml) for ready-to-drink soft drinks, and 7·5 kcal per fluid ounce (26 kcal/100 ml) for soft drinks for consumption after dilution.

In the formulation of low-calorie soft drinks, account must be taken of the contribution made to the caloric value of the product by the caloric value of the fruit/juice. A ready-to-drink orange drink containing 13·3% v/v of orange juice would have a caloric value equal to the maximum permitted, and no addition of sugar would be possible. The Soft Drinks Regulations[11] do not prohibit the addition of sugar to low-calorie soft drinks provided the caloric value limit is not exceeded. A ready-to-drink beverage could therefore contain a maximum of 1·4% w/v of added fructose, assuming that there is no other contribution to the caloric value from any other ingredient. This amount of fructose approximates to a sweetness equivalent to 2% sucrose.[112] In order to produce a sweetness match to a sugar-sweetened product at 10% sucrose content, the intense sweetener contribution must therefore equate to a sweetness equivalent of 8% sucrose, which exceeds the value of 5% recommended to minimise off-tastes.[112] The low-calorie limits therefore prevent the soft drink formulator from applying the principles developed for optimising sweetness values of mixtures of fructose and intense sweeteners.

Soft drinks for diabetics are also subject to the same fruit/juice requirements as 'normal' soft drinks but, unlike low-calorie soft drinks, they must not contain any added sugar. For the purposes of the Soft Drinks

Regulations[11] the word 'sugar' means any soluble carbohydrate sweetening matter, and the word 'carbohydrate' means a substance containing carbon, hydrogen and oxygen only, in which the hydrogen and oxygen occur in the same proportion as in water. Since soft drinks for diabetics have no calorie limits, it is therefore possible to add bulk sweeteners, such as sorbitol, to provide 'body' or 'mouthfeel' and calories in combination with intense sweeteners, which are necessary to make up the deficiency in sweetness because of the low sweetening power of sorbitol.

The claim that a soft drink is suitable for diabetics is restricted by the 1984 Food Labelling Regulations[23] which stipulate the following conditions:

1. A given quantity of the food must not have a higher energy content that the same quantity of a similar food in relation to which no diabetic claim is made.
2. A given quantity of the food must not have a higher fat content than the same quantity of a similar food in relation to which no diabetic claim is made.
3. A given quantity of the food must not have a readily absorbable carbohydrate content greater than 50% of the readily absorbable carbohydrate content of the same quantity of a similar food in relation to which no diabetic claim is made.
4. The food must not contain a greater quantity of mono- or di-saccharides, other than fructose, than the quantity that is technically necessary to retain the essential characteristics of the food while having regard to its claimed suitability for diabetics.

These conditions essentially bar low-calorie soft drinks, which are not made especially for diabetics and are therefore not labelled as diabetic soft drinks, from any claim as suitable for diabetics, even though they may contain no added sugar and have a zero caloric value.

### 9.2 Proposed Changes in the Soft Drinks Regulations

The present Soft Drinks Regulations[11] prevent the formulation of soft drinks containing less than $4\frac{1}{2}$% w/v of added sugar, on a ready-to-drink basis, unless they are diabetic soft drinks or low-calorie products containing less than 1·4% w/v total carbohydrate (equivalent to 5·3 kcal/ 100 ml). Between these two limits it is impossible to formulate soft drinks that are acceptable to the public. Although provision is made in the regulations for semi-sweet soft drinks, the addition of both sugar and saccharin is strictly controlled so that the sweetness of the resulting drink is approximately half that of the fully sweetened product. Such drinks do not appear to appeal to the consumer, and it is difficult to find any products of this type on the UK market.

This presents a conundrum to the soft drink formulator, and the availability of other intense sweeteners does not help. Whilst the Soft Drinks Regulations[11] do not prevent their addition to semi-sweet soft drinks to produce a more palatable product, the resulting drink would have to be labelled as semi-sweet because of its low sugar content. Since the product would not taste half as sweet, as was the intention of the regulations, it could be construed that such a description would be misleading. The legal acceptability of such a product is therefore questionable.

Proposals issued by the Ministry of Agriculture, Fisheries and Food[119] are currently under consideration by interested bodies to update the 1964 Soft Drinks Regulations.[11] One of the proposed changes is the abolition of the minimum added sugar requirement, but no changes are envisaged to the very restrictive low-calorie limits, or the limitations on the amounts of saccharin that can be added. The abolition of the minimum added sugar requirement will enable the formulation of soft drinks with a reduced calorie content, but the limit on saccharin will necessitate an increase in the use of the other intense sweeteners to achieve the desired sweetness.

It is also proposed to permit the addition of fructose to soft drinks that are suitable, or have been specially made, for diabetics.

### 9.3 Legislation in Other Countries

Legislation governing the use of intense sweeteners varies widely throughout the EEC and other countries. A comprehensive survey of the legal acceptability of intense sweeteners in soft drinks throughout the world is published by the Leatherhead Food Research Association,[57] and information is also available from the International Sweeteners Association (PO Box 768, CH-8026 Zürich, Switzerland).

## 10  MARKET SURVEY

A survey was made of the use of intense sweeteners in soft drinks, mainly confined to national brands, available at the time of writing on the UK market. Products sweetened entirely with carbohydrates were excluded.

### 10.1 Regular Soft Drinks

Intense sweeteners are used in these products as a partial replacement for carbohydrate sweeteners, for the technical and/or economic reasons previously outlined.

### 10.1.1 Ready-to-Drink

A total of 82 samples of soft drinks belonging to this category were examined. Despite the availability since 1983 of the new intense sweeteners acesulfame-K, aspartame and thaumatin, this segment of the market continues to be dominated by the use of mixtures of saccharin and carbohydrate sweeteners. In these products saccharin provides approximately 50% of the total sweetness at a very economic rate which cannot be matched by any of the alternative intense sweeteners legally permitted.

Many of these products are formulated with the maximum amount of saccharin allowed but, in order to attain the required sweetness for particular flavours, the amount of sugar added is also well in excess of the legal minimum requirement. There is therefore considerable potential in this category of soft drinks for the use of an intense sweetener additional to saccharin. During the period when cyclamate was also permitted in the UK, a mixture of cyclamate and saccharin, usually in the ratio 10:1, was extensively used for this purpose to produce a good quality sweetness.

### 10.1.2 Concentrates

A total of 68 samples of regular concentrated soft drinks were examined. Mixtures of saccharin and carbohydrate sweeteners also completely dominate this segment of the market, and the comments that were made for the ready-to-drink products apply equally to the concentrates.

In addition to fulfilling economic considerations, intense sweeteners for these products should be stable under market conditions for periods of 12–18 months in the pH range 2–3. Blends of saccharin and cyclamate exhibit exceptionally good sweetness stability under these conditions, and were extensively used in these products along with sugar in the period 1965–70 when cyclamate was permitted in the UK.

### 10.1.3 Superconcentrated Soft Drinks

All of the 25 samples examined were sweetened with a mixture of saccharin and sugar.

## 10.2 Low/Reduced-calorie Soft Drinks

Intense sweeteners are used in low-calorie soft drinks, usually as a complete replacement for carbohydrate sweeteners, to comply with the caloric value limits set by the 1964 Soft Drinks Regulations.[11]

### 10.2.1 Ready-to-Drink (Table 4)

The availability of a range of intense sweeteners since 1983 has revolutionised this segment of the market and has been a significant factor

TABLE 4
INTENSE SWEETENERS USED IN LOW/REDUCED-CALORIE SOFT DRINKS

| Intense sweetener(s) | Number of products | |
|---|---|---|
| | Ready-to-drink | Concentrated |
| Aspartame/saccharin | 47 | 11 |
| Aspartame/acesulfame-K | 11 | 9 |
| Acesulfame-K/saccharin | 6 | — |
| Saccharin | 4 | 6 |
| Aspartame | 2 | — |
| Acesulfame-K | 1 | — |

in the growth of low-calorie soft drinks, and in particular low-calorie carbonates. Almost 70% of the 71 samples examined were sweetened with an aspartame/saccharin mixture, 12% with an aspartame/acesulfame-K mixture and 7% with an acesulfame-K/saccharin combination. Only two products were sweetened solely with aspartame, one solely with acesulfame-K and four with saccharin only.

This is in contrast to the US market, where aspartame is now incorporated as the sole sweetening agent in 85% of all diet soft drinks, the remainder being sweetened with saccharin or an aspartame/saccharin combination.[120] No other intense sweetener is currently permitted in the USA. One of the incentives for the replacement of saccharin by aspartame in the USA is the removal from the soft drink label of the long mandatory warning statement that must accompany the use of saccharin.[108] Manufacturers have therefore enhanced the stability of aspartame by raising the pH of their products, including an overage (extra amount) to compensate for loss during storage, and they have instituted careful stock control to reduce shelf-life times. By these means it appears that aspartame-sweetened soft drinks can survive the normal distribution cycles and temperature environments experienced in the USA.[35]

Typical levels of use of aspartame in carbonated beverages in the USA are[48]

| | | |
|---|---|---|
| Cola | 550–680 mg/litre at pH 2·4–3·1 |
| Lemon/lime | 300–600 mg/litre at pH 3·0–3·1 |
| Orange | 550–900 mg/litre at pH 3·1–3·4 |

*10.2.2 Concentrates* (Table 4)

Of the products examined in this segment of the market, 76% were sweetened with a multiple intense sweetener system. Aspartame/saccharin combinations were used in 44%, and 32% used an aspartame/acesulfame-K mixture. The remaining 24% continued to use saccharin as the sole sweetener. None of the products used either aspartame or acesulfame-K as the sole sweetener and, in contrast to the diet carbonates, no product used the acesulfame-K/saccharin combination.

Low-calorie concentrated soft drinks require a longer shelf-life than carbonates, and the pH range (2–3) is lower, resulting in a more destructive environment. The use of saccharin or acesulfame-K with aspartame enhances the sweetness stability of the aspartame in these products, whilst the aspartame improves the quality of the sweetness.

### 10.2.3 Superconcentrates

Of the 10 samples examined, 7 were sweetened solely with saccharin, whilst the remainder used an aspartame/saccharin combination.

### 10.3 Soft Drinks for Diabetics

#### 10.3.1 Ready-to-Drink

Very few products especially made for diabetics could be found on the market, but those that were available were all sweetened with an aspartame/saccharin combination. A large number of low-calorie carbonates are also sugar-free and, since some 50–70% of all diabetics actually require a reduction in energy intake, such products are recommended for inclusion in the diet of the overweight diabetic.[121]

#### 10.3.2 Concentrates

Of the 9 products especially made for diabetics, 6 were sweetened with saccharin and the remainder with an aspartame/saccharin mixture. In addition to saccharin, sorbitol was also used in a number of products to improve flavour and mouthfeel, and in the case of opaque soft drinks it also aided their appearance by stabilising cloudiness.[122]

## 11 CONCLUSIONS

Intense sweeteners play a major role in the formulation of soft drinks, and are used extensively in regular, low-calorie and diabetic soft drinks. The abolition of the minimum sugar requirement will further increase their potential in regular soft drinks. As the diet segment of the market continues to expand, the demand for intense sweeteners will continue to grow.

No single intense sweetener is ideally suited to meet all the requirements of soft drinks. Only by blending the available intense sweeteners can a satisfactory substitute for carbohydrate sweeteners be produced.

The search for the ideal intense sweetener will continue but, in conformity with the concept of an 'acceptable daily intake', the consumption of a small quantity of a variety of intense sweeteners is preferable to the ingestion of a relatively large amount of a single intense sweetener. The use of multiple sweeteners is therefore likely to continue, although the consumer may not always appreciate the reasons for using such a system, particularly in the present anti-additive climate of opinion.

## REFERENCES

1. Heasman, M. *Food Process.* (1987) September, 17.
2. Jeffries, D. A. *Drink Market Updates.* No. 5. *Soft Drinks in the UK*, 2nd edn. Food Research Association, Leatherhead, Surrey, 1987.
3. Anon., *Sweeteners in Food Regulations 1983*, SI 1983, No. 1211. HMSO, London.
4. Wolkstein, M. *Food Eng. Int.* (1983) December, 33.
5. Hall, R. *Soft Drinks Manage. Int.* (1988) February, 34.
6. Beattie, G. B. *Perfum. Essent. Oil Rec.* (1961) November, 723.
7. Anon., *Soft Drinks Order 1943*, No. 838.
8. Anon., *Soft Drinks Order 1947*, No. 2756.
9. Food Standards Committee, *Report on Soft Drinks.* HMSO, London, 1959.
10. Anon., *Soft Drinks Regulations 1963*, SI 1963, No. 844. HMSO, London.
11. Anon., *Soft Drinks Regulations 1964*, as amended SI 1964, No. 760. HMSO, London.
12. Brook, M. *J. Roy. Soc. Health*, **89**(3) (1969) 140.
13. Anon., *Fed. Reg.*, **34**(202) (1969) 17063.
14. Anon., *Artificial Sweeteners in Food Regulations 1969*, SI 1969, No. 1817. HMSO, London.
15. Woodroof, J. G. & Phillips, G. F., *Beverages: Carbonated and Non-Carbonated.* AVI, Westport, CT, 1974, Ch. 3, p. 69.
16. Anon., *Fed. Reg.*, **35**(167) (1970) 13644.
17. Anon., *Fed. Reg.*, **39** (1974) 27317.
18. Anon., *Fed. Reg.*, **40** (1975) 56907.
19. Anon., *Fed. Reg.*, **46** (1981) 38284.
20. Anon., *Fed. Reg.*, **48** (1983) 31376.
21. Royal College of Physicians of London, Obesity. *J. Roy. Coll. Physicians*, **17** (1983) 3.
22. Porikos, K. P., Hesser, M. F. & van Itallie, T. B. *Physiol. Behav.*, **29** (1982) 293.
23. Anon., *Food Labelling Regulations 1984*, as amended SI 1984, No. 1305. HMSO, London.
24. Linke, H. A. B. In *Developments in Sweeteners*, Vol. 3, ed. T. H. Grenby. Elsevier Applied Science, London, 1987, Ch. 6, p. 151.
25. Ziesenitz, S. C. & Siebert, G. *Caries Res.*, **20** (1986) 498.

26. Linke, H. A. B. & Kohn, J. S. *Caries Res.*, **18** (1984) 12.
27. Grenby, T. H. & Saldanha, M. G. *Caries Res.*, **20** (1986) 7.
28. Mishiro, Y. & Kaneko, H. *J. Dent. Res.*, **56** (1977) 1427.
29. Siebert, G., Ziesenitz, S. C. & Lotter, J. *Caries Res.*, **21** (1987) 141.
30. Sicard, P. J. & Leroy, P. In *Developments in Sweeteners*, Vol. 2, ed. T. H. Grenby, K. J. Parker & M. G. Lindley. Applied Science Publishers, London, 1983, Ch. 1, p. 1.
31. Anon., *Martindale: Extra Pharmacopoeia*, 27th edn, ed. A. Wade. The Pharmaceutical Press, London, 1977, p. 64.
32. Von Rymon Lipinski, G. W. *Food Chem.*, **16** (1985) 259.
33. Hoppe, K. & Gassmann, B. *Nährung*, **24** (1980) 423.
34. Von Rymon Lipinski, G. W. *Soft Drinks* (1982) October, 434.
35. Franta, R., Beck, B., Katz, F., Primack, N., Varvil, R. D. & Voirol, F. A. *Food Technol.*, **40**(1) (1986) 116.
36. Von Rymon Lipinski, G. W. In *Alternative Sweeteners*, ed. L. O'B. Nabors & R. C. Gelardi. Marcel Dekker, New York, 1986, Ch. 5, p. 89.
37. Von Rymon Lipinski, G. W. & Huddart, W. E. *Chem. Ind.* (1983) 6 June, 427.
38. Joint FAO/WHO Expert Committee on Food Additives, 27th Report, Technical Report Series No. 696. WHO, Geneva, 1983.
39. Beck, C. I. In *Low Calorie and Special Dietary Foods*, ed. B. K. Dwivedi. CRC Press, West Palm Beach, FL, 1978, Ch. 5, p. 59.
40. Beck, C. I. In *Symposium: Sweeteners*, ed. G. E. Inglett. AVI Westport, CT, 1974, Ch. 15, p. 164.
41. Cloninger, M. R. & Baldwin, R. E. *Science*, **170** (1970) 81.
42. Cloninger, M. R. & Baldwin, R. E. *J. Food Sci.*, **39** (1974) 347.
43. Larson-Powers, N. & Pangborn, R. M. *J. Food Sci.*, **43**(1) (1978) 41.
44. Larson-Powers, N. & Pangborn, R. M. *J. Food Sci.*, **43**(1) (1978) 47.
45. Samundsen, J. A. *J. Food Sci.*, **50** (1985) 1510.
46. Baker, E. *Soft Drinks*, **38** (1984) 479.
47. Baldwin, R. E. & Korschgen, B. M. *J. Food Sci.*, **44** (1979) 938.
48. Vetsch, W. *Food Chem.*, **16** (1985) 245.
49. Mazur, R. H. *J. Toxicol. Environ. Health*, **2** (1976) 243.
50. Mazur, R. H. In *Symposium: Sweeteners*, ed. G. E. Inglett. AVI, Westport, CT, 1974, Ch. 14, p. 159.
51. Mazur, R. H. & Ripper, A. In *Developments in Sweeteners*, Vol. 1, ed. C. A. M. Hough, K. J. Parker & A. J. Vlitos. Applied Science Publishers, London, 1979, Ch. 5, p. 125.
52. Mazur, R. H., Schlatter, J. M. & Goldkamp, A. H. *J. Am. Chem. Soc.*, **91** (1969) 2684.
53. Goodburn, K. E., *Food Ingredient Guide. No. 2. A User's Guide to the New Intense Sweeteners—An Update*. Food Research Association, Leatherhead, Surrey, 1987.
54. Homler, B. E. *Food Technol.*, **38**(7) (1984) 50.
55. Ojima, T., Nagashima, N. & Ozawa, T. (Ajinomoto Co., Tokyo), European patent 0097950 (1983).
56. Anon., *Food Chem. News* (1988) 22 February, 14.
57. Marshall, J. P. & Pollard, J. A., *Food Legislation Surveys. No. 1. Sweeteners—International Legislation*, 4th edn. Food Research Association, Leatherhead, Surrey, 1987.

58. Joint FAO/WHO Expert Committee on Food Additives, 25th Report, Technical Report Series No. 669. WHO, Geneva, 1981.
59. Kunst, E. A. *Amer. Soft Drinks J.*, **125** (1971) 99.
60. Helgren, F. J., Lynch, M. J. & Kirchmeyer, F. J. *J. Am. Pharm. Assoc.*, **44** (1955) 353.
61. Rader, C. P., Tihanyi, S. G. & Zienty, F. B. *J. Food Sci.*, **32** (1967) 357.
62. Frisch, H. R. *Chem. Can.*, **2** (1950) 22.
63. UK patent 1 194 344 (*Chem. Abstr.*, **73**, 75882).
64. French patent 1 502 207 (*Chem. Abstr.*, **69**, 95264).
65. UK patent 1 239 518 (*Chem. Abstr.*, **75**, 117353).
66. US patent 3 773 526 (*Chem. Abstr.*, **80**, 69411).
67. US patent 3 743 518 (*Chem. Abstr.*, **79**, 64836).
68. Walter, G. J. & Mitchell, M. L. In *Alternative Sweeteners*, ed. L. O'B. Nabors & R. C. Gelardi. Marcel Dekker, New York, 1986, Ch. 2, p. 15.
69. Joint FAO/WHO Expert Committee on Food Additives, 28th Report, Technical Report Series No. 710. WHO, Geneva, 1984.
70. Stephens, J. P. *Food Flavour. Ingred. Process. Packag.*, **5**(3) (1983) 12.
71. Higginbotham, J. D. In *Alternative Sweeteners*, ed. L. O'B. Nabors & R. C. Gelardi. Marcel Dekker, New York, 1986, Ch. 6, p. 103.
72. Talres Development NV, US patent 4 228 198 (1980).
73. Talres Development NV, US patent 4 096 285 (1978).
74. San Ei Chem. Ind. KK, Japanese patent appl. 087 873 (1979).
75. Beck, K. M. *Food Technol.*, **11**(3) (1957) 156.
76. Vincent, H. C., Lynch, M. J., Pohley, F. M., Helgren, F. J. & Kirchmeyer, F. J. *J. Am. Pharm. Assoc.*, **44** (1955) 442.
77. John & E. Sturge Ltd, Technical data sheet for cyclamate, 1963.
78. Kasperson, R. W. & Primack, N. In *Alternative Sweeteners*, ed. L. O'B. Nabors & R. C. Gelardi. Marcel Dekker, New York, 1986, Ch. 4, p. 71.
79. Beck, K. M., US Society of Soft Drink Technologists, 6th Annual Meeting (1959) 113.
80. Renwick, A. G. & Williams, R. T. *Biochemistry*, **129** (1972) 869.
81. Grenby, T. H. *J. Roy. Soc. Health*, **105** (1985) 117.
82. Joint FAO/WHO Expert Committee on Food Additives, 26th Report, Technical Report Series No. 683. WHO, Geneva, 1982.
83. Guadagni, D. G., Maier, V. P. & Turnbaugh, J. H. *J. Sci. Food Agric.*, **25** (1974) 1199.
84. Horowitz, R. M. & Gentili, B. In *Alternative Sweeteners*, ed. L. O'B. Nabors & R. C. Gelardi. Marcel Dekker, New York, 1986, Ch. 7, p. 135.
85. Givaudan, L. & Cie, S. A., Netherlands patent appl. 7 412 072 (1975) (*Chem. Abstr.*, **83**, 204966).
86. Crosby, G. A. & Furia, T. E. In *CRC Handbook of Food Additives*, 2nd edn, Vol. II, ed. T. E. Furia. CRC Press, Boca Raton, FL, 1980, p. 204.
87. Inglett, G. E., Krbechek, L., Dowling, B. & Wagner, R. *J. Food Sci.*, **34** (1969) 101.
88. Berry, C. W. & Henry, C. A. Baylor College of Dentistry, Meeting of the American Association for Dental Research. *Food Chem. News* (1983) 21 March, 25.
89. Ishii, K., Toda, J., Aoki, H. & Wakabayashi, H., US patent 3 653 923 (1972).
90. Stevia Corporation Ltd, *Sato Stevia*. Technical Publication, 1985.

91. Ishima, N. & Kakayama, O. *Shokuhin Sogo Kenkyusho Kenkyo Hokoku*, 31 (1976) 80.
92. Chang, S. S. & Cook, J. M. *J. Agric. Food Chem.*, **31** (1983) 409.
93. Fujita, H. & Edahiro, T. *Shokuhin Kogyo*, **22**(20) (1979) 66.
94. Bakal, A. I. & Nabors, L. O'B. In *Alternative Sweeteners*, ed. L. O'B. Nabors & R. C. Gelardi. Marcel Dekker, New York, 1986, Ch. 14, p. 295.
95. Pfizer, *Alitame—A New High-Intensity Sweetener*. Technical Summary, 1987.
96. Jenner, M. R. This volume, Chapter 5, pp. 121–41.
97. Joint FAO/WHO Expert Committee on Food Additives, Summary and Conclusions of the 33rd Meeting (1988) ICS/88.19.
98. Schiffman, S. S., Crofton, V. A. & Beeker, T. G. *Physiol. Behav.*, **34** (1985) 369.
99. Askar, A., Hassanien, F. R., Abd El-Fadeel, M. G., El-Saidy, S. & El-Zoghabi, M. S. *Alimenta*, **24**(2) (1985) 37.
100. Helgren, F. J., US patent 2 803 551 (1957).
101. Harper, E. N., US Society of Soft Drink Technologists, 12th Annual Meeting (1965) 17.
102. Lavia, A. F. & Hill, J. A., French patent 2 087 843 (1972).
103. Bakal, A. I. In *Alternative Sweeteners*, ed. L. O'B. Nabors & R. C. Gelardi. Marcel Dekker, New York, 1986, Ch. 16, p. 325.
104. G. D. Searle & Company, Vol. 1, US Food Additive Petition No. 2A3661 (1982).
105. Weisberg, K. *Beverage World Int.*, **5**(4) (1987) 5.
106. UK patent application, GB 2 153 651A (1985).
107. Weickmann, F., Weickmann, H., Fincke, K., Weickmann, F. A. & Huber, B., German patent 1 961 769 (1969).
108. Houghton, H. W. In *Low Calorie Products*, ed. G. G. Birch & M. G. Lindley. Elsevier Applied Science, London, 1988, p. 11.
109. Lanton, B. *Food Ind. S. Afr.* (1988) February, 23.
110. Porter, A. B. *Chem. Ind.* (1983) 19 September, 696.
111. Sharrock, J. *Food Process.* (1987) June, 17.
112. Van Tornout, P., Pelgroms, J. & van der Meeren, J. *J. Food Sci.*, **50** (1985) 469.
113. Hyvonen, L., Koivistoinen, P. & Ratilainen, A., *J. Food Sci.*, **43** (1978) 1580.
114. Hyvonen, L., Research Report. University of Helsinki, 1981.
115. Johnson, L., Unpublished Report. Hoffmann–La Roche, Nutley, NJ, 1982.
116. Phillips, K. C. In *Developments in Sweeteners*, Vol. 3, ed. T. H. Grenby. Elsevier Applied Science, 1987, Ch. 1, p. 1.
117. Moskowitz, H. R. & Wehrly, T. *J. Food Sci.*, **37** (1972) 411.
118. Moskowitz, H. R., Wolfe, K. & Beck, C. *J. Food Qual.*, **2** (1978) 17.
119. Ministry of Agriculture, Fisheries and Food, *Proposals for Revised Soft Drink Regulations*, 18 May 1988.
120. Botma, Y. *Fluss. Obst.*, **2** (1988) 59.
121. Metcalfe, J. *Balance*, No. 86 (1985) April, 7.
122. Jackson, K. G., Howells, J. & Armstrong, J. In *Developments in Sweeteners*, Vol. 3, ed. T. H. Grenby. Elsevier Applied Science, London, 1987, Ch. 8, p. 213.

*Chapter 8*

# ARE SWEETENERS REALLY USEFUL TO DIABETICS?

N. Finer

*Division of Medicine, United Medical and Dental Schools,*
*Guy's Hospital, London, UK*

## SUMMARY

*Diabetes mellitus is a disease syndrome characterised by increased blood glucose levels and a tendency for patients to develop vascular complications affecting the retina, kidney, heart and nerves. Diet occupies a pivotal position in the treatment of both insulin-dependent and non-insulin-dependent diabetes mellitus. Although traditionally sucrose has been proscribed from the diabetic diet, recent evidence shows that small amounts of simple sugars, particularly if taken with meals, do not lead to a deterioration of blood glucose control. Although these findings do not preclude a case for including alternative sweeteners in the diabetic diet, there is still a lack of evidence that either nutritive or non-nutritive sweeteners are helpful in improving control of the disease or altering the long-term outcome. Non-nutritive sweeteners may be useful, particularly to children with diabetes mellitus, in allowing them to consume sweet-tasting drinks and snacks between meals. However, if they are used as a supplement to sucrose, rather than as a substitute, they will not lead to the desired goal.*

## 1 INTRODUCTION

That people suffering from diabetes mellitus must not take sugar is one of the medical 'facts' most widely held by the public. Although diabetes physicians no longer consider this to be an absolute truth, it is certainly still true that diabetics are nearly always advised to restrict the amount of simple sugars in their diet. The most important of these simple sugars quantitatively is sucrose, which in the Western world provides about 14%

215

TABLE 1
'METABOLIC' CLASSIFICATION OF SWEETENERS

| Nutritive | | Non-nutritive[a] |
|---|---|---|
| Insulin-requiring and metabolised to glucose | Non-insulin-requiring | |
| Sucrose | Fructose | Aspartame |
| Glucose | Sorbitol | Saccharin |
| Lactose | Mannitol | Cyclamate |
| High-fructose corn syrup | Xylitol | Acesulfame K |
| | Maltitol | Thaumatin |

[a] Some intense sweeteners (e.g. aspartame) have a caloric value but are used in such small quantities that they are effectively non-caloric.

of the total energy in the diet.[1] Sucrose tastes sweet. It is possible to replace sweetness from sucrose with a variety of alternative sweeteners, some of which are a source of energy (nutritive sweeteners) and some of which provide no energy or virtually none (non-nutritive sweeteners) (Table 1). Sucrose (and glucose) require insulin for their metabolism, while other sweet nutritive sugars and polyols (such as fructose, sorbitol, xylitol and maltitol) do not. Sweeteners that do not require insulin for their metabolism might be thought to be particularly useful for diabetics, and this has led to the development of a wide range of specially formulated foods aimed at the diabetic market. However, if diabetics can take sucrose in their diet, there would be no particular need for such foods incorporating alternative sweeteners. This review considers whether it really is necessary to proscribe diabetics from taking sucrose and, if so, whether, and which, alternative sweeteners are helpful to diabetics in managing their disease.

## 2 WHAT IS DIABETES?

Diabetes mellitus covers a group of disorders characterised by hyperglycaemia (elevated blood glucose levels) and a propensity to develop specific clinical complications with retinal, renal, neurological and cardiovascular damage. These complications result mainly from micro- and macro-vascular disease and occur with greater frequency with increasing duration of diabetes. Diabetes mellitus can be categorised broadly into two main diseases: insulin-dependent diabetes mellitus

(IDDM) or type I and non-insulin-dependent diabetes mellitus (NIDDM) or type II.[2]

## 2.1 Insulin-dependent Diabetes Mellitus

Established IDDM is characterised by a total lack of endogenous insulin production from the pancreatic $\beta$-cell. It is commonest in the young, with a peak incidence at about 12 years of age; it was previously known as 'juvenile diabetes'. The prevalence in the United Kingdom is about 0·2%, with an incidence of about 0·01%, which is increasing.[3] Patients with IDDM must take insulin; without it they soon become ill and die from the acute metabolic complications of ketoacidosis. The object of treatment is to try and maintain blood glucose levels as close to normal as possible. In order to achieve this goal the diabetic must understand his illness and its treatment, and try to balance the amount of insulin injected against the intake of carbohydrate. For this reason, as well as the need to avoid dietary changes that add to the risk of developing various complications of diabetes (such as raised dietary saturated fat and the risk of athero-sclerosis[4]), it is clear that diet is an important component of the treatment of IDDM.

## 2.2 Non-insulin-dependent Diabetes Mellitus

NIDDM is the commoner form of diabetes mellitus by far, and affects about 1–2% of adults in Western Europe and the USA.[5] It is strongly associated with obesity, which is thought to act permissively by allowing an underlying, genetically determined, metabolic disorder to manifest itself.[6,7] In NIDDM insulin production may be either reduced or increased, with varying degrees of tissue resistance to endogenously produced insulin. Thus, initially, the obese non-insulin diabetic may have hyperglycaemia associated with high serum insulin levels and insulin resistance. With time insulin levels can fall as pancreatic $\beta$-cell function fails, and the patient may go on to require insulin. Initial treatment of NIDDM is dietary. Overweight patients should reduce weight, and this can restore insulin sensitivity. Improved insulin sensitivity, in conjunction with a diet avoiding foods that rapidly supply glucose to the liver and systemic circulation, will often control the disease without the need for drugs. Increasing the total carbohydrate content of the diet may increase insulin sensitivity and therefore ameliorate the disease in some individuals.

From this brief introduction it can be seen that the day-to-day life of the diabetic is constrained by a continuous diet. Until recently it was a diet that proscribed, above all else, simple sugars. It was assumed that, since these sugars require insulin for their metabolism, including them in the diet would inevitably lead to excessive rises in blood glucose. Most foods and drinks that taste sweet contain such simple sugars, and as a result it was taken for granted by the patient, the physician and the public that diabetics were to be forbidden to eat anything sweet. Is this still true? Can sweeteners like glucose and sucrose be included in the diabetic diet? If not, can alternative sweeteners, either nutritive or non-nutritive, usefully replace them to improve the quality of life for diabetics and even, perhaps, to ameliorate the disease?

## 3 IS SUGAR HARMFUL?

It is necessary to show that sucrose is harmful to diabetics to justify removing it from their diet. However, it is also important to consider that removing any one nutrient from the diet could also lead to its substitution by another nutrient that could be equally or more harmful. The very low carbohydrate diets recommended to diabetics for many years were inevitably high fat diets. Such diets are likely to contribute to the development of vascular disease.[4] The specific place of sucrose in the aetiology of various diseases, including diabetes, has recently been the subject of intensive consideration by many national food policy organisations.[1,8] They have been unanimous in their conclusions that, while sucrose contributes to excessive energy intake in the obese, along with dietary fat, it is not a specific or independent cause of any disease other than perhaps dental caries.

### 3.1 Does Sugar Cause Diabetes?

Based largely on epidemiological data, Yudkin[9] and Cohen[10] causally related sugar consumption to the later incidence of diabetes mellitus in different populations. They suggested that sugar was a specific factor in causing diabetes mellitus. Both earlier[11] and more recent evidence has failed to support this hypothesis.[12] Dietary survey data from three population groups in the UK confirmed the association of obesity with glucose intolerance, but failed to find a relationship between glucose tolerance and sucrose consumption; in one of the subgroups it was inversely correlated to sucrose intake.[13] Enigmatically, this study found that increasing degrees of obesity were associated with smaller energy intakes, which casts some doubt on the validity of the dietary records. However, others have also failed to find evidence implicating sucrose as a specific agent in causing diabetes.[14,15]

There is also no evidence that sucrose leads to a deterioration in glucose tolerance (i.e. an impairment of glucose utilisation, more minor than that in diabetes mellitus, and which only rarely proceeds to the development of diabetes[1]). In one study, normal male volunteers were fed for 10 days with 45% or 65% of their energy as either sucrose or corn syrup;[16] glucose tolerance improved in the group receiving the higher carbohydrate diet, in agreement with an earlier study.[17] A long-term study also failed to find any benefit from replacing sucrose with other sugars. Two years of total replacement of sucrose by an average of 60 g of either fructose or xylitol had no effect on glucose tolerance in human volunteers.[18]

The recent US Sugars Task Force concluded their extensive review of both animal and human evidence by stating '...there is no persuasive evidence ... that current use levels of sugars or sweeteners directly contribute to the development of abnormal glucose tolerance in the general US population'.[1]

### 3.2 Does Sugar Cause Obesity?

Obesity is a major determinant of NIDDM,[19] although more recent evidence suggests that the distribution of fat deposits is more important than the total quantity.[20,21]

The evidence against a specific role of sugar in causing obesity parallels that against sugars causing diabetes mellitus.[12,13,22,23] Notwithstanding, sugar is an easily identified source of energy in the diet, and as a result is often one of the first things that obese patients are instructed to give up to achieve a caloric deficit and hence weight loss. There is a disappointing lack of evidence for the usefulness of low-calorie sweeteners in weight control.[24,25] Despite widespread use of non-nutritive sweeteners, there is little evidence that they are effective in producing weight loss or preventing weight gain.[26-31] In an analysis of 78 694 women who were part of the cohort of volunteers enrolled in the large American Cancer Society study, 21·6% were found to be long-term users of artificial sweeteners. There was no significant difference between the age-adjusted percentages of artificial sweetener users and non-users who lost weight at any initial weight level; on the contrary, users were more likely to gain weight, across the whole range of obesity (Fig. 1).[26] However, most of these negative data precede the availability of newer sweeteners, such as aspartame, that are devoid of a bitter aftertaste and preferred by most consumers. It remains to be shown whether this new generation of sweeteners will be more useful in this respect. Whether non-nutritive sweeteners have any particular place in helping diabetics adhere to their diet will be discussed later.

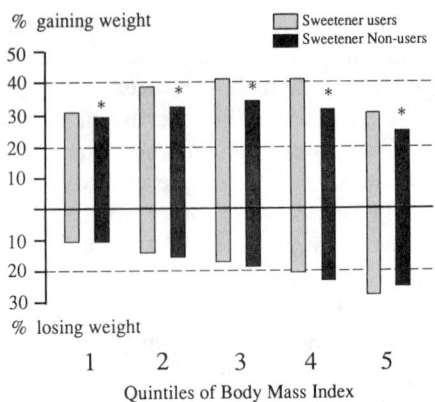

FIG. 1.   Artificial sweetener use and weight change over 1 year in 78 694 women (data from Ref. 26). The bars show the percentage of women either gaining or losing weight over a period of 6 years, by quintile of body weight and separated by whether or not they were users of artificial sweeteners. An asterisk (*) indicates a significant difference ($p < 0.05$) between users and non-users. Significantly more users of artificial sweeteners gained weight; there was no significant difference between them for weight losers.

### 3.3  Does Sugar Impair Blood Glucose Control?

Control of blood glucose levels is the cornerstone of diabetic treatment. There is increasing evidence that tight control of blood glucose towards physiological levels can delay the development of microangiopathic complications of diabetes (nephropathy, retinopathy and neuropathy),[32-34] and improve the outcome of diabetic pregnancies.

Before the discovery of insulin, the only method of altering blood glucose levels in diabetes was by dietary means. Recommendations have included both a high carbohydrate intake (in Egyptian prehistory) and a low carbohydrate diet (by Rollo at the turn of the 18th century). In the 20th century extremely low energy (fat and carbohydrate restricted) diets,[35] which were used to prolong life in insulin-dependent diabetics, were suddenly made unnecessary by the discovery and manufacture of insulin. Carbohydrate continued to be restricted, and refined sugar eliminated, in the belief that this would minimise hyperglycaemia.[36] Adherence to such diets was poor.[37] Only in the past 20 years have diabetes physicians and national diabetes associations relaxed their advice on limiting carbo-hydrate. There has in fact been a reversal towards recommending diets high in complex carbohydrates. The problem of these diets is that they are often unpalatable and patient compliance is often poor. For example, a group of highly motivated diabetics who had increased their carbohydrate intake from 32% to 60% of total calories for a 6-week study were reassessed 2 years later. Their carbohydrate intake had fallen on average to 41%.[38] Now, in the past few years, the outright ban on including sucrose in the diabetic diet has been officially relaxed.[12,39,40]

One of the main reasons for advocating complex carbohydrate, as opposed to simple sugars, was the assumption that they would be more slowly absorbed from the gut, more slowly digested, and so less likely to cause unwanted rises in blood glucose levels. This belief may have been in part suggested by assumptions arising out of the use of the oral glucose tolerance test as a diagnostic test for diabetes mellitus or impaired glucose tolerance.[16] In this 'stress' test, an unphysiologically large amount of glucose (usually 75 g) is given to a patient who has fasted overnight. An excessive rise in blood glucose is the hallmark of diabetes. This stimulus with a concentrated glucose solution is, however, far removed from the normal dietary 'stresses' on glucose metabolism.

The fact first noted by Moskovitz 50 years ago, that blood glucose responses were not proportional to the carbohydrate content of food eaten,[41] has been extensively investigated by several groups, including those of Crapo and Jenkins, and has led to the development of the concept of the glycaemic index.[42] This gives the increment in blood glucose after a standard quantity of a particular food is eaten, compared with a standard food. Initially glucose was used as the standard, then wholemeal bread and cottage cheese, and most recently white bread.[43] On this latest basis sucrose in non-diabetic individuals has a lower glycaemic index (85–92%) than wholemeal bread (93–106%) or Weetabix (109%).[43]

Multiple factors are involved in determining the glycaemic indices of various foods. Adding fibre to a meal has complex effects on the rate of carbohydrate digestion within the gut and on absorption from it. The fibre may also increase the stimulus for the release of gut hormones, such as glucose-dependent insulinotropic polypeptide (GIP), that enhance pancreatic insulin secretion. Clearly, post-absorptive metabolism (i.e. whether or not the carbohydrate is metabolised to glucose) will also influence the glycaemic response. The use of the glycaemic index in formulating diabetic diets has been criticised on the grounds that there is such a large variation in response between individuals and laboratories, and, perhaps more important, testing individual components of a meal gives little useful information about the effect of a normal mixed meal. In fact most mixed meals give very similar glycaemic responses. Despite these reservations, the concept of incorporating low-glycaemic index foods into the diabetic diet will often improve blood glucose levels in patients with NIDDM, and also give patients with IDDM more stable responses to intermittent insulin injections.

### 3.3.1 Patients with Insulin-dependent Diabetes Mellitus

Control of blood glucose in patients with IDDM is determined mainly by the amount and type of insulin injected, which is usually tailored to suit dietary needs and social habits. A study in poorly controlled insulin-dependent diabetic children showed that the effects of a breakfast containing high-glycaemic-index foods, which raised post-prandial blood glucose levels compared with a breakfast with low-glycaemic-index foods, could be corrected for by appropriate adjustments to the insulin given before the meal. Flexibility in dosing with insulin is gaining widespread popularity amongst diabetics and their physicians, and has been made considerably more practical by the introduction of more convenient delivery systems such as pen injectors. These devices allow doses of short-acting (soluble) insulin to be given just before each meal, while overnight and background basal insulin needs are met by once-daily long-acting insulin (e.g. ultratard). These regimens can improve blood glucose control compared with conventional once- or twice-daily regimens. When combined with monitoring of capillary blood glucose by the patient at home using test strips, and with suitable education in how to alter insulin dosage, near-normoglycaemia can often be achieved. Several studies have shown that, even without such intensive self-management, incorporating moderate amounts of sucrose into the diet of patients with IDDM has no significant deleterious effect on blood glucose levels.[44-47]

### 3.3.2 Patients with Non-insulin-dependent Diabetes Mellitus

Evidence exists to show that sucrose taken with meals does not necessarily aggravate post-prandial hyperglycaemia in NIDDM. Ten non-obese patients with NIDDM (and 10 with IDDM) ate a breakfast containing milk, rice crispies, eggs, bacon and coffee. The carbohydrate content provided nearly 50% of total calories and included 42 g of either glucose, fructose, sucrose or potato or wheat starch. In the patients with IDDM, sucrose produced similar blood glucose values, in terms of peak and area of increment, to each of the test meals, and also the lowest values, equal to those with fructose, in patients with NIDDM[46] (see Fig. 2). The authors concluded that there was 'no reason for diabetics to be denied foods containing sucrose ... provided that sucrose is consumed in controlled amounts in nutritionally balanced meals that also contain protein and fat'.

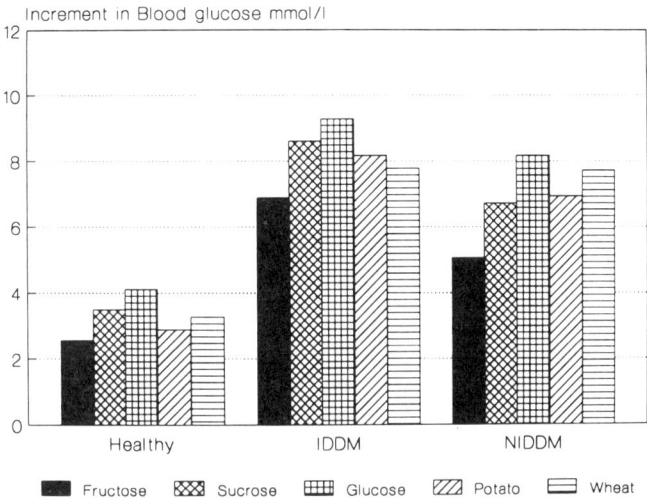

FIG. 2.   The glycaemic response to breakfasts containing various carbohydrates
(data from Ref. 46).

A French group came to similar conclusions, based on their findings that a mixed meal of meat, vegetables, cheese and cake containing 70 g carbohydrate gave similar blood glucose curves in patients with well controlled IDDM and NIDDM no matter whether the cake contained either 20 g of sucrose or saccharin to sweeten it.[47] On the other hand, 11 patients with NIDDM who received 50% of their energy as carbohydrate with either 16% or 1% of calories from sucrose, each for two 14-day periods, showed worse glycaemic control on the sucrose-containing diet.[48] The differences were small (1–2 mmol/litre) and, surprisingly, were only apparent for a few hours after the midday meal. A study of patients with NIDDM and overt hypertriglyceridaemia revealed no benefit in glycaemic control from restricting sucrose and carbohydrate below 120 g daily.[49]

Thus it is hard to find strong evidence against including moderate amounts of sugar in the diabetic diet, provided it is taken as part of a mixed meal. This does not exclude a place for alternative sweeteners, particularly if they are used to flavour beverages or snacks taken on their own, between meals. Non-nutritive sweeteners could also contribute to reducing energy intake in the overweight or obese diabetic.

### 3.4 Sugar and Lipid Disorders

Although the metabolic disorder of diabetes is characterised by hyperglycaemia, other metabolic features may be just as, or more, important as causes of diabetic complications. Diabetics are at increased risk from vascular disease, in particular ischaemic heart disease, compared with the non-diabetic population.[4] For this reason an important goal of treatment has been to formulate diets that reduce the risk of coronary heart disease. Replacing saturated fat by carbohydrate lowers blood levels of low-density lipoprotein (LDL) cholesterol, one of the major risk factors. On the other hand, very high carbohydrate diets can have unwanted effects. They can lead to higher blood levels of triglycerides, very low density lipoproteins (VLDL) and decreased levels of high-density lipoprotein (HDL),[50,51] all factors that have been implicated as increasing the risk of developing coronary heart disease.[52,53] A recent dietary modification shown to improve lipid levels[54] is partially to replace some of the complex carbohydrates in the diet by mono-unsaturated fat.

Increasing sucrose intake leads to higher levels of plasma cholesterol and triglycerides in non-diabetic patients,[55-57] but this effect is not easily separable from that of a general increase in total dietary carbohydrate. Mildly hypertriglyceridaemic patients with NIDDM given 220 g sucrose a day with a high (60–65%) carbohydrate diet developed raised fasting, but not post-prandial, triglyceride levels, while a trace-sucrose and low-carbohydrate diet did not affect fasting, but did increase post-prandial triglyceride levels.[49] When total carbohydrate and fat in the diet were kept constant, a 75-fold variation in dietary sucrose (from 3 to 220 g sucrose daily) had no effect on plasma triglycerides over a 1-month period.[58]

This area remains contentious,[59] and there is insufficient evidence to say whether alternative sweeteners alter overall carbohydrate and fat intake and make any significant contribution, either way, to the risks of coronary heart disease.

## 4 HOW ARE SWEETENERS METABOLISED?

The quantities of intense, or non-nutritive, sweeteners required in foods are so small that they make a negligible contribution to energy intake, although they or their metabolites may have significant metabolic effects in terms of toxicity. On the other hand, the extent to which nutritive sweeteners, often used as bulking agents, are made available for metabolism is determined by their rate of digestion, including their transit through the gut and the speed with which they are absorbed through the gut mucosa.[60] The metabolic effects of parenterally administered alternative sweeteners will not be considered here, except to say that adverse effects are common unless quantities are restricted to less than 0·25 g/kg/h.[61]

## 4.1 Nutritive Sweeteners

The potential value of nutritive sweeteners to the diabetic is in their ability to provide sweetness and bulk in foods at a lower energy and glucose load than sucrose (see Fig. 3). Quantitating these properties is not simple. Nutritive sweeteners have both a physical energy value (derived from complete combustion) and a physiological one that depends upon many 'host' factors. The methodology for measuring physiological or 'metabolisable energy' is complex and contentious, since energy losses in the faeces and urine from incomplete digestion are compounded by losses from fermentation to carbon dioxide and methane lost as flatus.[62]

These theoretical problems are not without importance to diabetics. For insulin-treated patients, foods are usually classified on the basis of carbohydrate equivalents to allow choice in foods which 'balance' injected insulin. Patients with IDDM therefore need to know not just the caloric value of a food but also whether it will count towards their prescribed carbohydrate allowance. In the case of fructose, the most commonly used alternative to sucrose, early suggestions that a factor of 70% should be used in calculating its carbohydrate equivalence were quickly dismissed as impracticable. It was decided that it should be regarded as a 'not readily absorbable carbohydrate' and excluded from dietary calculations, provided that the amount consumed was restricted to less than 25 g daily.[63] Defining a carbohydrate equivalence for the various polyol sweeteners is more difficult still. With the growing variety of nutritive sweeteners, and the ubiquity with which they are now appearing in foods,[64] it is easy for a diabetic consuming a variety of 'diet' foods to exceed the recommended limits.

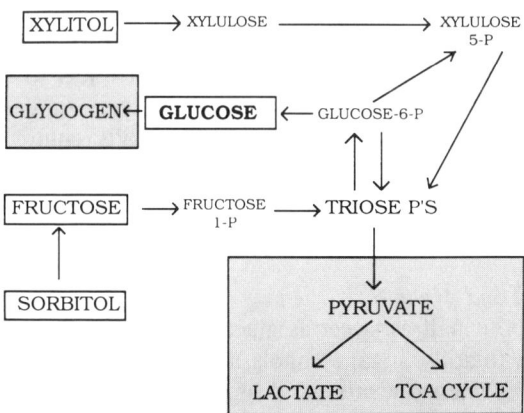

FIG. 3.    Metabolic pathways of some caloric sweeteners; the shaded areas represent pathways that require insulin.

### 4.1.1 Fructose

'High fructose corn syrup' (derived by incomplete degradation of corn starch, and subsequent partial enzymic conversion of glucose to fructose) is now the main source of fructose in the diet and used extensively in place of sugar cane-based sweeteners in soft drinks.[65] However, since these HFCSs usually contain about 40–50% glucose they must be regarded as metabolically similar to sucrose.

Fructose alone is extensively used as a bulk sweetener in diabetic diet foods. Because it is sweeter than sucrose,[66] it allows a significant reduction in caloric density in foods traditionally sweetened with sucrose. Gastric emptying of fructose solutions is faster than that of glucose solutions.[67] Fructose is actively absorbed from the jejunum,[68] but about 1 in 3 people show evidence of incomplete absorption after 50 g given as a 10% solution (equivalent to two cans of soft drink).[69] Gastrointestinal symptoms are unusual with doses of less than 50 g.[70] Fructose enters the liver cell independently of insulin, and is metabolised to triose phosphates (Fig. 3). The subsequent metabolism of these 3-C compounds through pyruvate and the tricarboxylic acid (Krebs) cycle is dependent upon insulin and, in its presence, can lead to fat synthesis. However, in the absence of adequate insulin, or in poorly controlled diabetics, fructose-derived trioses enter the gluconeogenic pathway, contributing to overall hepatic glucose output, and may in these circumstances raise blood glucose levels.[71]

Rapid initial phosphorylation of fructose can deplete ATP levels, increase nucleotide catabolism and lead to hyperuricaemia.[72,73] although clinically significant alterations in serum uric acid do not seem to occur with oral fructose, even when it provides 20% of dietary carbohydrate calories.[74] On the other hand, subjects fed diets containing 7·5% or 15% of calories as fructose had increased uric acid responses to a sucrose tolerance test compared with those fed starch. Uric acid levels were higher still (although still within the 'normal' range) in hyperinsulinaemic individuals.[75] These findings suggest that further caution is required in including excessive quantities of fructose in the diet of patients with NIDDM, since hyperuricaemia is associated with increased risk of coronary heart disease.[76]

### 4.1.2 Sorbitol and Mannitol

Sorbitol is about half as sweet as glucose.[66] The human intestine has a limited ability to utilise sugar alcohols, and only 50–75% of an oral dose is bioavailable.[77] The slow rate of absorption of sorbitol from the small intestine results in it reaching the colon, where it exerts an osmotic effect causing diarrhoea. In the colon sorbitol is fermented by commensal bacteria to produce hydrogen, carbon dioxide and methane, causing flatulence and abdominal discomfort if daily intake exceeds 20–30 g.[61] Symptoms can arise in some individuals with as little as 10 g.[45] Diarrhoea, often severe and intractable, is a feature of autonomic neuropathy, which can occur as a complication of long-standing diabetes. The acceptability of polyol sweeteners in such patients has not been specifically studied.

Following absorption, sorbitol is rapidly metabolised by the liver to fructose, and passes into the metabolic pathways already described (Fig. 3). Dietary sorbitol does not appear in plasma and, even when infused directly into the circulation, plasma sorbitol does not diffuse across membranes, and does not therefore contribute to cataract development, which is another complication of diabetes.[78]

Mannitol has similar biochemical properties to sorbitol, and is also metabolised to fructose.

### 4.1.3 Xylitol

Xylitol is a naturally occurring polyol and is also produced commercially from vegetable xylose. It has a caloric value of about 4 kcal/g, but can be used in smaller quantities than sucrose since it is sweeter. It is slowly absorbed and therefore creates similar problems to sorbitol in causing flatulence and diarrhoea. Symptoms are observed after doses of 30–40 g/day but, if gradually introduced to the diet, up to 90 g/day may be tolerated.[79] It has mainly been used in chewing-gum to replace sucrose and reduce dental caries, and is seldom used in diabetic products.[80] It does not appear to have any significant unwanted metabolic effects.[18]

### 4.1.4 Maltitol

Maltitol is not yet approved for use in the United Kingdom, although it is a major component of Lycasin (a hydrogenated glucose syrup) which is in use. Maltitol is only partially absorbed from the gut and undergoes extensive bacterial fermentation there, so it is not surprising that it also causes diarrhoea and flatulence.[81]

### 4.2 Non-nutritive Sweeteners

The metabolism of non-nutritive sweeteners should be considered separately from that of nutritive sweeteners. Because of their intense sweetness they are taken only in milligram quantities, so the main concern about their use by diabetics should be any possible toxicity rather than their contribution to intermediary metabolism.[82,83]

### 4.2.1 Do They Stimulate Insulin Release?

Concern has been expressed that stimulating the sweet taste may promote insulin release in some way, leading to hypoglycaemia in normal individuals and so perhaps fostering appetite. Additionally, hyper-insulinaemia itself (albeit at pharmacological levels) can stimulate hunger, increase the palatability of sucrose and increase food intake.[84] In high doses, acesulfame K stimulated insulin release in rats, but did not lead to hypoglycaemia.[85] Somatostatin, which suppresses glucose-induced pancreatic insulin release, did not abolish this effect in studies on isolated rat islet cells.[86] The implications of these findings are profound, and if confirmed in man they could suggest that the use of acesulfame K might lead to hypoglycaemia, with subsequent stimulation of appetite and counter-regulatory responses, such as release of growth hormone and catecholamines. Tordoff, in a series of experiments, showed a cephalic-mediated stimulation of appetite by saccharin in rats. This was brought about not by hyperinsulinaemia but probably by a rise in serum tonicity leading to a stimulation of drinking.[87] Blundell has used non-nutritive sweeteners to study the effect of uncoupling sweet taste from its caloric consequences in man. He has produced evidence that artificial sweeteners such as saccharin, aspartame and acesulfame K have opposite effects to those of nutritive sweeteners, and can increase hunger and appetite ratings.[88]

If this paradoxical stimulation of appetite by non-nutritive sweeteners were due to inappropriate insulin-release, it is unlikely to be of significance in patients with IDDM (who have no pancreatic $\beta$-cell insulin secretion) or in the hyperinsulinaemic patient with NIDDM who is insulin-resistant. Further work is needed urgently in this area but, as Blundell has pointed out,[88] the results of such research appear to be judged often on their commercial implications, and the questioning of the motives of research workers can serve as a disincentive to become involved in this area of research.

### 4.2.2 Saccharin

Saccharin was first synthesised in 1879; it is several hundred times as sweet as sucrose.[89] About 80% of saccharin taken by mouth is absorbed and then excreted either unchanged or as 2-sulphamoylbenzoic acid.[77,90] The controversy over its safety, and in particular whether or not its use was associated with bladder cancer, has been reviewed elsewhere.[29,90] One important study of diabetics who were shown to be above-average consumers of saccharin found that they were no more likely to die from bladder or other cancers than non-diabetics.[91] Saccharin remains the most commonly used alternative sweetener worldwide, despite its removal from GRAS (generally recognised as safe) status in the United States in 1971. A proposed ban on its use in the United States was deferred owing to public (and industry) pressure. It is of interest that a poll in 1978 found that most health professionals thought saccharin to be essential or of great value in the treatment of obesity and diabetes.[92] However, consumer preference for other sweeteners which do not have its bitter aftertaste is leading to its replacement in many food and drink products.

### 4.2.3 Cyclamate

Cyclamate is about 30 times as sweet as sucrose, and was widely used until withdrawn in the United States and the United Kingdom in 1970. Unabsorbed cyclamate is metabolised in the colon to form cyclohexyl-amine, a compound with the potential to stimulate the sympathetic nervous system, but not in sufficient quantities to cause clinical effects.[77]

### 4.2.4 Aspartame

Aspartame is not strictly a non-nutritive sweetener, since it is a dipeptide which is metabolised like other peptides, providing 4 kcal/g energy. However, the intensity of its sweet taste (200 times that of sucrose) allows calorically insignificant quantities of it to be used to sweeten foods and drinks. Its popularity is attested to by the rapidity with which it has replaced saccharin in soft drinks and milk-based products despite its greater cost and shorter shelf-life. Aspartame is the methyl ester of the dipeptide L-aspartyl-L-phenylalanine and is hydrolysed in the gut to form the dipeptide and methanol. The dipeptide is taken up by mucosal cells and metabolised to its constituent amino acids—aspartic acid and phenyl-alanine. Concern has been raised about possible adverse effects from all three of these metabolites but it has not been substantiated.[77,93,94]

Methanol is a potent toxin causing both metabolic acidosis and blindness when consumed in quantity.[95] About 10% by weight of aspartame is absorbed into the circulation as methanol, but this is undetectable even at doses of 34 mg/kg body weight of aspartame (nearly double the total likely daily intake).[96] Larger quantities do acutely raise blood methanol levels, but to no higher levels than those found after drinking a double martini; blood formate levels (thought to be the metabolite of methanol responsible for its toxicity) do not increase.[97]

Plasma aspartate concentrations did not rise in adults given 10 mg/kg body weight aspartame as a carbonated drink,[94] and changes in plasma phenylalanine were similar to those seen after a high-protein meal.[98] Other concerns about the possible neurochemical consequences of aspartame on brain serotonin metabolism[99,100] have been widely refuted.[94,101,102] Aspartame has been extensively tested for safety in diabetics.[103,104]

### 4.2.5 Acesulfame K

Acesulfame K has been available for use in the United Kingdom since 1983. It is less intensely sweet than saccharin and has the advantage over aspartame of being heat stable; it can therefore be incorporated into baked products.[105] It is not yet widely used.

## 5 ARE ALTERNATIVE SWEETENERS USEFUL TO DIABETICS?

If the case against sucrose is largely unproven, are there any positive advantages in using alternative sweeteners? Is there evidence that the insulin-independent metabolism of fructose or sugar alcohols can improve glycaemic control, or other factors that are thought to be important in mediating diabetic complications? Do non-nutritive sweeteners, by satisfying the hedonic desire for sweet taste, improve life for diabetics? There is a dearth of information with which to answer these questions.[61]

### 5.1 Do They Improve Diabetic Control?

The absence of data on this results in part from the difficulty in designing appropriate clinical trials, as well as the difficulty in funding such trials. The development of techniques such as home blood glucose monitoring (on finger-prick samples of blood using glucose test strips) and measurement of glycosylated proteins[106] means that clinical research methods are now available for measuring glycaemic control in free-living diabetics. Food manufacturers have not approached the evaluation of their products in the same way as pharmaceutical companies have to. The cynic might speculate that this is because foods for diabetics may be sold with or without any proof of efficacy.

The majority of single meal studies in IDDM[46,107,108] and NIDDM[109] do not suggest that substitutes for sucrose improve diabetic glycaemic control. Longer-term studies are no more supportive for the use of alternative sweeteners.[110,111] Early introduction of insulin treatment to newly diagnosed diabetics is thought to preserve pancreatic $\beta$-cell function and can lead to a 'honeymoon' period of lower insulin requirements and easier control. It is not known whether the reduced stimulus to insulin secretion by fructose and polyol sweeteners could be useful in a similar way.

## 5.2 How are Alternatively Sweetened Products Controlled?

For food additives to be approved for use in the UK the Food Advisory Committee of the Ministry of Agriculture, Fisheries and Food, and the Department of Health and Social Security, have to be convinced of a need for them. The terms for this are extremely wide and do not depend upon a medical or therapeutic need being defined. Information on the toxicology of the sweeteners is necessary to satisfy the Committee on Toxicity of Chemicals in Food, Consumer Products and the Environment, and the likely intake of a sweetener is taken into account in considering its safety and before it is approved; no data need be provided on its efficacy as anything other than the provider of a sweet taste. Regulations regarding claims for foods as therapeutic agents or quasi-drugs come under the Food Labelling Regulations. A physician faced with as little knowledge about the efficacy of a new drug as there is about diet foods would be unlikely to prescribe it. Perhaps it is time that 'diet products' were evaluated as therapeutic agents. The 1955 Food and Drugs Act did incorporate the principle that foods designed as 'therapy' should be used under medical advice,[112] but this is a view that has become increasingly eroded. The public certainly thinks of diet products as medical products. A survey by the Food and Drugs Administration in the United States found that 9% of those interviewed thought that 'diet foods' were for diabetics, and 28% thought the word 'dietetic' on a food label meant it was for use by persons with diabetes.[113]

Studies are urgently needed to assess the value of sweeteners in the diabetic diet as well as in weight control. Protocols which will allow the place of alternative sweeteners in the long-term management of diabetes to be assessed independently of any concomitant changes in other dietary components are difficult but not impossible to devise.[61] Physicians and patients need reliable data to confirm or refute the sort of claim made by manufacturers that 'fructose offers many advantages to ... the clinically obese, and to non-insulin-dependent diabetics'.[114] Data are needed to confirm or refute the sort of tautologous claim that a non-nutritively sweetened product can help weight loss if 'included as part of a calorie-controlled diet'. Even if sweeteners were shown to be of no specific value in treating diabetes, this would not preclude their use for the enjoyment of their taste alone.

# 6 DO DIABETICS HAVE A 'SWEET TOOTH'?

Diabetics have a higher threshold than normal individuals for detecting a sweet taste from glucose and sucrose[115-117] but perhaps not from fructose.[118] By contrast, there is no evidence for altered sweet taste sensitivity in obesity.[119] Whether a change in taste sensitivity should lead to an increased or decreased desire for, and consumption of, sweet foods has not been studied. Adding sucrose to butter beans (as in South African and South American cuisine) improved their palatability without any deterioration in glycaemic and insulin responses. Those prepared without sucrose were least liked, while the most popular were those prepared with 30% sucrose. Interestingly, in this study unsweetened beans were rated more highly by diabetics than normal subjects (73% of normal subjects compared with only 37% of diabetics rejected the unsweetened beans), although, on the other hand, twice as many diabetics compared with control subjects rejected the beans with 30% sucrose addition.[120]

In 1977, 91% of all diabetics were thought to be using saccharin.[29] Only 17% of 500 diabetics in a survey reported no cravings for sweetness or alternative sweeteners.[121] Most mothers of diabetic children believe that artificially sweetened foods will help their children to adhere to a sugar-free diabetic diet.[122] It has been suggested that sweetness may improve the palatability of unpopular foods that are encouraged in the diabetic diet, such as beans, but there is little direct evidence to support this.

A survey of 100 women attending a diabetic clinic was carried out in 1963 to assess whether saccharin and cyclamates, the only non-caloric sweeteners then available, improved adherence to a carbohydrate-restricted diet.[123] It found no evidence that the use of non-caloric sweeteners improved dietary compliance, and suggested that some patients used them as a substitute for sugar, while others used them as a dietary supplement.

# 7 SHOULD DIABETICS USE ALTERNATIVE SWEETENERS?

There seems to be no reason to restrict the use of non-nutritive sweeteners in the diabetic diet, and they do allow greater flexibility in choosing convenience foods, desserts and soft drinks. This may be particularly useful in helping diabetic children feel less socially stigmatised by having to avoid sweet snacks between meals. The balance of medical opinion is that nutritive sweeteners only have a limited use in the diabetic diet. They are mostly less sweet than sucrose or non-nutritive sweeteners, have roughly the same caloric value as sucrose, and may cause unwanted effects on the gut if taken in anything other than moderate quantities. Furthermore, they detract from the current emphasis of diabetic dietary advice, which focuses on 'healthy eating'. The modern diabetic diet is a prescription for an overall attitude to eating patterns, of which sucrose restriction is only one small part.

## REFERENCES

1. Glinsmann, W. H., Irausquin, H. & Park, Y. K., Evaluation of health aspects of sugars contained in carbohydrate sweeteners. *J. Nutr.*, **116** (1986) S1–S216.
2. World Health Organisation Expert Committee on Diabetes Mellitus, Second Report. WHO Technical Reports, WHO, Geneva, 1980.
3. Calnan, M. & Peckham, C. S., Incidence of insulin-dependent diabetes in the first sixteen years of life. *Lancet*, **i** (1977) 589–90.
4. WHO Expert Committee on the Prevention of Coronary Heart Disease, WHO Technical Report Series No. 678. WHO, Geneva, 1982.
5. Mann, J. I., Pyorala, K. & Teuscher, A., *Diabetes in Epidemiological Perspective*. Churchill Livingstone, Edinburgh, 1983.
6. Bennett, P. H., Rushforth, H. B., Miller, M. & LeCompte, P. M., Epidemiologic studies of diabetes in the Pima Indians. *Rec. Progr. Horm. Res.*, **32** (1976) 333–76.
7. Zimmett, P. & Whitehouse, S., Bimodality of fasting and two-hour glucose tolerance distributions in a Micronesian population. *Diabetes*, **27** (1978) 793–9.
8. National Advisory Council on Nutrition Education, *Proposals for Nutritional Guidelines for Health Education in Britain*. Health Education Council, London, 1983.
9. Yudkin, J., Dietary fat and dietary sugar in relation to ischaemic heart disease and diabetes. *Lancet*, **ii** (1964) 4–5.
10. Cohen, A. M., Bavly, S. & Poznanski, R., Change of diet of Yemenite Jews in relation to diabetes and ischaemic heart disease. *Lancet*, **ii** (1961) 1399–401.
11. Himsworth, H. P., Diet and the incidence of diabetes mellitus. *Clin. Sci. Mol. Med.*, **2** (1935) 117–48.
12. Nutrition Sub-Committee of the British Diabetic Association's Medical Advisory Committee. Dietary recommendations for the 1980's—a policy statement by the British Diabetic Association. *Hum. Nutr. Appl. Nutr.*, **36A** (1982) 378–86.
13. Keen, H., Thomas, B. J., Jarrett, R. J. & Fuller, J. H., Nutrient intake, adiposity, and diabetes. *Br. Med. J.*, **1** (1979) 655–8.
14. Kahn, H. A., Herman, J. B., Medalie, J. H., Neufeld, H. N., Riss, E. & Goldbourt, U., Factors related to diabetes incidence: a multivariate analysis of two years observation on 10,000 men. *J. Chronic Dis.*, **23** (1971) 617–29.
15. Baird, J. D., Diet and the development of clinical diabetes. *Acta Diabetol. Lat.*, **9** (1972) Suppl. 1, 621–37.

16. Thompson, R. G., Hayford, J. T. & Danney, M. M., Glucose and insulin responses to diet. *Diabetes*, **27** (1978) 1020–6.
17. Anderson, J. W., Herman, R. H. & Zakim, D., Effect of high glucose and sucrose diets on the glucose tolerance of normal men. *Am. J. Clin. Nutr.*, **26** (1973) 600–7.
18. Huttunen, J. K., Makinen, K. K. & Scheinin, A., Turku sugar studies. XI. Effects of sucrose, fructose and xylitol diets on glucose, lipid and urate metabolism. *Acta Odont. Scand.*, **34** (1976) Suppl. 70, 345–51.
19. West, K. M., *Epidemiology of Diabetes Mellitus and its Vascular Lesions*. Elsevier, New York, 1978.
20. Lapidus, L., Bengtsson, C., Larsson, B., Pennert, K., Rybo, E. & Sjorstrom, L., Distribution of adipose tissue and risk of cardiovascular disease and death: a 12-year follow-up of participants in the population study of women in Gothenburg, Sweden. *Br. Med. J.*, **298** (1984) 1257–61.
21. Larsson, B., Bjorntorp, P. & Tibblin, G., The health consequences of moderate obesity. *Int. J. Obesity*, **5** (1981) 97–116.
22. Richardson, J. F., The sugar intake of businessmen and its inverse relationship with relative weight. *Br. J. Nutr.*, **27** (1972) 449–60.
23. Walker, A. R. P., Studies on sugar intake and overweight in South African black and white schoolchildren. *S. Afr. Med. J.*, **48** (1974) 1650.
24. Finer, N., Sugar substitutes in the treatment of obesity and diabetes mellitus. *Clin. Nutr.*, **4** (1985) 207–14.
25. Booth, D. A., Evaluation of the usefulness of low-calorie sweeteners in weight control. In *Developments in Sweeteners*, Vol. 3, ed. T. H. Grenby. Elsevier Applied Science, London, 1987, p. 287.
26. Stellman, S. D. & Garfinkel, L., Artificial sweetener use and one-year weight change among women. *Prev. Med.*, **15** (1986) 195–202.
27. McCann, M. B., Trulson, M. F. & Stulb, S. C., Non-caloric sweeteners and weight reduction. *J. Am. Diet. Assoc.*, **32** (1954) 327–30.
28. Foltin, R. W., Fischman, M. W., Emurian, C. S. & Rachlinkski, J. J., Compensation for caloric dilution in humans given unrestricted access to food in a residential laboratory. *Appetite*, **10** (1988) 13–24.
29. Committee for a Study on Saccharin and Food Safety Policy, *Saccharin: Technical Assessment of Risks and Benefits.* Assembly of Life Sciences/ Institute of Medicine/National Research Council/National Academy of Sciences, Washington, DC, 1978.
30. Alexander, M. M., Have formula diets helped? *J. Am. Diet. Assoc.*, **40** (1968) 538.
31. Parham, E. S. & Parham, A. R., Saccharin use and sugar intake by college students. *J. Am. Diet. Assoc.*, **76** (1980) 560–3.
32. Alberti, K. G. M. & Hockaday, T. D. R., Diabetes mellitus. In *Oxford Textbook of Medicine*, ed. D. J. Weatherall, J. G. G. Ledingham and D. A. Warrell. Oxford University Press, Oxford, 1987, p. 51.
33. Tchoubroutsky, G., Relation of diabetic control to development of microvascular complications. *Diabetologia*, **15** (1978) 143–52.
34. Skyler, J. S., Complications of diabetes mellitus: relationship to metabolic dysfunction. *Diabetes Care*, **2** (1979) 499–509.
35. Allen, F. M., Stillman, E. & Fitz, R., Total dietary regulation in the treatment of diabetes. *Monogr. Rockefeller Inst. Med. Res.*, **11** (1919) 1–78.

36. Joslin, E. P., *Treatment of Diabetes Mellitus.* Lea and Febiger, Philadelphia, 1928.
37. West, K. M., Diet therapy of diabetes: an analysis of failure. *Ann. Intern. Med.,* **79** (1973) 425–34.
38. Geekie, M., Eaton, J., Simpson, H. & Mann, J. I., Will diabetics accept an increase in dietary carbohydrate? *Diabetologia,* **21** (1981) 507–10.
39. Committee on Food and Nutrition of the American Diabetes Association, 1979, Principles of nutrition and dietary recommendations for individuals with diabetes mellitus. *Diabetes,* **28** (1979) 1027–30.
40. Special Report Committee of the Canadian Diabetes Association, Guidelines for the nutritional management of diabetes mellitus, 1980. Special report from the Canadian Diabetes Association. *J. Can. Diet. Assoc.,* **42** (1981) 110–18.
41. Moskovitz, E., Der Einfluss vegetabiler Nährungsmittel auf den Blutzucker bei Diabetikern. *Z. Klin. Med.,* **131** (1937) 648–59.
42. Jenkins, D. J. A., Wolever, T. M. S., Taylor, R. H. *et al.,* Glycemic index of foods: a physiological basis for carbohydrate exchange. *Am. J. Clin. Nutr.,* **34** (1981) 362–6.
43. Jenkins, D. J. A., Wolever, T. M. S. & Jenkins, A. L., Starchy foods and glycemic index. *Diabetes Care,* **11** (1988) 149–59.
44. Erkelens, D. W., Stofkooper, A., Van der Bogaard, E. & Van der Snoek, B. E., Glycaemic effect of mono-, di- and poly-saccharides in a mixed meal in diabetic patients. *Neth. J. Med.,* **28** (1985) 157–63.
45. Crapo, P. A., Use of alternative sweeteners in diabetic diet. *Diabetes Care,* **11** (1988) 174–82.
46. Bantle, J. P., Laine, D. C., Castle, G. W., Thomas, J. W., Hoogwerf, B. J. & Goetz, F. C., Postprandial glucose and insulin responses to meals containing different carbohydrates in normal and diabetic subjects. *N. Engl. J. Med.,* **309** (1983) 7–12.
47. Slama, G., Haardt, M. J., Jean-Joseph, P. *et al.,* Sucrose taken during mixed meal has no additional hyperglycaemic action over isocaloric amounts of starch in well-controlled diabetics. *Lancet,* **ii** (1984) 122–5.
48. Coulston, A. M., Hollenbeck, C. B., Donner, C. C., Williams, R., Chiou, Y.-A. M. & Reavan, G. M., Metabolic effects of added dietary sucrose in individuals with non-insulin-dependent diabetes mellitus. *Metabolism,* **34** (1985) 962–6.
49. Jellish, S., Emanuele, M. A. & Abraira, C., High sucrose carbohydrate diets versus sucrose restricted diets in overt hypertriglyceridemic diabetics. *Am. J. Med.,* **77** (1984) 1015–22.
50. Coulston, A. M., Hollenbeck, C. B., Swislocki, A. L. M., Chen, Y.-D. & Reavan, G. M., Deleterious effects of high carbohydrate sucrose-containing diets in patients with non-insulin-dependent diabetes mellitus. *Am. J. Med.,* **82** (1987) 213–20.
51. Sestoft, L., Karup, T., Palmvig, B., Meinertz, H. & Faergeman, O., High-carbohydrate, low-fat diet: effect on lipid and carbohydrate metabolism, GIP and insulin secretion in diabetics. *Dan. Med. Bull.,* **32** (1985) 64–9.
52. Castelli, W. P. & Anderson, K., A population at risk: prevalence of high cholesterol levels in hypertensive patients in the Framingham study. *Am. J. Med.,* **80** (1986) 23–32.
53. Carlson, L. A., Bottiger, L. E. & Ahfeldt, P. E., Risk factors for myocardial

infarction in the Stockholm prospective study: a 14-year follow-up focusing on the role of plasma triglycerides and cholesterol. *Acta Med. Scand.*, **206** (1979) 351–60.

54. Garg, A., Bonanome, A., Grundy, S. M., Zhang, Z.-J. & Unger, R. H., Comparison of a high-carbohydrate diet with a high mono-unsaturated-fat diet in patients with non-insulin-dependent diabetes mellitus. *N. Engl. J. Med.*, **319** (1988) 829–34.

55. Nikkila, E. A. & Kekki, M., Effects of dietary fructose and sucrose on plasma triglyceride metabolism in patients with endogenous hypertriglyceridemia. *Acta Med. Scand.*, **542** (1972) Suppl., 221–7.

56. Hayford, J. T., Danney, M. M. & Thompson, R. G., Triglyceride integrated concentrations: effects of variation of source and amount of carbohydrate. *Am. J. Clin. Nutr.*, **32** (1979) 1670–8.

57. Reiser, S., Hallfrisch, J., Michaelis, O. E., IV, Lazar, F. L., Martin, R. E. & Prather, E. S., Isocaloric exchange of dietary starch and sucrose in humans. I. Effects on fasting blood lipids. *Am. J. Clin. Nutr.*, **32** (1979) 1659–69.

58. Abraira, C. & Derler, J., Large variations of sucrose in constant carbohydrate diets in Type II diabetes. *Am. J. Med.*, **84** (1988) 193–200.

59. Reavan, G. M., Dietary therapy for non-insulin-dependent diabetes mellitus. *N. Engl. J. Med.*, **319** (1988) 862–4.

60. Zeisnitz, S. L. & Siebert, G., The metabolism and utilization of polyols and other bulk sweeteners compared with sugar. In *Developments in Sweeteners*, Vol. 3, ed. T. H. Grenby. Elsevier Applied Science, London, 1987, p. 109.

61. Talbot, J. M. & Fisher, K. D., The need for special foods and sugar substitutes by individuals with diabetes mellitus. *Diabetes Care*, **1** (1978) 231–40.

62. Hobbs, D. C., Methodology in the measurement of caloric availability. In *Low Calorie Products*, ed. G. G. Birch & M. G. Lindley. Elsevier Applied Science, London, 1988, p. 245.

63. Anon., *Food Labelling Regulations 1984*. HMSO, London, 1984.

64. Jackson, K. G., Howells, J. & Armstrong, J., Sweeteners in special foods for diabetic, slimming and medical purposes. In *Developments in Sweeteners*, Vol. 3, ed. T. H. Grenby. Elsevier Applied Science, London, 1987, p. 213.

65. Bujake, J. E. In *Alternative Sweeteners*, ed. L. O'B. Nabors & R. C. Gelardi (*Food Science and Technology*, Vol. 17). Marcel Dekker, New York, 1986, p. 277.

66. Moskowitz, H. R., The sweetness and pleasantness of sugars. *Am. J. Psychol.*, **84** (1971) 387–405.

67. Moran, T. H. & McHugh, P. R., Distinctions among three sugars in their effects on gastric emptying and satiety. *Am. J. Physiol.*, **241** (1981) R25–R30.

68. Cook, G. C., Absorption products of D-fructose in man. *Clin. Sci.*, **37** (1969) 675–87.

69. Ravich, W. J., Bayless, T. M. & Thomas, M., Fructose: incomplete intestinal absorption in humans. *Gastroenterology*, **84** (1983) 26–9.

70. Anderson, D. E. H. & Nygren, A., Four cases of long-standing diarrhea and colic pains cured by fructose-free diet; a pathogenic discussion. *Acta Med. Scand.*, **203** (1978) 87–92.

71. Felber, J. P., Renold, A. & Zahno, G. R., The comparative metabolism of glucose, fructose and sorbitol in normal subjects and disease states. *Med. Prob. Pediatr.*, **4** (1959) 482–7.

72. Maenpaa, P. H., Raivio, K. O. & Kekornaki, M. P., Liver adenine nucleotides: fructose-induced depletion and its effect on protein synthesis. *Science*, **161** (1968) 1252–4.
73. Fox, I. H. & Kelly, W. N., Studies on the mechanism of fructose-induced hyperuricaemia in man. *Metabol. Clin. Exp.*, **23** (1972) 713–21.
74. Turner, J. L., Bierman, E. L., Brunzell, J. D. & Chait, A., Effect of dietary fructose on triglyceride transport and glucoregulatory hormones in hypertriglyceridemic men. *Am. J. Clin. Nutr.*, **32** (1979) 1043–50.
75. Hallfrisch, J., Ellwood, K., Michaelis, O. E., IV, Reiser, S. & Prather, E. S., Plasma fructose, uric acid, and inorganic phosphorus responses of hyperinsulinemic men fed fructose. *J. Am. Coll. Nutr.*, **5** (1986) 61–8.
76. Jacobs, D., Hyperuricemia as a risk factor in coronary heart disease. *Adv. Exp. Med. Biol.*, **76B** (1977) 231–7.
77. Caballero, B., Absorption and metabolism of sweetening agents. *Clin. Nutr.*, **3** (1984) 65–70.
78. Gabbay, K. H., The sorbitol pathway and the complications of diabetes. *N. Engl. J. Med.*, **288** (1973) 831–6.
79. Makinen, K. K., Long-term tolerance of healthy human subjects to high amounts of xylitol and fructose; general and biochemical findings. *Int. J. Vitam. Nutr. Res.*, **15** (1976) 92–104.
80. Govindji, A., Special products—a close-up on ingredients. *Balance*, **106** (1988) 46–9.
81. Secchi, A., Pontirolli, A. E., Cammelli, L., Bizzi, A., Cini, M. & Pozza, G., Effects of oral administration of maltitol on plasma glucose, plasma sorbitol, and serum insulin levels in man. *Klin. Wochenschr.*, **64** (1986) 265–9.
82. Anon., Synthetic sweeteners: cyclamates, saccharin, aspartame. *Med. Lett. Drugs Ther.*, **17** (1975) 61–2.
83. Macdonald, I., The therapeutic potential of artificial sweeteners. *Practitioner*, **204** (1970) 268–70.
84. Rodin, J., Wack, J., Ferrannini, E. & DeFronzo, R. A., Effect of insulin and glucose on feeding behavior. *Metabolism*, **34** (1985) 826–31.
85. Liang, Y., Steinbach, G., Maier, V. & Pfeiffer, E. F., The effect of artificial sweetener on insulin secretion. I. The effect of acesulfame K on insulin secretion in the rat (studies in vitro). *Horm. Metab. Res.*, **19** (1986) 233–8.
86. Liang, Y., Maier, V., Steinbach, G., Lalic, L. & Pfeiffer, E. F., The effect of artificial sweetener on insulin secretion. II. Stimulation of insulin release from isolated rat islets by acesulfame K (in vitro experiments). *Horm. Metab. Res.*, **19** (1987) 285–9.
87. Tordoff, M. G., Saccharin and food intake. In *Low Calorie Products*, ed. G. G. Birch & M. G. Lindley. Elsevier Applied Science, London, 1987, p. 127.
88. Blundell, J. E., Rogers, P. J. & Hill, A. J., Artificial sweeteners and appetite in man. In *Low Calorie Products*, ed. G. G. Birch & M. G. Lindley. Elsevier Applied Science, London, 1988, p. 147.
89. Moskowitz, H. R., The psychology of sweetness. In *Sugars in Nutrition*, ed. H. L. Sipple & K. W. McNutt. Academic Press, New York, 1974, p. 38.
90. Kalkhoff, R. K. & Levin, M. E., The saccharin controversy. *Diabetes Care*, **1** (1978) 211–22.
91. Armstrong, B., Lea, A. J., Adelstein, A. M., Donovan, G. C., White, G. C. &

Ruttle, S., Cancer mortality and saccharin consumption in diabetes. *Br. J. Prev. Soc. Med.*, **29** (1975) 73–81.

92. Market Facts Inc., Professional assessment of the physiological and psychological benefits of saccharin. In *A Report to the Calorie Control Council.* Medical Services Group, Chicago, 1978.

93. Horwitz, D. L. & Bauer-Nehrling, J. K., Can aspartame meet our expectations? *J. Am. Diet. Assoc.*, **83** (1983) 142–6.

94. Stegink, L. D., Filer, L. J. & Baker, G. L., Plasma amino acid concentrations in normal adults ingesting aspartame and monosodium-L-glutamate as part of a soup/beverage meal. *Metabolism*, **36** (1987) 1073–9.

95. Tephly, T. R. & McMartin, K. E., Methanol metabolism and toxicity. In *Aspartame: Physiology and Biochemistry*, ed. L. D. Stegink & L. J. Filer, Jr. Marcel Dekker, New York, 1984, p. 111.

96. Stegink, L. D., Aspartame metabolism in humans: acute dosing studies. In *Aspartame: Physiology and Biochemistry*, ed. L. D. Stegink & L. J. Filer, Jr. Marcel Dekker, New York, 1984, p. 509.

97. Stegink, L. D., Brummell, M. C., McMartin, K. E. *et al.*, Blood methanol concentrations in normal adult subjects administered abuse doses of aspartame. *J. Toxicol. Environ. Health*, **7** (1981) 218–90.

98. Stegink, L. D., Filer, L. J., Baker, G. L. & McDonnell, J. E., Effect of aspartame loading upon plasma and erythrocyte amino acid levels in phenylketonuric heterozygotes and normal subjects. *J. Nutr.*, **609** (1979) 708–17.

99. Wurtman, R. J., Aspartame: possible effect on seizure susceptibility. *Lancet*, **ii** (1985) 1060.

100. Yokogoshi, H., Roberts, C. H., Caballero, B. & Wurtman, R. J., Effects of aspartame and glucose administration on brain and plasma levels of large neutral amino acids and brain 5-hydroxyindoles. *Am. J. Clin. Nutr.*, **40** (1984) 1–7.

101. Fernstrom, J. D., Reply to letter by R. J. Wurtman. *Am. J. Clin. Nutr.*, **44** (1986) 801–3.

102. Fernstrom, J. D., Fernstrom, M. H. & Grubb, P. E., Effects of aspartame ingestion on the carbohydrate-induced rise in tryptophan hydroxylation rate in rat brain. *Am. J. Clin. Nutr.*, **44** (1986) 195–205.

103. Visek, W. J., Chronic ingestion of aspartame in humans. In *Aspartame: Physiology and Biochemistry*, ed. L. D. Stegink & L. J. Filer, Jr. Marcel Dekker, New York, 1984, p. 495.

104. Horwitz, D. L., Aspartame use by persons with diabetes. In *Aspartame: Physiology and Biochemistry*, ed. L. D. Stegink & L. J. Filer, Jr. Marcel Dekker, New York and Basel, 1984, p. 633.

105. Von Rymon Lipinski, G. W., Acesulfame K: properties, physiology and applications in calorie-reduced and low-calorie products. In *Low Calorie Products*, ed. G. G. Birch & M. G. Lindley. Elsevier Applied Science, London, 1988, p. 101.

106. Gonen, B. & Rubenstein, A. H., Haemoglobin A1 and diabetes mellitus. *Diabetologia*, **15** (1978) 1–8.

107. Arvidsson-Lenner, R., Studies of glycemia and glycosuria in diabetics after breakfast meals of different composition. *Am. J. Clin. Nutr.*, **29** (1976) 716–25.

108. Akerblom, H. K., Siltanen, I. & Kallio, A. K., Does dietary fructose affect the control of diabetes in children? *Acta Med. Scand.*, **542** (1972) Suppl., 197–202.
109. Arvidsson-Lenner, R., Specially designed sweeteners and food for diabetics— a real need. *Am. J. Clin. Nutr.*, **29** (1976) 726–33.
110. Steinke, J., Wood, F. C., Domenge, L., Marble, A. & Renold, A. E., Evaluation of sorbitol in the diet of diabetic children at camp. *Diabetes*, **10** (1961) 218–27.
111. Pelkonnen, R., Aro, A. & Nikkila, E. A., Metabolic effects of dietary fructose in insulin-dependent diabetes of adults. *Acta Med. Scand.*, **542** (1972) Suppl., 187–93.
112. Frazer, A. C., Health aspects of artificial sweeteners. *Rep. Soc. Health*, **3** (1969) 133–6.
113. Food and Drug Administration, Special dietary foods: label statements. *Fed. Register*, **42** (1977) 19203.
114. Finnsugar Xyrofin (UK) Ltd, Fructose: Natural Sweetness and Health (advertising brochure), 1988.
115. Hardy, S. L., Brennand, C. P. & Wyse, B. W., Taste thresholds of individuals with diabetes mellitus and of control subjects. *J. Am. Diet. Assoc.*, **79** (1981) 286–9.
116. Porte, D., Jr, Robertson, R. P., Halter, J. B., Kulkosky, P. J., Makous, W. L. & Woods, S. C. In *Food Intake and Chemical Senses*, ed. Y. Katsuki, M. Sato, S. F. Takagi & Y. Oomura. University of Tokyo Press, Tokyo, 1977, p. 331.
117. Lawson, W. B., Zeidler, A. & Rubenstein, A., Taste detection and preferences in diabetics and their relatives. *Psychosom. Med.*, **41** (1979) 219–27.
118. Schelling, J.-L., Tetreault, L., Lasagna, L. & Davis, M., Abnormal taste threshold in diabetes. *Lancet*, **i** (1965) 508–12.
119. Grinker, J., Obesity and sweet taste. *Am. J. Clin. Nutr.*, **31** (1978) 1078–87.
120. Vorster, H. H., van Tonder, E., Kotze, J. P. & Walker, A. R. P., Effects of graded sucrose additions on taste preference, acceptability, glucemic index, and insulin response to butter beans. *Am. J. Clin. Nutr.*, **45** (1987) 575–9.
121. Mehnert, H., Advantages and disadvantages of artificial sweeteners and sugar substitutes. In *Health and Sugar Substitutes*, Proceedings of ERGOB Conference. Karger, Basel, 1978, p. 262.
122. Court, J. M., Diet in the management of diabetes: why have special foods? *Med. J. Aust.*, **1** (1976) 841–3.
123. Farkas, C. S. & Forbes, C. E., Do non-caloric sweeteners aid patients with diabetes to adhere to their diets? *J. Am. Diet. Assoc.*, **46** (1965) 482–4.

*Chapter 9*

# ASPECTS OF THE ENERGY VALUE ASSESSMENT OF THE POLYOLS

Pierre Würsch & Gillian Anantharaman

*Nestlé Research Centre, Nestec Ltd, Lausanne, Switzerland*

## SUMMARY

*The evaluation of the digestion, absorption and metabolic utilization of polyols has been the subject of a large number of studies, in which various methodologies have been applied. Each method provides specific but often incomplete information. This review describes and comments on some of the direct and indirect methods of assessing the digestibility and the metabolizability of the polyols by animals and humans.*

## 1 INTRODUCTION

Many alternative bulk sweeteners have been proposed to replace acidogenic sugars as a means of reducing the incidence of dental caries. Among them the polyols take a predominant position. They have approximately the same heat of combustion or gross energy content (GE) as common dietary sugars but they are neither digested, absorbed nor metabolized to the same extent as sugars, and hence provide less energy to the consumer.

In practice, the energy that a food will provide (metabolizable energy, ME) is estimated by determination of its protein, fat, carbohydrate and ethanol content, to which appropriate metabolizability factors are then assigned (4, 9, 4 and 7 kcal/g respectively). The sum of the individual values for the four macronutrients gives the approximate ME of the food. Information about the digestion and metabolism of sugar alcohols can be obtained in various ways involving animals or humans. The objective of this review is not to define the true energy value of each sugar alcohol, but to present the numerous types of methodology applied to polyols and to discuss the results thus obtained. Each method will provide valuable but partial information on the digestibility, absorption and metabolism of the polyol or of its subunits, and assembling this information should allow a reasonably accurate assessment of the energy value of the polyols. This should be in agreement with results obtained from energy balance studies conducted in animals or humans.

## 2 FATE OF POLYOLS ALONG THE INTESTINAL TRACT

The energy value of the polyols or of any carbohydrate depends on

—the amount of carbohydrate that is actually absorbed, after hydrolysis if it is not a monosaccharide;
—the extent to which the body metabolizes the absorbed carbohydrate;
—the amount of unabsorbed carbohydrate that reaches the large intestine and is subsequently fermented by the colonic bacteria, generating metabolites which in turn could be absorbed and to a certain extent salvaged by the host.

These different aspects are presented schematically in Fig. 1.

In a recent review, Ziesenitz and Siebert[1] described in detail the digestive and absorptive process of the polyols, with special emphasis on Palatinit; the data are summarized in Section 4. The fermentation, in the large intestine, of food that escapes digestion and absorption in the small intestine is a fairly common process in animals and man. Colonic bacteria use this food as a substrate for growth and maintenance. In addition to cell mass, they produce short-chain fatty acids (SCFA), occasionally lactic acid, carbon dioxide and hydrogen, sometimes methane and heat. Establishing the quantity of energy lost by the bacteria in this process is important in the case of the polyols since energy in the form of the metabolites can be absorbed by the colonic epithelial cells and utilized by the host. However, the quantification of this energy gain for the host is very difficult to estimate owing to the complexity of the environment.

### 2.1 Large Intestine Fermentation

The wet weight of bacterial residues in the human colon has been estimated to be between $175 \, g$[2] and $1.5 \, kg$,[3] containing $10^{11}–10^{12}$ bacteria/g, practically all of which are anaerobes.[4] Most of these bacteria derive their energy primarily from carbohydrates and their derivatives.[5] These are

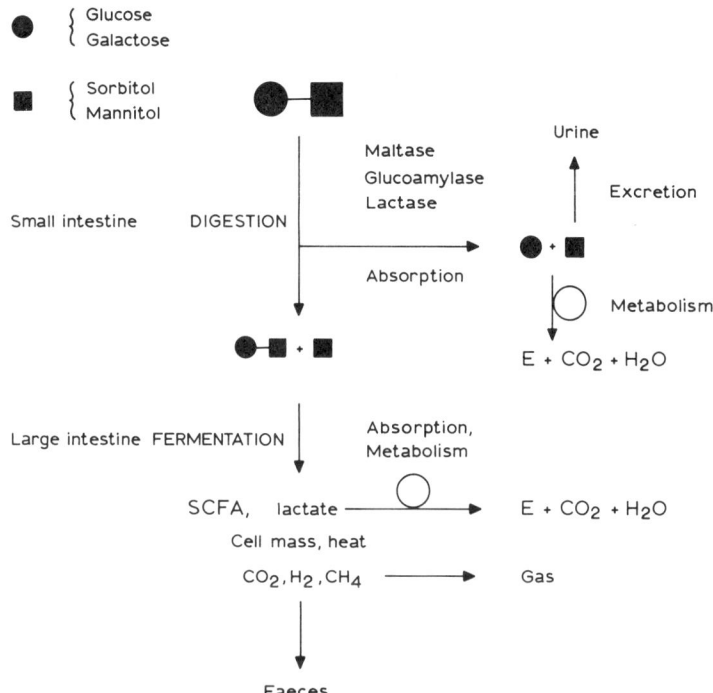

FIG. 1.   The metabolic fate of disaccharide alcohols.

mainly the polysaccharide fraction of plant cell walls referred to collectively as dietary fibre, together with unabsorbed sugars and starch.

The amounts of starch and sugars passing the terminal ileum have been estimated using various direct and indirect methods on ileostomized patients, by perfusion experiments and by measurement of hydrogen excretion. Unabsorbed, ingested starch was found to range from 0·5% for rice to 20% for beans,[6] corresponding to 5–10 g/day. This fermentable carbohydrate, together with the dietary fibre (daily intake of 15–25 g), stimulates microbial growth and increases microbial cell excretion in faeces. McNeil[7] calculated that the daily faeces of people on a Western diet contain on average 15–20 g bacteria (dry weight basis) and that 1·5–2 mol ATP are needed to replace the energy lost. This amount of ATP can be produced by the anaerobic fermentation of 50–65 g hexose, which roughly corresponds to twice the daily intake of dietary fibre plus unabsorbed carbohydrate entering the colon. Besides the ATP, the fermentation of these carbohydrates yields short-chain fatty acids (SCFA), which represent about two-thirds of the energy of the substrate. The molar distribution and yield of SCFA, however, depend on the type and concentration of fermentable product, on the nature of the diet and on the environment, e.g. the type of microflora responsible for the fermentation.

Miller and Wolin[5] have proposed a quantification of the human colonic fermentation which yields SCFA, methane, carbon dioxide, water and ATP. They calculated that SCFA account for approximately 72% of the energy content of the carbohydrate, excluding the substrate used by the micro-organisms for their growth.

Isaacson et al.,[8] investigating fermentation in vitro of glucose by mixed populations of anaerobic bacteria from the rumen of cows, found that the yield of microbial material was 23–46% of the carbohydrate fermented. Salyers et al.,[9] using single strains of human colonic bacteroides, obtained a yield of 30–40% with either glucose or arabinogalactan as substrate. These values are in agreement with those reported by McNeil.[7] Fermentation in vitro of [U-$^{14}$C]sorbose by homogenates of human faeces was found to yield 64·5% SCFA.[10] Thus, increased provision of foods which are to a large extent metabolized by bacterial fermentation stimulates bacterial growth and hence increases the apparent weight of the host. In so doing, even to only a minor extent, the host maintenance requirement will rise primarily as a result of an increase in the cost of movement.

Short-chain fatty acids (SCFA), also named volatile fatty acids (VFA), have a chain length ranging from one to six carbon atoms. They occur ubiquitously as free fatty acids in the gastrointestinal tract of mammals, where they are the end-products of the microbial fermentation of carbohydrates.[11,12] The SCFA provide up to 70–80% of the energy requirements of ruminants.[13] Imoto and Namioka,[14] working on pigs, showed that 0·7–0·8 mol of SCFA is absorbed daily, providing 10% of their maintenance energy requirements. Production and absorption of SCFA also take place in a similar manner in the large intestine of other monogastrics, including man. The rates of absorption of acetate, propionate and butyrate in humans were found to be in the range 6–12 $\mu$mol/cm$^2$/h;[15] these values are similar to those in the large intestine of the pig[16] and in rumen epithelium.[17] Ruppin et al.,[18] using a perfusion method, confirmed that propionate is absorbed by the human large intestine. The same authors estimated that the human colon has the capacity to absorb more than 500 kcal/day in the form of SCFA.

The rate of propionate absorption was related linearly to initial concentration, and no difference has been found between rates of absorption of acetate, propionate and butyrate.[18,19] SCFA are rapidly metabolized in the digestive epithelium, and their complete oxidation to $CO_2$ is often a predominant catabolic pathway. As a consequence, the rate of transport of SCFA from the epithelial cells to the blood may be considerably lower than the absorption from the intestinal lumen, especially when the capacity of epithelial cells for metabolizing SCFA is not saturated,[20] and the oxidative metabolism is therefore an important energy source for the epithelium itself. Bergman and Wolff[21] showed that approximately 30%, 50% and 90% of acetate, propionate and butyrate respectively do not reach the portal blood of ruminants, but no precise data exist for man. However, these SCFA, in particular acetate produced by fibre or by lactulose fermentation, have been detected in the peripheral blood.[22] Acetate is then metabolized in the peripheral tissues, whereas propionate and butyrate are rapidly metabolized by the liver. The molar ratio acetate/butyrate in portal blood was found to be around 10 but a ratio of 3–4 is generally found in whole stools or faecal dialysates.[12,23]

The extensive metabolism of SCFA in the epithelium has been confirmed by numerous studies *in vitro*.[24] Rat everted caecal sacs[25] and guinea-pig isolated colonocytes[26] have been shown to metabolize SCFA in the order butyrate > propionate > acetate. The catabolic rate of labelled acetate, which is the most abundant of the three SCFA, has been compared *in vitro* with its utilization rate for lipid biosynthesis. Although acetate is the major precursor for lipid biosynthesis, it was found that the oxidation rate of acetate to $CO_2$ was 5–50 times higher than the rate of fatty acid synthesis in rat small intestine. Moreover, the rate of acetate oxidation to methane gradually increases down the length of the rat small intestine.[27-29]

The metabolism of higher SCFA is less well documented, but it is known that propionate and the higher chain lengths are potentially glycogenic in ruminants and other mammals. Lactate is likely to be derived from this glycolysis, a process that is known to occur in the digestive epithelium.[30-31] Nevertheless, it is assumed that $CO_2$ is the major metabolite of propionate and butyrate in the digestive tract epithelium.[31,32] It has been pointed out, however, that butyrate is also readily metabolized into ketone bodies by the mucosa of the rumen, the caecum and the colon, but not that of the small intestine.[33] This conversion into ketone bodies is important in maintaining a constant supply of energy to the mucosal cell and to other tissues. It appears that, even in the presence of glucose, butyrate is an important energy source for the colonocytes of rats and humans.[34] This acid seems to have a glucose-sparing effect. Numerous other metabolites, such as free amino acids, acids of the tricarboxylic acid cycle and sugars have been found in the large intestinal wall, especially in rabbits.[35] However, there are few data on the metabolism and nutritional contribution of SCFA in the epithelium of the large intestine of most mammals, in particular of humans.

The colonic epithelial cells, like all other mammalian cells, must grow, differentiate and be renewed if they are to remain healthy and serve their intended functions. Lupton *et al.*[36] showed that supplementing a basal fibre-free diet of rats with either 10% pectin or guar significantly increased the mucosal mass, cell number and proliferation, with the greatest differences occurring in the caecum, compared with supplementation by 20% oat bran or cellulose. Both pectin and guar gum were completely fermented, unlike the bran. Stimulation of colonic cell division was also observed with lactulose and sorbitol.[37] Sakata[38] showed that daily infusion of SCFA in the caecum stimulates epithelial cell production and leads to increased mucosal growth. The effect of SCFA was dose-dependent, and varied among acids (butyrate > propionate > acetate). The SCFA, and particularly butyrate, seem therefore to be salvaged to a great extent by the epithelial cells, and used for maintenance and cell proliferation.

Thus the SCFA spare circulating substrates such as glucose, glutamine and ketone bodies. It is still unknown, however, whether such a sparing effect is significant to the host when the amount of fermented carbohydrate is low, as is the case in a standard Western diet, or whether a more rapid cell turnover would compensate for this apparent sparing effect.

## 3 ASSESSING THE ENERGY VALUE OF POLYOLS: METHODOLOGY AND LIMITATIONS

In order that the energy present within a food be harnessed by its consumer, digestion, absorption and metabolism must take place. Each step in these processes is itself energetically demanding. Hence the nutritive value of the food will be dependent on (1) its energy content, (2) the degree to which it can be digested by intestinal enzymes, (3) the ability of the host to metabolize the products of digestion, and (4) the intricacy of the pathways involved in the derivation of a utilizable form of energy. Each of these points will be briefly considered in relation to the types of methodology and limitations currently encountered in ascribing a nutritive value to the polyols.

### 3.1 The Energy Content of Polyols

The energy content of a food (gross energy, GE) is determined by bomb calorimetry in which the food undergoes complete combustion in an oxygen-rich atmosphere, and the heat liberated is measured and expressed in calories or joules. The technique has no ambiguity in the analysis of nitrogen-free food constituents such as the polyols since it yields the end products $CO_2$ and water.[39] GE values for the polyols are similar to that of sucrose at 16·5 kJ/g.

## 3.2 Assessment of Digestibility and Absorption of Polyols

Blaxter[40] stated that 'faecal loss of energy is by far the most important single factor determining the relative nutritive value of different foods as sources of energy'. The energy lost in faeces does not arise directly from unabsorbed food but from enzymic secretions, sloughed cells from the gastrointestinal tract and both dead and living bacteria. Thus, in the case of the polyols, faecal energy will increase not necessarily as the result of the presence of malabsorbed food in the faeces but as the result of increased faecal bacterial mass. Considerable research by various techniques has been conducted in rats, mice, pigs and man to evaluate the relative importance of (a) digestion and absorption in the small intestine, and (b) utilizable products of bacterial fermentation to the digestible value of the polyols.

### 3.2.1 Germ-free Animals

To avoid interference by the intestinal flora, germ-free rats and mice have been used to assess the degree of malabsorption of substrates. Maltitol[41] and Palatinit[1] were fed to germ-free rats. Between 19% and 26% of the administered Palatinit was recovered in the faeces, essentially as hexitol. Of a 10 g/kg load of maltitol, only 10·5% was recovered in the faeces, as sorbitol only. As explained by Ziesenitz and Siebert,[1] the amount of polyol found in the faeces tends to underestimate the digestion and absorption in the small intestine, mostly because the intestinal contents are in contact with the mucosa of the small and the large intestine, whereas in conventional animals the product is utilized by the bacteria in the caecum in the large intestine. This prolonged period of contact with the mucosa allows the digestion and absorption of the hexitol to continue. To prevent this phenomenon taking place, Würsch et al.[42] measured the intestinal contents of germ-free mice 3 h after the administration of either 140 mg/kg maltitol or 70 mg/kg sorbitol. In both cases most of the unabsorbed product reached the caecum and only 5% of the administered maltitol was recovered intact (Fig. 2). The recovery of sorbitol and maltitol, together with its sorbitol moiety, reached means of 51% and 39% respectively. This latter value is much higher than that found in the faeces of germ-free mice

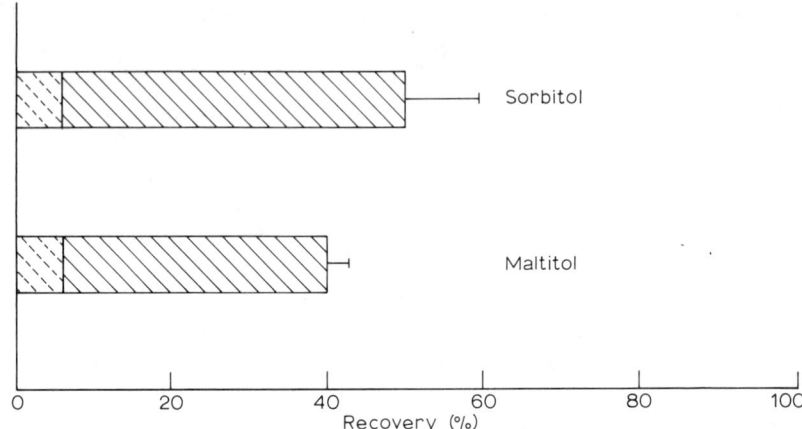

FIG. 2.   Recovery of sorbitol and maltitol in the small intestine (shaded) and the caecum (hatched) 3 h after oral administration of 140 mg/kg maltitol or 70 mg/kg sorbitol to germ-free mice (adapted from Ref. 42).

collected over 48 h (28·4%). These results on germ-free animals suggest that they are able to digest maltitol extensively as well as Palatinit in the small intestine. More interestingly, in comparing the absorption of sorbitol with that of maltitol, it seems that the limited digestibility of the disaccharide alcohols is to a large extent a consequence of the slow absorption rate of sorbitol and mannitol, and therefore the digestion rate seems to have only a limited effect on carbohydrate malabsorption.

### 3.2.2   Ileal Recovery
The quantitative direct measurement of malabsorbed carbohydrates in humans and large animals has been attempted by various means, e.g. intubation of healthy volunteers and analysis of ileal aspirates. However, this invasive technique delays gastric emptying and shortens small intestinal transit time.[43]

Beaugerie[44] quantified the unabsorbed sugars remaining after oral intake of sorbitol, maltitol and Lycasin by healthy volunteers, using an aspiration technique. The results presented in Fig. 3 show that the recoveries were quite low: 20·6% (range 9–33%) for sorbitol; 24·3% (range 15–38%) for maltitol; and 15·5% (range 10–25%) for Lycasin. Maltitol was digested to a very large extent, and only one-third of the freed sorbitol remained unabsorbed. The levels of recovery were, however, low compared with what was estimated with the hydrogen excretion method described below. Using the jejunal perfusion technique, Dharmaraj et al.[45] showed

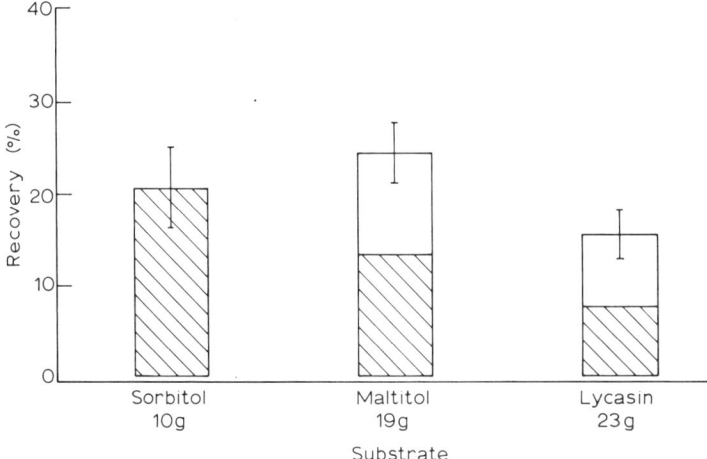

FIG. 3.    Terminal ileum recovery of sorbitol (hatched) and maltitol (block) after ingestion of 10 g sorbitol, 10 g maltitol or 23 g Lycasin 8055 by 5 healthy volunteers (% based on ingested polyol) (adapted from Ref. 44).

that lactitol hydrolysis and absorption in the small intestine were insignificant.

Elegant, direct measurement of malabsorbed starch in ileostomy effluents was introduced by Sandberg and co-workers in 1981,[46] and since then has often been used to quantify unabsorbed starch from various foods. This convenient method has, however, been rarely used for the measurement of the malabsorbed polyols. Around 55% of Palatinit was recovered in terminal ileum, partly intact and partly digested to sorbitol and mannitol.[47] In fistulated pigs the recovery of Palatinit in a mixed diet was 19–34%.[48] By measuring the quantity of glucose and sorbitol reaching the portal vein of pigs after oral administration of maltitol, Rérat et al.[49] concluded that maltitol was completely digested, and that all glucose and a small fraction of sorbitol were absorbed.

### 3.2.3 Hydrogen Breath Test

The measurement of hydrogen ($H_2$) in breath has been widely used in the investigation of carbohydrate malabsorption, in particular of lactose and more recently of starch in foods.[50,51] This method is based on the fact that carbohydrates not absorbed by the small intestine are fermented by bacteria in the colon with the evolution of hydrogen gas. The hydrogen is in turn rapidly absorbed and excreted through the lungs. This malabsorption of carbohydrate can even be quantified by comparing the areas under the breath hydrogen excretion curves after the test carbohydrate versus a non-absorbable carbohydrate. Lactulose is most frequently utilized as the reference compound.

However, large inter- and intra-individual variations are observed, so any quantification requires a large group of subjects. Furthermore, adaptation of the microflora to the non-digestible carbohydrate can result in a great change in hydrogen breath excretion,[52] while in some cases the colonic flora may lack bacteria capable of fermenting carbohydrate to produce hydrogen gas.[53]

Several methods of measuring hydrogen have been proposed. Bond and Levitt[54] measured the breath excretion rate of hydrogen in a rebreathing closed system and estimated the amount excreted during the 2 h after increased hydrogen production. Another method consists of collecting at intervals end-expiratory air and analysing (a) hydrogen using an electrochemical cell or (b) hydrogen and methane by gas chromatography.[50,54-56] Several authors have shown that the amounts of starch or glucose metabolized in the colon, as calculated from the excretion of hydrogen in breath, were close to the actual load, i.e. the carbohydrates produced equivalent volumes of hydrogen.[50] Furthermore, the volume of hydrogen excreted was found to be proportional to the carbohydrate fermented.[50,54] This linearity of response was also observed by Fritz et al.,[57] who measured the $H_2$ excreted after consumption of 10–40 g of Palatinit. The ability to produce $H_2$ can also be assessed by measuring the $H_2$ liberated during incubation of a faecal homogenate with test carbohydrates.[50,58] This technique has shown that lactulose, glucose, sucrose, lactose and maltose produce the same volume of hydrogen;[50,54,58] however, it seems that the fermentation of sorbitol, which comprises one sugar moiety of most disaccharide alcohols, produces more hydrogen than the dietary sugars and lactulose.[59,60] Würsch et al.[60,61] determined the area under the excreted hydrogen curve for 6 volunteers after consumption of 7 polyols or related materials. Results in Fig. 4 show that the mean areas for sorbitol, mannitol, maltitol and Palatinit were not significantly different from each other, suggesting that the totals of unabsorbed sugars were similar in each case. The digestibility of lactitol is known to be very low, and therefore most if not all of the disaccharide will ultimately be fermented. Consequently the hydrogen excretion presented in Fig. 4 would be caused by the fermentation of 20 g of lactitol. The mean areas for maltitol and Palatinit were 67% and 60% respectively of that for lactitol, suggesting that there is only a 50% absorption of these two disaccharides in the small intestine.

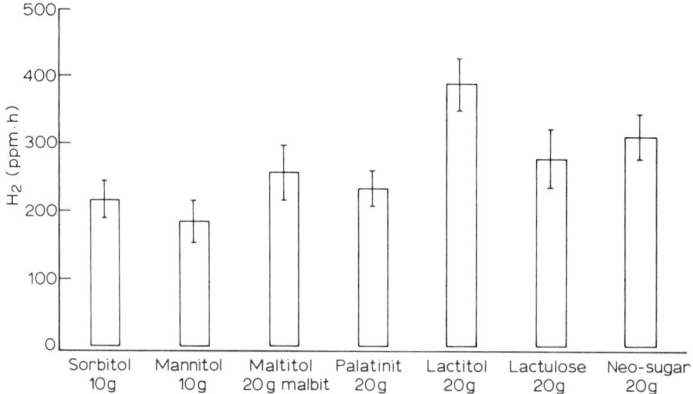

FIG. 4. Cumulative breath hydrogen excretion produced after consumption of various sugars by 6 healthy volunteers, for 5 h, following its initial increase ($\pm$ SEM).[61]

### 3.2.4 Indirect Calorimetry

The conversion of food into utilizable energy results in the production and loss of heat from the body. This is a process of oxidation, and therefore results in oxygen consumption and carbon dioxide production. The change in concentration of these two gases in inhaled and expired air is proportional to the heat produced by the body. Thus these gases may be measured by indirect calorimetry, from which heat production can be derived. In addition, the method yields information on overall carbohydrate and lipid utilization since the respiratory quotient (RQ) or ratio of oxygen consumed to carbon dioxide expired is dependent on the substrate used. For glucose the RQ is 1 and for fat 0·7. During fasting the RQ will approach 0·7 since body fat stores will be mobilized.[62] The administration of a carbohydrate load to a fasting subject will result in an increase in RQ which theoretically should be dependent on the rate and extent of absorption of load.[63] However, it must be noted that the RQ of polyols calculated using stoichiometric equations is not equal to that of glucose (xylitol 0·91, sorbitol 0·92, maltitol 0·96).

When xylitol was administered orally the total increase in carbohydrate oxidation during the subsequent 2·5 h amounted to some 25% of that induced by glucose.[64] On the other hand, the carbohydrate oxidation rate after intravenous infusion of sorbitol rose to 27 mg/min as compared with 39 mg/min for xylitol and 101 mg/min for glucose.[65] This low oxidation rate of the polyols suggests that part of the polyol was used for the synthesis of liver glycogen, as shown by Froesch et al.,[66] with secondary release of glucose as a fuel in normoglycaemia. However, whether the polyol is oxidized as it is or whether it is converted first into glucose and then oxidized, the oxygen consumed and the respiratory quotient remain the same.

Dietary sugars like glucose, fructose and sucrose have been reported to induce different thermic effects, i.e. the total increment in heat production measured for 3–4 h after ingestion was highest with sucrose, lowest for glucose and intermediate for fructose.[67,68] These differences have been attributed to the different metabolic pathways and rate of metabolism for glucose and fructose. By measuring the dietary-induced thermogenesis over a much longer period, the difference found would be smaller. However, if the duration of measurement increases, the metabolic state of the subject invariably changes and any low increase in metabolic rate induced by a slowly metabolizable sugar would no longer be quantifiable. Consequently the indirect calorimetry method does not allow measurement specifically of the amount of carbohydrate absorbed in the small intestine.

It has been repeatedly shown that ingestion of the polyols, maltitol and Palatinit produces small increases in blood glucose and insulin response, suggesting a slow absorption rate of the glucose moiety. The increase in metabolic rate after maltitol and L-sorbose (a ketohexose which has been shown to be only partially absorbed[10]) was found to be closely similar for both sugars, and approximately 50% of that obtained following glucose ingestion.[69] In the same way, the proportions of carbohydrate oxidized during the 6 h following ingestion of maltitol[70] or Palatinit[71] were found to be approximately 40% and 43% respectively of that following a sucrose load. Consequently the decrease in lipid oxidation was also significantly less after ingestion of the polyols.

### 3.3 Metabolizability of Polyols

On absorption, the monosaccharide alcohols have in general a similar metabolizability to that of glucose[72] and are completely available for use by the body. However, considerable quantities of mannitol appear in the urine if large doses are administered. For the disaccharide and oligosaccharide polyols, it has been shown in animals and man that small quantities can be absorbed intact without cleavage in the intestinal tract, and therefore remain unmetabolizable and are excreted in the urine.[73] The products of bacterial polyol metabolism (lactic acid and volatile fatty acids) are readily metabolized but are said to yield 5–20% less energy than glucose or fructose.[74]

An alternative approach to the problem of evaluating the metabolizable energy value of polyols is the global method of examining their effect on energy balance. Generally, such studies are conducted in laboratory animals for reasons that will later become apparent.

### 3.3.1 Growth Studies
Rapidly growing small animals such as rats have to be employed. A basal diet is prepared, into which a reference standard carbohydrate source or the test material is incorporated to yield equal levels of gross energy. The diet is fed to weanling animals during their phase of linear growth, and the weight gain per unit of energy consumed is evaluated.[75,76] For test materials which vary greatly in metabolizability from the standard, this will provide a reasonable general assessment of the energy value of the material. However, if the difference is small and the level of inclusion of the material in the diet is low, then the error associated with the variation in growth among the animals is likely to mask an effect or give rise to an erroneous conclusion. In the case of the polyols the problem may be complicated by an increase in bacterial mass within the colon, minimizing body weight differences. Further, since it is known that rats will increase their intake to keep up maximal growth when the metabolizable energy density of the diet is reduced,[77] activities associated with feeding could increase, resulting in greater energy expenditure and reduced efficiency. In this instance pair-feeding might be considered to ensure identical gross energy intakes, but this in itself may bring about an increase in food-seeking behaviour, and thus raise expenditure.[78] The technique may show maximum advantage and greater reliability if large numbers of animals are used and energy retention is measured using a comparative slaughter technique.[79] In this instance empty carcass weights, i.e. after removal of all intestinal residues, could then be compared. A further improvement would be to compare the protein content of the gut (intestine) and remaining carcass, since the polyols could possibly cause a hypertrophy of the colon and caecum which would not be representative of the animal's growth. Even with these precautions the extrapolation of such studies to adult man must be treated with caution in view of such major differences as the relative sizes of caecum in rodents and in man.

### 3.3.2 Maintenance Studies
As in the growth studies, a basal diet into which the standard or test carbohydrate is incorporated is fed to adult animals at a level allowing weight maintenance. Each animal should be fed on both diets, and for each regimen a sufficient period of adaptation must be permitted. Therefore

TABLE 1
METABOLIZABLE ENERGY OF DISACCHARIDE ALCOHOLS ASSESSED IN
GROWTH OR MAINTENANCE STUDIES ON RATS, PIGS AND GUINEA-PIGS

| | |
|---|---|
| Palatinit | 1·5–3·0 kcal (7 studies)[73,79–81,84,85] |
| Maltitol | 2·8 kcal (1 study)[86] |
| Lactitol | 2·0–2·8 kcal (4 studies)[87,88] |

these studies are long-term and, although feasible, are impracticable to conduct in man. However, they may be applied to large animals such as pigs.[80,81]

There are numerous points to be taken into account in interpreting such tests. First, the environment must be maintained stable throughout, since variation in temperature is an important determinant of food intake and energy balance.[40] Second, bacterial fermentation in the colon results in heat production. This could be beneficial to an animal housed in a cold environment since its needs for thermoregulation would be reduced. Thus the energy value of a diet may be dependent on the temperature at which it is fed, so all experiments should be conducted at thermoneutrality.[74,82] Third, where possible the use of activity meters or monitors should verify that this component of expenditure remains constant.[83] Finally, the maintenance of weight does not necessarily indicate the maintenance of body energy. For example, it is feasible that an increase in bacterial mass could take place simultaneously with a reduction in body fat. In terms of weight the change might be minimal but in terms of energy it might be significant.

The application of these techniques to disaccharide alcohols is illustrated by the summary of results from several authors shown in Table 1. These assessments of the metabolizable energy were performed with high dietary levels of polyols ranging from 10 to 20 g/kg body weight. For purposes of comparison, the daily intake by humans, based on an average intake of 15–30 g/daily, would be below 1 g/kg. Extrapolation of data from a high to low dose range should therefore be done with great caution.

### 3.3.3 Complete Energy Balance in Man

This approach has been used in a recent study[83] in which subjects ate a basal diet containing either sucrose or lactitol in a cross-over design. For the test period they passed 24 h in a calorimetric chamber and collected all excreta. This is probably the most thorough human experiment ever, yet even with highly motivated subjects the uncontrolled variations became evident. Minor changes in energy balance were found between the two regimens, but activity, assessed by an actometer, was reduced on lactitol. Variation between subjects was high and, as the authors state, the results must be considered with some caution. This illustrates the problems involved in determining the effect of a very small change in the metabolizability of a diet due to the inclusion or omission of one element— rather like looking for a needle in a haystack.

Palatinit was also submitted to a diet study with two groups of volunteers who received a 1500-kcal diet in which 24 g of sucrose was given to one group and 48 g of Palatinit to the other. After 3 weeks, the mean loss of weight was 1·4 kg in both groups, implying that Palatinit provided 50% of the energy of sucrose.[1,108]

### 3.3.4 Metabolism of Carbon-14 Radiolabelled Polyols

This method of assessing the metabolic fate of substances has been applied to sorbitol,[89] mannitol,[90] maltitol[91] and lactitol.[92] It provides valuable information on the rate of metabolism in the body by following (a) the radiorespirometric (radioactivity in expired gas) pattern of $^{14}CO_2$ excreted in the breath after consumption of the labelled substrate, (b) the level of substrate, (c) the degradation products excreted in the faeces and urine, and (d) the level of incorporation in the organs. This method does not provide quantitative data on the fraction metabolized by the large intestinal flora, however, because the $^{14}CO_2$ recovered from expired breath arises from the metabolism of the polyol by the host and by the intestinal flora, and from the metabolism of the short-chain fatty acids by the host.

Rennhardt and Bianchine[91] administered 11 g maltitol enriched in $^{14}C$-maltitol to 4 human subjects. The faecal and urinary recoveries of radioactivity in 12 h were less than 6% and 4% respectively. The radiorespirometric pattern presented a broad peak at 3–7 h, suggesting that a substantial amount of labelled material was absorbed from the lower intestinal tract after fermentation of the sugar alcohol. A radiotracer study was performed recently on conventional rats and mice and on germ-free mice.[42] The radiorespirometric patterns for the conventional animals were similar in profile and were characterized by a rapid (0–2 h) increase followed by a plateau until the fifth hour (Fig. 5). The germ-free mice produced a rapid excretion of $^{14}CO_2$ which peaked at the second hour. The faecal recovery of radioactivity after 48 h reached around 15% in conventional animals against 28% in germ-free mice. In another study on germ-free rats, the recovery was 27·4%, as compared with 35·7% after administration of sorbitol.[93] However, as mentioned in Section 3.2, the faecal recovery in germ-free animals does not represent the polyols which

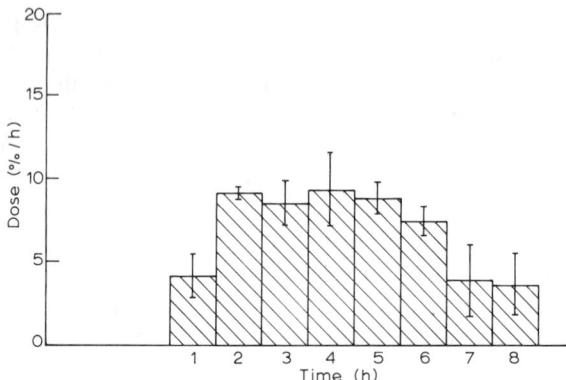

FIG. 5.    Recovery of excreted $^{14}CO_2$ following a single dose of 140 mg/kg U-$^{14}$C-maltitol to rats ($n = 3$; $\pm$ SEM) (adapted from Ref. 42).

were not absorbed at the small intestinal level, because sorbitol is very likely also absorbed, although much more slowly, in the caecum and in the colon.

The case of lactitol is interesting since this disaccharide alcohol is barely hydrolysed in the small intestine and consequently most, if not all, reaches the lower intestinal tract. In a metabolic study of $^{14}$C-lactitol in rats, this absence of small intestinal absorption resulted in the start of $^{14}CO_2$ exhalation only 2–4 h after administration of the polyol.[94] In humans[92] the cumulative $^{14}CO_2$ reached 3% within 3 h and peaked at 5–8 h, indicating a negligible intestinal absorption of sorbitol and galactose, as the production of $CO_2$ from their hepatic metabolism would have peaked at 4 h.[89] The $^{14}CO_2$ excreted accounted for 63% of isotope given, whilst 6·5% and 2% of the label was recovered in the faeces and urine respectively. The remaining 28·5% was assumed to be incorporated in the host tissues. Using a series of assumptions, the authors calculated that the energy yield of absorbed SCFA was 54·5% of the energy content of lactitol, which is in agreement with the values obtained by energy balance studies.[83]

## 4 PROPERTIES OF THE POLYOLS

### 4.1 Sorbitol

Using sucrose as a standard, Staub[95] reported the energy availability of sorbitol to be 80% in the rat growth assay technique of Rice et al.[76] Sorbitol absorption has been shown to be passive and follows zero-order kinetics; for example, from 1 $\mu$M to 200 mM concentration in the lumen, absorption of a fairly constant low percentage of the initial content was observed per time unit.[96] In duodeno-jejunal loops the rate of sorbitol absorption was about 6 times lower than that of the glucose analogue 3-*O*-methylglucose, which is known to be absorbed at a rate somewhat lower than that of glucose.[97,98]

Ertel *et al.*[99] administered $^{14}$C-sorbitol to rats and measured the residual $^{14}$C activity in the gastrointestinal contents at different times. After 3 h, 55% of sorbitol was still present. Würsch *et al.*[42] repeated this experiment on germ-free mice and found that 51% (range 16–67%) of the administered sorbitol was recovered, mostly in the caecum (Fig. 2). These results conflict with those of Grossklaus,[100] who used conventional rats and recovered only around 10% of the polyol. Probably part of the polyol had been metabolized by the intestinal flora.

Only a few studies have shown the slow and partial absorption of sorbitol in humans.[101] Beaugerie[44] recently measured the non-absorbed sorbitol in human volunteers by ileal perfusion; 10 g of polyol was ingested after a meal and 21% was recovered (Fig. 3). However, great individual variation was observed, with the recovery ranging from 7% to 33%. Malabsorption of sorbitol in man has been shown indirectly by measuring the pulmonary hydrogen resulting from the colonic fermentation of the polyol. Sorbitol ingested at 10 or 33 g produced almost as much hydrogen as the same dose of lactulose, which is a non-digestible disaccharide.[59,102,103] Knowing that sorbitol is partially absorbed, one can conclude that sorbitol produces during fermentation more hydrogen than the same amount of lactulose, dietary disaccharides and glucose, since it has been found that the volume of hydrogen liberated from faecal fermentation is similar with lactulose, glucose, sucrose, lactose and maltose.[50]

## 4.2 Mannitol

Several studies have shown that mannitol is poorly absorbed from the small intestine and is practically unmetabolized by humans. Nasrallah and Iber[90] administered $^{14}$C-mannitol orally and recovered 17·5% of it in the urine and 31·5% of the radioactivity in the faeces. The cumulative radioactivity in the expired $CO_2$ during the first 6 h was less than 51%, and the peak was at 10 h, suggesting that the absorbed polyol was not oxidized and part of it was fermented in the colon. Alternatively, close to 100% of intravenously administered polyol was recovered intact in the urine.[90,104–106] Saunders and Wiggins[107] reported recently that 74% of a 10-g oral load of mannitol passed unabsorbed through the small intestine of patients with ileostomies, indicating that 26% was absorbed, a value close to the 20% of urinary excretion reported by Elia *et al.*[105]

## 4.3 Palatinit

Palatinit is an equimolecular mixture of 6-$O$-D-glucopyranosyl-D-sorbitol and 1-$O$-D-glucopyranosyl-D-mannitol obtained by catalytic hydrogenation of isomaltulose. Both of these disaccharide alcohols are digested at a very slow rate by the small intestinal enzymes of most mammals.[1]

The disaccharide alcohol has been submitted to numerous studies in animals and humans in order to assess its digestibility, tolerance and energy value.[1] Several of these studies were undertaken on humans and indicated a reduced physiological digestibility and energy value of the Palatinit. Recovery from ileostomized patients[47] and comparative breath hydrogen measurements[61] indicated that half of the substrate was absorbed after hydrolysis. A comparative 1500-kcal diet study of 3 weeks, with alternatively Palatinit or sucrose, by healthy volunteers showed that the product was only partially utilized, yielding 50% of the energy of sucrose.[108]

## 4.4 Maltitol

Maltitol, 4-$O$-$\alpha$-D-glucopyranosyl-D-sorbitol, is commonly prepared by hydrogenation of maltose, which in turn is obtained by controlled hydrolysis of starch. In both normal and diabetic humans, maltitol causes a lower rise in blood sugar and insulin levels than a corresponding oral dose of glucose or sucrose.[70,109]

This reduced response has been attributed to the slow digestion rate of maltitol by the small intestine enzymes, which was repeatedly confirmed by kinetic measurements of the maltitol hydrolytic activity of intestinal sucrase-isomaltase and maltase-glucoamylase in crude extracts or when purified from rat and human sources.[1,110,111]

Furthermore, the digestion of maltitol has been found to be incomplete, as shown in Figs 2 and 3. Beaugerie[44] recovered at the terminal ileum 3·7–21·5% of intact maltitol besides 9–16% of free sorbitol, after administration of 19 g maltitol. This extensive digestion was also observed in germ-free rats and mice[23,41,42] and in pigs,[49] although in these species the total intestinal maltase activity is fairly high.[1,42] Even so, 39% and 42% of the administered sugar alcohol reached the large intestine in germ-free mice and pigs respectively.

The great variability in the absorption of glucose and sorbitol was also observed by Rennhardt and Bianchine,[91] who measured the radiorespirometric profile of $^{14}CO_2$ after administration of $^{14}C$-maltitol. The comparable breath hydrogen excretion measured after consumption of sorbitol, mannitol, maltitol and Palatinit shows convincingly that maltitol and Palatinit are both absorbed after digestion to a similar extent in man (Fig. 4).

The limiting steps in the digestion of maltitol in the human small intestine are the hydrolysis of maltitol and the absorption of sorbitol.

### 4.5 Lactitol

Lactitol is poorly digested in the human small intestine. The rate of digestion by small intestinal enzymes in man is reported to be 1–3% of that of lactose, which is itself around 25% that of sucrose.[112] As shown by the jejunal perfusion technique on human subjects, most of the product reaches the large intestine intact.[45] Expired $^{14}CO_2$ by rats receiving a single oral dose of $^{14}C$-lactitol reached only 16·5% of the administered dose within 5 h and 48·5% after 24 h. Radioactivity first appeared between 2 and 3 h after administration, and peaked between 5 and 6 h, indicating that practically no polyol was hydrolysed and metabolized by the rat, but it was fermented by the intestinal flora.[94] In humans the radioactivity recovered in breath $CO_2$, after administration of 20 g of radiolabelled lactitol, was 63%, and the metabolizable energy was estimated to be 54·5% of the energy content of lactitol, provided by the metabolism of the absorbed short-chain fatty acids by the host.[92] When a significant part of the diet of guinea-pigs or rats,[88] e.g. 10–20%, was replaced by lactitol, the metabolizable energy of the polyol ranged between 56% and 70%. The metabolizable energy of lactulose which derived from caecal fermentation was found to be comparable to that of lactitol, indicating that the polyol is very poorly digested, but it still produced at least 2 kcal/g via bacterial flora fermentation.[88] In a recent human study in which 8 subjects received 50 g lactitol or 49 g sucrose daily, it was found that 60% more metabolizable energy was required from lactitol than from sucrose to maintain energy equilibrium.[83]

## 5 CONCLUSION

Despite the great quantity of work that has been done, considerable ambiguity still surrounds the energy value of the polyols. This may be a point of academic interest rather than practical importance when looking at the contribution of polyols to daily energy requirements, since they are unlikely to contribute more than 5–10% of the gross energy intake in view of their limited tolerance. It is, however, important to consider in relation to specific polyol-containing products and dietetic applications, as well as to legislational demands. It therefore remains an area of research which warrants further attention. Models and techniques should be carefully chosen so as to be close to real situations, and to allow comparisons of the data from different research centres and groups to be made without any problems.

## REFERENCES

1. Ziesenitz, S. C. & Siebert, G., The metabolism and utilization of polyols and other bulk sweeteners compared with sugar. In *Developments in Sweeteners*, Vol. 3, ed. T. H. Grenby. Elsevier Applied Science, London, 1987, pp. 109–54.
2. Banwell, J. G., Branch, W. & Cummings, J. H., The microbial mass in the human large intestine. *Gastroenterology*, **80** (1981) 1104A.
3. Hill, M. J. & Drasar, B. S., The normal colonic bacterial flora. *Gut*, **16** (1975) 318–23.
4. Moore, W. E. C., Cato, E. P. & Holderman, L. U., Some current concepts in intestinal bacteriology. *Am. J. Clin. Nutr.*, **31** (1978) S33–S42.
5. Miller, T. L. & Wolin, M. J., Fermentation by saccharolytic intestinal bacteria. *Am. J. Clin. Nutr.*, **32** (1979) 164–72.
6. Würsch, P., Starch in human nutrition. *World Rev. Nutr. Diet.*, **60** (1989) 199–256.
7. McNeil, N. I., The contribution of the large intestine to energy supplies in man. *Am. J. Clin. Nutr.*, **39** (1984) 338–42.
8. Isaacson, H. R., Hinds, F. C., Bryant, M. P. & Owens, F. N., Efficiency of energy utilization by mixed rumen bacteria in continuous culture. *J. Dairy Sci.*, **58** (1975) 1645–59.
9. Salyers, A. A., Vercellotti, J. R., West, S. E. H. & Wilkins, T. D., Fermentation of mucin and plant polysaccharides by strains of bacteroides from the human colon. *Appl. Environ. Microbiol.*, **33** (1977) 319–22.
10. Würsch, P., Welsch, C. & Arnaud, M. J., Metabolism of L-sorbose in the rat and the effect of the intestinal microflora on its utilisation both in the rat and in the human. *Nutr. Metab.*, **23** (1979) 145–55.
11. Wolin, M. J., Fermentation in the rumen and human large intestine. *Science*, **213** (1981) 1463–8.
12. Cummings, J. H. & Branch, W. J., Fermentation and the production of short-chain fatty acids in the human large intestine. In *Dietary Fiber: Basic and Clinical Aspects*, ed. G. V. Vahouny & D. Kritchevsky. Plenum Press, New York, 1986, pp. 131–50.
13. Hayssen, V. & Lacy, R. C., Basal metabolic rates in mammals: taxonomic differences in the allometry of BMR and body mass. *Comp. Biochem. Physiol.*, **81A** (1965) 741–54.
14. Imoto, S. & Namioka, S., VFA production in the pig large intestine. *J. Anim. Sci.*, **47** (1978) 467–78.
15. McNeil, N. I., Cummings, J. H. & James, W. P. T., Short chain fatty acid absorption by the human large intestine. *Gut*, **19** (1978) 819–22.

16. Argenzio, R. A. & Southworth, M., Site of organic acid production and absorption in the gastrointestinal tract of the pig. *Am. J. Physiol.*, **228** (1975) 454–60.

17. Stevens, C. E. & Stettler, B. K., Transport of fatty acid mixtures across rumen epithelium. *Am. J. Physiol.*, **211** (1966) 264–71.

18. Ruppin, H., Bar-Meir, S., Soergel, K. H., Wood, C. M. & Schmitt, M. G., Absorption of short-chain fatty acids by the colon. *Gastroenterology*, **78** (1980) 1500–7.

19. McNeil, N. I., Human large intestinal absorption of short-chain fatty acids. In *Colon and Nutrition*, ed. H. Kasper & H. Goebell. MTP Press, Lancaster, UK, 1982, pp. 55–8.

20. Engelhardt, W. & Rechkemmer, G., Absorption of inorganic ions and short-chain fatty acids in the colon of mammals. In *Intestinal Transport: Fundamental and Comparative Aspects*, ed. M. Gilles-Baillien & R. Gilles. Springer-Verlag, Berlin, 1983, pp. 26–45.

21. Bergman, E. N. & Wolff, J. E., Metabolism of volatile fatty acids by liver and portal-drained viscera in sheep. *Am. J. Physiol.*, **221** (1971) 586–92.

22. Pomare, E. W., Branch, W. J. & Cummings, J. H., Carbohydrate fermentation in the human colon and its relation to acetate concentration in venous blood. *J. Clin. Invest.*, **75** (1985) 1448–54.

23. Dankert, J., Zijlstra, J. B. & Wolthers, B. G., Volatile fatty acids in human peripheral and portal blood: quantitative determination by vacuum distillation and gas chromatography. *Clin. Chim. Acta*, **110** (1981) 301–7.

24. Weekes, T. E. C. & Webster, A. J. F., Metabolism of propionate in the tissues of the sheep gut. *Br. J. Nutr.*, **33** (1975) 425–38.

25. Mottaz, P. & Worbe, J. F., Transfert des acides gras volatils dans la paroi du caecum isolé de rat. *Compt. Rend. Soc. Biol.*, **171** (1977) 375–80.

26. Engelhardt, W. V. & Rechkemmer, G., The physiological effects of short-chain fatty acids in the hind gut. *Bull. Roy. Soc. N.Z.*, **20** (1983) 149–55.

27. Dietschy, J. M., The role of bile salts in controlling the rate of intestinal cholesterogenesis. *J. Clin. Invest.*, **47** (1968) 286–300.

28. Sallee, V. L. & Dietschy, J. M., Determinants of intestinal mucosal uptake of short- and medium-chain fatty acids and alcohols. *J. Lipid Res.*, **14** (1973) 475–84.

29. Miguel, S. G., Fatty acid and sterol synthesis by rat small intestine *in vitro*. *Lipids*, **12** (1977) 1080–3.

30. Giesecke, D. & Strangassinger, M., Lactic acid metabolism. In *Digestive Physiology and Metabolism in Ruminants*, ed. Y. Ruckebusch & P. Thivend. AVI, Westport, CT, 1980, pp. 523–39.

31. Watford, M., Lund, P. & Krebs, H. A., Isolation and metabolic characteristics of rat and chicken enterocytes. *Biochem. J.*, **178** (1975) 589–96.

32. Hood, R. L., Thompson, E. H. & Allen, C. E., The role of acetate, propionate and glucose as substrates for lipogenesis in bovine tissues. *Int. J. Biochem.*, **3** (1972) 598–606.

33. Henning, S. J. & Hirt, F. J. R., Ketogenesis from butyrate and acetate by the caecum and the colon of rabbits. *Biochem. J.*, **130** (1972) 785–90.

34. Roediger, W. E. W., Utilization of nutrients by isolated epithelial cells of the rat colon. *Gastroenterology*, **83** (1982) 424–9.

35. Marty, J. & Vernay, M., Absorption and metabolism of the volatile fatty acids in the hindgut of the rabbit. *Br. J. Nutr.*, **51** (1984) 265–77.
36. Lupton, J. R., Coder, D. M. & Jacobs, L. R., Long-time effects of fermentable fibres on rat colonic pH and epithelial cell cycle. *J. Nutr.*, **118** (1988) 840–5.
37. Lupton, J. R., Coder, D. M. & Jacobs, L. R., Influence of luminal pH on rat large bowel epithelial cell cycle. *Am. J. Physiol.*, **249** (1985) G382–8.
38. Sakata, T., Stimulatory effect of short-chain fatty acids on epithelial cell proliferation in the rat intestine; a possible explanation for trophic effect of fermentable fibre, gut microbes and luminal trophic factors. *Br. J. Nutr.*, **58** (1987) 95–103.
39. Davidson, S., Passmore, R. & Brock, J. F., *Human Nutrition and Dietetics*, 5th edn. Churchill Livingstone, Edinburgh, 1972, pp. 8–9.
40. Blaxter, K. L., *The Energy Metabolism of Ruminants*, 1st edn. Hutchinson, London, 1962, p. 186.
41. Lian-Loh, R., Birch, G. G. & Coates, M. E., The metabolism of maltitol in the rat. *Br. J. Nutr.*, **48** (1982) 477–81.
42. Würsch, P., Koellreutter, B., Gétaz, F. & Arnaud, M. J., Metabolism of $^{14}$C-maltitol in rats, mice and germ-free mice and comparative digestibility between maltitol and sorbitol in germ-free mice. *Br. J. Nutr.* (in press).
43. Read, N. W., Aljancki, M. N., Bates, T. E. & Barber, D. C., Effect of gastrointestinal intubation on the passage of a solid meal through the stomach and small intestine in humans. *Gastroenterology*, **84** (1983) 1568–72.
44. Beaugerie, L., Contribution à l'étude du transport intestinal du sorbitol chez l'homme. Thesis, Hôpital St. Lazare, Paris, 1987.
45. Dharmaraj, H., Patil, H., Grimble, G. K. & Silk, D. B. A., Lactitol, a new hydrogenated lactose derivative: intestinal absorption and laxative threshold in normal human subjects. *Br. J. Nutr.*, **57** (1987) 195–9.
46. Sandberg, A. S., Andersson, H., Hallgren, B., Hasselblad, K., Isaksson, B. & Hulten, L., Experimental model for *in vivo* determination of dietary fibre and its effect on small bowel absorption of nutrients. *Br. J. Nutr.*, **45** (1981) 283–94.
47. Kronenberg, H.-G., Spengler, M. & Strohmeyer, G., Zur Resorption von Palatinit, dem äquimolecularen Gemisch von α-D-Glucopyranosido-1,6-sorbit und α-D-Glucopyranosido-1,6-mannit im Dünndarm von colectomierten Patienten. Unpublished report, 1979.
48. Zinner, P. M., Kirchgessner, M., Ascherl, R. & Erhardt, W., Zur präcäcalen Absorption von Palatinit beim ausgewachsenen Schwein. *Z. Tierphysiol. Tierernährg. Futtermittelkde.*, **53** (1985) 79–83.
49. Rérat, A., Vaugelade, P. & Vaissade, P., Etude de la digestion d'un sirop de glucose hydrogéné riche en maltitol chez le porc éveillé. *Bull. Acad. Nat. Méd.*, **171** (1987) 183–7.
50. Flourié, B., Florent, C., Etanchaud, F., Evard, D., Franchisseur, C. & Rambaud, J.-C., Starch absorption by healthy man evaluated by lactulose hydrogen breath test. *Am. J. Clin. Nutr.*, **47** (1988) 61–6.
51. Levitt, M. D., Hirsh, P., Fetzer, C. A., Sheahan, M. & Levine, A. S., $H_2$ excretion after ingestion of complex carbohydrates. *Gastroenterology*, **92** (1987) 383–9.

52. Florent, C., Flourié, B., Leblond, A., Rautureau, M., Bernier, J.-J. & Rambaud, J.-C., Influence of chronic lactulose ingestion on the colonic metabolism of lactulose in man (an *in vivo* study). *J. Clin. Invest.*, **75** (1985) 608–13.

53. Levitt, M. D. & Donaldson, R., Use of respiratory hydrogen ($H_2$) excretion to detect carbohydrate malabsorption. *J. Lab. Clin. Med.*, **75** (1970) 937–45.

54. Bond, J. H. & Levitt, M. D., Quantitative measurement of lactose absorption. *Gastroenterology*, **70** (1976) 1058–62.

55. Solomons, N. W., Viteri, F. E. & Hamilton, L. H., Application of a simple gas chromatographic technique for measuring breath hydrogen. *J. Lab. Clin. Med.*, **90** (1977) 856–62.

56. Rumessen, J. J., Kokholm, G. & Gudmand-Høyer, E., Methodological aspects of breath hydrogen ($H_2$) analysis. Evaluation of a $H_2$ monitor and interpretation of the breath $H_2$ test. *Scand. J. Lab. Invest.*, **47** (1987) 555–60.

57. Fritz, M., Siebert, G. & Kasper, H., Dose dependence of breath hydrogen and methane in healthy volunteers after ingestion of a commercial disaccharide mixture Palatinit®. *Br. J. Nutr.*, **54** (1985) 389–400.

58. Burbige, E. J., Lewis, D. R. & Chin, C. K., Hydrogen liberated by fecal homogenates incubated with dietary fiber supplements *in vitro*. *Gastroenterology*, **84** (1983) 117.

59. Schnell-Dompert, E. & Siebert, G., Metabolism of sorbitol in the intact organism. *Hoppe-Seyler's Z. Physiol. Chem.*, **361** (1980) 1069–75.

60. Würsch, P. & Schweizer, T., Sugar substitutes and their energy value for the human body. *Deutsch. Zahnärtzl. Z.*, **42** (1987) s151–3.

61. Würsch, P., Koellreuter, B. & Schweizer, T., Hydrogen excretion after ingestion of five different sugar alcohols and lactulose. *Eur. J. Clin. Nutr.*, (in press).

62. Lusk, G., *The Elements of the Science of Nutrition*, 4th edn. Johnson Reprint Corporation, London, 1976, pp. 61–74.

63. Felber, J. P., Magnenat, G., Castellaz, M. *et al.*, Carbohydrate and lipid oxidation in normal and diabetic subjects. *Diabetes*, **26** (1977) 693–9.

64. Mueller-Hess, R., Geser, C. A., Bonjour, J. P., Jéquier, E. & Felber, J. P., Effects of oral xylitol administration on carbohydrate and lipid metabolism in normal subjects. *Infusion Therapie*, **2** (1975) 247–52.

65. Pellaton, M., Acheson, K., Maeder, E., Jéquier, E. & Felber, J. P., The comparative oxidation of glucose, fructose, sorbitol and xylitol in normal man. *J. Parent. Ent. Nutr.*, **2** (1978) 627–33.

66. Froesch, E. R., Zapf, J., Keller, U. & Oelz, O., Comparative study of the metabolism of U-$^{14}$C-fructose, U-$^{14}$C-sorbitol and U-$^{14}$C-xylitol in the normal and in the streptozotocin-diabetic rat. *Eur. J. Clin. Invest.*, **2** (1971) 8–14.

67. Benedict, F. G. & Carpenter, T. M., Food ingestion and energy transformation with special reference to the stimulating effect of nutrient. *Carnegie Inst. Wash. Publ.*, **261** (1913) 47–250.

68. Sharief, N. N. & Macdonald, I., Differences in dietary-induced thermogenesis with various carbohydrates in normal and overweight men. *Am. J. Clin. Nutr.*, **35** (1982) 267–72.

69. Pittet, P., Acheson, K., Raman, K. & Jéquier, E., Effect of the sugars maltitol and sorbose on carbohydrate metabolism in man. *Experiencia*, **31** (1975) 213.

70. Felber, J. P., Tappy, L., Vouillamoz, D., Randin, J. P. & Jéquier, E., Comparative studies of maltitol and sucrose by means of continuous indirect calorimetry. *J. Parent. Ent. Nutr.*, **11** (1987) 250–4.

71. Thiébaud, D., Jacot, E., Schmitz, H., Spengler, M. & Felber, J.-P., Comparative study of isomalt and sucrose by means of continuous indirect calorimetry. *Metabolism*, **33** (1984) 808–13.

72. Dutch Nutrition Council, *The Energy Value of Sugar Alcohols.* Voedingsraad, The Hague, 1987.

73. Grupp, U. & Siebert, G., Metabolism of hydrogenated palatinose, and equimolar mixture of α-D-glucopyranosido-1,6-sorbitol and α-D-glucopyranosido-1,6-mannitol. *Res. Exp. Med.* (Berl.), **173** (1978) 261–78.

74. Van Es, A. J. H., Energy utilization of low digestibility carbohydrates. In *Low Digestibility Carbohydrates*, ed. D. C. Leegwater, V. J. Feron & R. J. J. Hermus. Pudoc, Wageningen, 1987, pp. 121–7.

75. Karimzadegan, E., Clifford, A. J. & Hill, F. W., A rat bioassay for measuring the comparative availability of carbohydrate and its application to legume foods, pure carbohydrate and polyols. *J. Nutr.*, **109** (1979) 2249–59.

76. Rice, E. E., Warner, W. D., Mone, P. E. & Poling, C. E., Comparison of the metabolizable energy contribution of foods by growth under conditions of energy restriction. *J. Nutr.*, **61** (1957) 253–66.

77. Adolph, E. F., Urges to eat and drink in rats. *Am. J. Physiol.*, **151** (1947) 110–25.

78. Morrison, S. D., The constancy of the energy expended by rats on spontaneous activity and the distribution of activity between feeding and non-feeding. *J. Physiol.*, **197** (1986) 305–23.

79. Berschauer, F. & Spengler, M., Energetische Nutzung von Palatinit. *Deutsch. Zahnärztl. Z.*, **42** (1987) s145–50.

80. Zinner, P. M. & Kirchgessner, M., Zur energetischen Verwertung von Palatinit. *Z. Ernährungswiss.*, **21** (1982) 272–8.

81. Kirchgessner, M., Zinner, P. M. & Roth, H.-P., Energiestoffwechsel und Insulinaktivität bei Ratten nach Palatinitfütterung. *Int. J. Vit. Nutr. Res.*, **53** (1983) 86–93.

82. Swift, R. W., The effect of feed on the critical temperature of the albino rat. *J. Nutr.*, **28** (1944) 359–64.

83. Van Es, A. J. H., Degroot, L. & Vogt, J. E., Energy balances of eight volunteers fed on diets supplemented with either lactitol or saccharose. *Br. J. Nutr.*, **56** (1986) 545–54.

84. Musch, K., Siebert, G., Schiweck, H. & Steinle, G., Ernährungsphysiologische Untersuchungen mit Isomalt an der Ratte. *Z. Ernährungswiss. Suppl.*, **15** (1973) 3–16.

85. Février, C. & Pascal, G., Utilisation énergétique du Palatinit et du saccharose chez le porc en finition. Unpublished report, INRA, Station de Recherches sur l'Elevage des Porcs, 1985.

86. Maranesi, M., Gentili, P. & Garenini, G., Nutritional studies on maltitol. *Acta Vitaminol. Enzymol.*, **6** (1984) 3–9.

87. Van Beek, L., Unpublished report, CIVO–TNO, Zeist, The Netherlands, 1977.

88. Bird, S. P., Hewitt, D. & Gurr, M. I., Digestible and metabolizable energy values of lactitol and lactulose for the rat and miniature pig. *Proc. Nutr. Soc.* (1985) 40A.

89. Adcock, L. H. & Gray, C. H., The metabolism of sorbitol in the human subject. *Biochem. J.*, **65** (1957) 554–60.
90. Nasrallah, S. M. & Iber, F. L., Mannitol absorption and metabolism in man. *Am. J. Med. Sci.*, **258** (1969) 80–8.
91. Rennhard, H. H. & Bianchine, J. R., Metabolism and caloric utilization of orally administered maltitol-[14]C in rat, dog and man. *J. Agric. Food Chem.*, **24** (1976) 287–90.
92. Grimble, G. K., Patil, D. H. & Silk, D. B. A., Assimilation of lactitol, and 'unabsorbed' disaccharide in the normal human colon. *Gut*, **29** (1988) 1666–71.
93. Anon., [14]C-Maltitol, [14]C-sorbitol and [14]C-glucose: metabolic utilization in the rat. Internal Report, Instituto di Recerche Biomediche, 'Antoine Marxer' Ivrea, Italy, 1987.
94. Leegwater, D. C., Studies on the metabolic fate of orally administered [14]C-lactitol in the rat. Unpublished report, CIVO–TNO, Zeist, The Netherlands, 1978.
95. Staub, H. W., Caloric availability of dietary polyols. *Fed. Proc.*, **37** (1978) 678 (abstr.).
96. Lauwers, A.-M., Daumerie, C. & Henquin, J. C., Intestinal absorption of sorbitol and effects of its acute administration on glucose homeostasis in normal rats. *Br. J. Nutr.*, **53** (1985) 53–62.
97. Honegger, P. & Semenza, G., Multiplicity of carriers for free glucalogues in hamster small intestine. *Biochem. Biophys. Acta*, **315** (1973) 390–410.
98. Semenza, G. & Mühlhaupt, E., Studies on intestinal sucrase and sugar transport. *Biochem. Biophys. Acta*, **173** (1969) 104–12.
99. Ertel, N. M., Algun, S., Kemp, F. W. & Mittler, J. C., The metabolic fate of exogenous sorbitol in the rat. *J. Nutr.*, **113** (1983) 566–73.
100. Grossklaus, R., Dosisabhängigkeit der energetischen Nutzung von Zuckeraustauschstoffen. *Deutsch. Zahnärztl. Z.*, **42** (1987) s154–8.
101. Mehnert, H., Stuhlfauth, K., Mehnert, B., Lausch, R. & Seitz, W., Vergleichende Untersuchungen zur Resorption von Glucose, Fructose und Sorbitol beim Menschen. *Klin. Wochenschr.*, **37** (1959) 1138–42.
102. Beaven, J., Bjørneklett, A., Jensen, E., Blomhoff, J. P. & Skrede, S., Pulmonary hydrogen and methanol and plasma ammonia after the administration of lactulose or sorbitol. *Scand. J. Gastroenterol.*, **18** (1983) 343–7.
103. Hyams, J. S., Sorbitol intolerance: an unappreciated cause of functional gastrointestinal complaints. *Gastroenterology*, **84** (1983) 30–3.
104. Cobden, I., Hamilton, I., Rothwell, J. & Axon, A. T. R., Cellobiose/mannitol: physiological properties of probe molecules and influences of extraneous factors. *Clin. Chim. Acta*, **148** (1985) 53–62.
105. Elia, M., Behrens, R., Northrop, C., Wraight, P. & Neale, G., Evaluation of mannitol, lactulose and [51]Cr-labelled ethylenediaminetetra-acetate as markers of intestinal permeability in man. *Clin. Sci.*, **73** (1987) 197–204.
106. Laker, M. F., Bull, J. H. & Menzies, I. S., Evaluation of mannitol for use as a probe marker of gastrointestinal permeability. *Eur. J. Clin. Invest.*, **12** (1982) 485–91.
107. Saunders, D. R. & Wiggins, H. S., Conservation of mannitol, lactulose and raffinose by the human colon. *Am. J. Physiol.*, **241** (1981) G397–402.

108. Spengler, M. & Boehme, K., Körpergewichtsverlauf gesunder Probanden unter 3-wöchiger Gabe von Tagesdosen von 48 g Isomalt, bzw. 24 g Saccharose im Rahmen einer 1500 kcal Diät. Unpublished report, Bayer A, 1985.

109. Kearsley, M. W., Birch, G. G. & Lian-Loh, R. H. D., The metabolic fate of hydrogenated glucose syrups. *Stärke*, **34** (1982) 279–83.

110. Würsch, P. & Del Vedovo, S., Inhibition of human digestive enzymes by hydrogenated malto-oligosaccharides. *Int. J. Vit. Nutr. Res.*, **51** (1981) 161–5.

111. Rosiers, C., Verwaerde, F., Dupas, H. & Bouquelet, S., New approach to the metabolism of hydrogenated starch hydrolysate: hydrolysis by maltase/glucoamylase complex of the rat intestinal mucosa. *Ann. Nutr. Metab.*, **29** (1985) 76–82.

112. Nilsson, U. & Jägerstad, M., Hydrolysis of lactitol, maltitol and Palatinit by human intestinal biopsies. *Br. J. Nutr.*, **58** (1987) 199–206.

*Chapter 10*

# EVALUATION OF THE INFLUENCE OF INTENSE SWEETENERS ON THE SHORT-TERM CONTROL OF APPETITE AND CALORIC INTAKE: A PSYCHOBIOLOGICAL APPROACH

PETER J. ROGERS & JOHN E. BLUNDELL

*BioPsychology Group, Psychology Department, University of Leeds, UK*

## SUMMARY

*Intense sweeteners provide a means of uncoupling sweet taste and caloric content; with appropriate experimental designs they can be used as tools to investigate the contribution of these factors to the control of appetite. In addition, intense sweeteners are widely perceived as aids to dietary control by consumers pursuing ideals of slimness. Therefore understanding the influences of intense sweeteners on human appetite has both theoretical and practical significance. When this issue is considered with regard to the psychobiological system controlling appetite, two major themes emerge: the stimulating effect of sweetness and the suppression of hunger by calories.*

*A number of laboratory studies have compared the effect of covertly substituting an intense sweetener for sucrose or glucose in foods and beverages. Two principles are involved: the 'addition' principle, in which the food is made sweeter (but caloric content does not change), and the 'substitution' principle, in which the level of sweetness is maintained but caloric content is reduced. Many studies have failed to distinguish between the logical consequences of such experimental manipulations. When a substitution is involved, almost always some increase in voluntary intake is observed which, at least partially, compensates for the lower energy value of the items containing intense sweeteners. It is important to note that nothing can be concluded about the role of sweetness from this experimental design, because the level of sweetness is not systematically varied. All the studies, including a comparison between different sweetness levels (same caloric value—the addition principle), have revealed a stimulatory effect of sweetness on hunger and/or food intake.*

*A further issue concerns possible differences within the family of intense sweeteners. Although all may be expected to exert similar initial effects through the generation of sweet taste in the mouth, post-ingestive and post-absorptive effects may vary according to the biological handling of these substances in the body. Recent evidence indicates, for example, that saccharin promotes medium-term increases in food intake, which may be due to its action(s) on post-ingestive mechanisms. It should not be assumed that intense sweeteners will all have equivalent effects on appetite.*

*Field studies comparing users and non-users of intense sweeteners have produced conflicting findings. One difficulty is that it is not possible to conclude that differences in food intake and body weight reported in these surveys were due to the use (or non-use) of intense sweeteners per se, because users and non-users may differ in other respects, such as the extent to which they attempt to exert conscious restraint over eating. Nonetheless, these studies are representative of the 'natural' use of intense sweeteners, whereas in laboratory studies subjects are usually selected to preserve uniformity. The further development of both these approaches will strengthen current understanding of appetite control and the role played by intense sweeteners.*

## 1 SWEETENERS IN THE LATE TWENTIETH CENTURY

The ways in which foods may increase the willingness to eat or satisfy our desire for further foods are issues of great theoretical importance and considerable practical significance in these closing decades of the 20th century. Products with raised palatability to promote consumption may have an over-nutritive potential. In addition, new types of foods and additives are constantly being added to the food supply, although often little is known about their effects on appetite. Consequently there is a considerable need to provide information about the appetite-enhancing and satiating capacity of different foods. First, a knowledge of how the composition of food alters energy intake and food selection throws light upon the mechanisms of appetite control. Food itself can be used as an experimental tool to investigate the mode of operation of appetite mechanisms. Second, this knowledge can be used to develop a coherent strategy for nutritional intake for everyday use in the home, at work and in the clinic. Information on the effects on appetite control exerted by particular components of food can help industry to provide appropriate foods for specific requirements, and can allow the consumer to select rationally a suitable diet. Despite the importance of these issues, only a little is known about the satiating properties (or appetite-enhancing capacities) of foods in general,[1] and only a few studies have examined the effects of individual food components.[2] However, various research strategies and experimental designs are available for investigating these matters.[1,3]

The above concerns are particularly relevant to the powerful psychobiological phenomenon of sweetness.[4] Why has sweetness become particularly important in the last quarter of the 20th century? There are two prominent reasons. First, advances in food technology have made it possible to develop and produce foods with precisely defined sensory properties. These properties are designed to make foods particularly attractive to the consumer (the eater and the purchaser). It is clear that the sensory characteristics of foods can be manipulated quite independently of their nutrient or caloric content. This disengagement, or uncoupling, of the sensory and nutritional components of food is likely to have effects upon the control of appetite and the pattern of ingestion.

Second, in economically well developed cultures such as in North America and Europe, the past 10 years have seen increased attention directed to the problem of obesity and the effects of excess weight on health.[5,6] This concern has been reflected in the extent to which the media have promoted the doctrine of slimness, particularly among women. Developing from these circumstances, a variety of eating disorders have been observed, ranging from mild but uncontrolled dieting to the disabling disorder of bulimia nervosa.[7] A sizeable proportion of the population is greatly concerned with body shape and weight, and is actively attempting to undereat. This type of behaviour, and its associated eating patterns, bring to importance the way in which foods influence hunger and provide for bodily requirements.

Taken together these two factors—increased sophistication in food product development and the prevalence of dietary concern among consumers—suggest that mechanisms of appetite control are being subjected to considerable pressures.

It is therefore appropriate to examine the role of intense sweeteners—agents that satisfy a requirement for producing highly palatable foods, and that are likely to be consumed not only for their taste characteristics but also for their reputed value in the control of calorie intake. Accordingly, the effects of intense sweeteners on appetite should be assessed in comparison with what is known about the operation of other characteristics of foods.

Does sweetness alone exert unique effects not shared by other physical properties of foods, and what effect does sweetness have in combination with other food characteristics? Traditionally the effects of food upon appetite have been examined through the concept of 'satiating efficiency', and this seems particularly relevant since subjective satiety or 'fillingness' of food is one of the three major variables determining food intake.[8] Moreover, the notion of satiating power lends itself to an understanding of the short-term effects of sweeteners.

## 2 THE SATIETY CASCADE

Eating food has the capacity to take away hunger. After satiation, further eating is inhibited for a while. What mechanisms are responsible? It is likely that the mechanisms involved in terminating eating and in maintaining inhibition range from those that operate when food is initially sensed to the effects of metabolites following digestion and absorption. By definition, satiety is not instantaneous but comes on over a considerable time period; it is therefore useful to distinguish between different phases of satiety which can be associated with different mechanisms. This concept is illustrated in Fig. 1. Four mediating processes are identified: sensory, cognitive, post-ingestive and post-absorptive. These maintain inhibition of eating (and hunger) during the early and late phases of satiety. Sensory effects are generated through the smell, taste, temperature and texture of foods, and it is likely that these factors help to bring eating to a halt and to inhibit the eating of foods with similar sensory properties in the short term. Such a

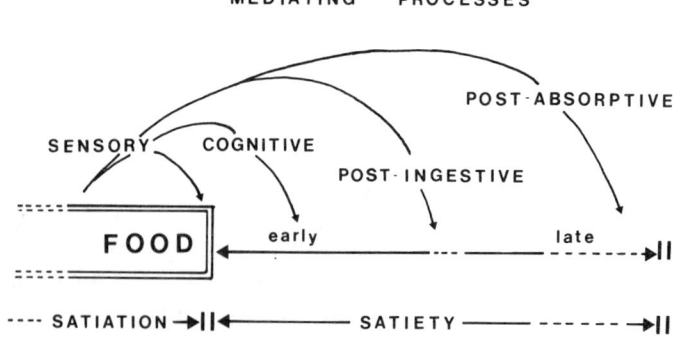

FIG. 1. The satiety cascade. Conceptualisation of the contributions of sensory, cognitive, post-ingestive and post-absorptive stimuli to the time course of satiety.

mechanism is embodied in the idea of sensory-specific influences first disclosed by Le Magnen.[9] Cognitive effects represent the beliefs held about the properties of foods and their presumed effect upon the eater.[10] The category identified here as post-ingestive processes includes a number of possible modes of action, including gastric distension and rate of gastric emptying, the release of hormones such as cholecystokinin from the duodenum, and the stimulation of physicochemically specific receptors along the gastro-intestinal tract.[11] The post-absorptive phase of satiety includes those mechanisms arising from the action of metabolites after absorption across the intestine and into the blood system. This category embraces the action of metabolites such as glucose and the amino acids, which may act directly upon the brain after crossing the blood-brain barrier, or which may influence the brain indirectly via neural inputs following stimulation of peripheral chemoreceptors.

The approximate expected stage of action of these mediating processes is shown in Fig. 1, but the mechanisms will overlap and their effects will be integrated to produce a joint effect. It should also be kept in mind that the psychobiological system for appetite control has the capacity to learn, i.e. to form associations between the sensory and post-absorptive properties of foods.[12,13] Therefore the sensory aspects of a food may come to predict the absorptive consequences, thereby allowing the sensory-mediating processes to exert an augmented effect. This effect will be weakened or distorted when there is uncoupling of the sensory and nutritional properties of foods, as in the case of the non-caloric sweeteners. All these factors add to the importance of measuring the strength of satiety at various times after the end of ingestion in order to throw light upon the effects of the separate mediating processes. Good experimental designs will be required to analyse the operations of the satiety 'cascade'. However, it is clear that the effects of foods upon this sequence have important consequences for appetite control, for the subjective feelings associated with eating and for food acceptability.

## 3 ROLE OF HUNGER IN FOOD CONSUMPTION

It is clear that food itself is a potent and natural anorexic substance; eating food takes away our hunger and being deprived of food generally leads to a build-up of hunger, although there are certain special circumstances when this is not so. The processes responsible for the inhibition of hunger are shown in Fig. 1. Since the intensity of hunger feelings is a major component of satiety, hunger is influenced by the processes identified. However, a dilemma arises when foods are produced to meet a popular demand for low-calorie items arising from an obsession with dieting and slimming. It can be seen from Fig. 1 that reducing the amount (or nature) of food eaten will weaken those processes (post-ingestive and post-absorptive) that maintain hunger at a low level and postpone further eating. Changing food composition will alter the capacity of food to influence the satiety cascade. It is well known that both casual and planned dieting lead to chronic low levels of hunger and acute intense hunger episodes. This illustrates the biological function of the hunger sensation, which is to motivate eating. Hunger reminds us that the body requires food. The importance of this is reflected in the nature of hunger; it is a nagging, irritating feeling, the presence of which constantly serves to stimulate thoughts of food and eating. These attributes of hunger are one reason why casual dieting almost always ends in failure. It appears to be very difficult to cope with unregulated hunger. It also draws attention to the fact that reducing the calorie content of foods in order to reduce overall energy intake will probably only be effective as a weight-controlling strategy if the individual has some capacity for the management of hunger.

## 4 STIMULATORY AND INHIBITORY EFFECTS OF FOODS

The satiety cascade in Fig. 1 illustrates the processes that mediate the inhibitory effects of food eaten upon further consumption. The pre-absorptive responses involve the mouth, stomach, duodenum and lower intestine, together with related endocrine glands. Additionally, some attributes of food act before eating commences. In man the sight and smell of food are sufficiently potent to trigger cephalic phase insulin responses,[14,15] the response in obese subjects being 4 times that in lean people. The stimulus-induced cephalic phase responses are accompanied by an augmentation of the desire to eat. Consequently one action of salient sensory components of food is to provoke physiological responses which anticipate ingestion and which may stimulate appetite.[16,17] Therefore, in assessing the overall effect of individual foods on appetite control, it is necessary to take into account both stimulatory and inhibitory actions. Moreover, it is worth noting that this dual action of food may have important implications when a potent sensory dimension, e.g. sweetness, is combined with a weak (low energy) inhibitory component (such as in a low-calorie food or beverage sweetened with an intense sweetener).[18]

Accordingly, an evaluation of the effect of intense sweeteners on appetite and food intake can be carried out against the background of the satiety cascade, keeping in mind both the stimulatory and inhibitory actions of food, and the critical role played by hunger sensations.

## 5 UNCOUPLING SWEETNESS AND CALORIES: A RESEARCH STRATEGY

The intense sweeteners provide exquisite experimental tools for disengaging the sensory phenomenon of sweetness from the metabolic effect of calorie intake. It is readily recognised that almost all sweet foods are associated with moderate to high caloric content. Naturally sweet items such as fruits contain sweet carbohydrates, while many manufactured sweet products contain carbohydrate and fat. Consequently most sweet-tasting foods provide a combination of the sensation of sweetness and energy value. It is readily seen from the satiety cascade (Fig. 1) that both of these factors contribute to the control of appetite. How important is the role played by each? This theoretical issue is given practical significance through the use of intense sweeteners. As experimental tools they can be used to determine the effects of uncoupling sweetness and energy value, and the consequences of this should have implications for the effect of these substances on appetite under normal circumstances. However, before any inferences can be drawn about an action upon appetite control, the logic of the experimental methods should be clearly set out. In such a controversial area of research it is relatively easy for a combination of rhetoric, poor experimental design and weak statistical analysis to create a false impression. The relationship of validity and experimental design is summarised below; it has been discussed fully elsewhere.[19]

## 6 METHODOLOGICAL CONSIDERATIONS

Although, to date, relatively few studies investigating the effects of intense sweeteners on appetite control have been published in detail, this is developing into a very active field of research. The remainder of this review examines methodology, as well as describing findings of the work and drawing some interim conclusions.

The central methodological issue concerns internal validity, i.e. the

**CALORIES**

|  | present/<br>high | absent/<br>low |
|---|---|---|
| **present/<br>high** | **A**<br>bulk<br>sweeteners | **B**<br>intense<br>sweeteners |
| **SWEETNESS** |  |  |
| **absent/<br>low** | **D**<br>starch<br>or sweetness<br>inhibitors | **C**<br>vehicle |

FIG. 2.    Balanced design for investigating the effects of calories and sweetness
using a preloading procedure.

capacity to identify the causal factor underlying any experimental outcome.
In research on sweetness, every food or beverage used in an experimental
treatment will possess, in addition to sweetness, other attributes, including
its calorie and nutrient content, volume, texture, flavour and visual
characteristics; together these will contribute to its acceptability to the
consumer. Obviously, in order to assign any experimental outcome to the
action of sweetness itself, all other factors, including acceptability, must be
held constant or systematically varied in a controlled manner. Internal
validity therefore rests upon good experimental design. As noted above,
one key issue is the identification of effects attributable to sweetness or
calories.

Figure 2 presents an experimental design in which sweetness and caloric
content are varied independently. C might usually be the food or drink
vehicle for A and B, with calories and sweetness (in the form of e.g. sucrose
or glucose) being added to make A, and sweetness alone (in the form of an
intense sweetener) to make B. D can be produced by adding calories (in the
form of e.g. starch) to C, or by suppressing the sweetness in A. At the same
time the other attributes of these foods or drinks must be held constant. In
practice the above requirements are difficult, but not impossible, to fulfil.
Yoghurt, for example, provides an excellent medium for these experimental
manipulations,[20] since it is commonly eaten in sweetened and unsweetened
forms, and the addition of starch or a derivative such as a low-DE
maltodextrin is well disguised.

In fact, only conditions equivalent to A and B have been included in
many of the studies. What inferences can be drawn when this design is used?
Since the level of sweetness is not varied, nothing can be concluded about

the effects of sweetness. However, the effects of A and B will reveal the extent to which caloric content influences appetite. Accordingly, these studies are relevant to the issue of caloric compensation. As will become apparent from the review presented below, some degree of caloric compensation invariably took place. Nevertheless, in some studies the increase in voluntary food intake was insufficient to compensate fully for the energy 'saving' achieved by substituting an intense sweetener for sugar in the food. Note, however, that it cannot be concluded that therefore sweetness (or other sensory properties) contributed substantially to the satiating effect of the food; sweetness was not varied systematically. The knowledge that something has been consumed equally might account for the observed reduction in appetite, i.e. an action upon food intake of the cognitive factors involved in the satiety cascade. The evidence indicates that addition of sweetness to a food or drink preload (i.e. comparing the effects of A and D or B and C in Fig. 2) *stimulates* subsequent intake (see below).

The above discussion considers only the capacity of intense sweeteners to uncouple sweetness from caloric content. However, in addition to providing a sweet taste, these substances may have post-ingestive effects with significant consequences for the control of appetite. This clearly complicates the interpretation of experimental findings. One solution is to control for possible post-ingestive influences by examining the effects of intense sweeteners without tasting. This can be accomplished simply and conveniently by requiring volunteers to swallow capsules containing the appropriate amount of intense sweetener. A second approach is to conduct studies that compare directly the effects of the various intense sweeteners. Depending upon their chemical formulation, there are potentially important differences in the ways these substances are metabolised. The identification of any post-ingestive effect of intense sweeteners on appetite will be of considerable interest irrespective of how this affects the interpretation of the evidence on the role of sweetness and caloric content.

## 7 REVIEW OF STUDIES

Two different approaches have been adopted in investigations of the effects of intense sweeteners on appetite control. First, laboratory-based studies have used a preload procedure. Intense sweeteners are administered in a food or drink preload, followed by the tracking of subjective motivational changes (e.g. hunger, fullness and food preferences) and the measurement of voluntary food intake, usually in a 'test' meal. In contrast to these acute studies, the second approach has involved monitoring the longer-term effects of consuming products containing intense sweeteners within particular dietary regimes.

## 7.1 Acute Studies

Under certain conditions intense sweeteners have been found to stimulate subjective hunger.[18,20-22] Compared with the effect of water, consumption of intense sweetener solutions resulted in increases in hunger ratings, while solutions of sucrose and glucose of equivalent sweetness reduced hunger. However, while glucose also reduced subsequent (voluntary) food intake in a lunchtime test meal, increases in intake were not observed following acesulfame-K, aspartame or saccharin.[22] On the other hand, Guss et al.[23] have reported data showing increases in food consumption following aqueous preloads containing aspartame (1% glucose and 1% fructose solutions sweetened with aspartame, compared with the effect of the same volume of water). It is possible that the procedure adopted by Rogers et al.[22] was relatively insensitive to the detection of overeating. In particular, there was a high 'baseline' food intake following the water preload. This interpretation is supported by the outcome of a subsequent study using yoghurt as the preload vehicle. Baseline intake was lower and saccharin produced a moderate but significant increase in test meal intake (Fig. 3), consistent with its stimulatory effects on hunger ratings observed in the same experiment.[20] An additional and striking feature of the results shown in Fig. 3 is the further increase in food intake associated with saccharin ingestion that endured well beyond the test meal. Preliminary data from a replication of this study indicate similar effects of consuming an aspartame-sweetened yoghurt on food intake in the short term (i.e. in the test meal) but not in later intervals.[24]

Brala and Hagen[25] examined the effect of sucrose and aspartame in milkshakes. The experimental procedure was lengthy and the design involved a complex interaction of variables, including the suppression of sweetness with *Gymnema sylvestre* extracts. Importantly, however, a control (the unsweetened vehicle) was also included. Disregarding for the moment the effects of gymnemic acid, which brought about only a partial and transient suppression of sweetness perception, the data indicate that subjects receiving the aspartame-sweetened preload ate more (10·6% or 60 kcal), in a test meal 1 hour later, than subjects receiving the equicaloric vehicle. The size of the test meal was reduced following the sucrose-containing milkshake; however, the lower intake did not compensate fully for the higher energy value of this preload, and subjects ingested more

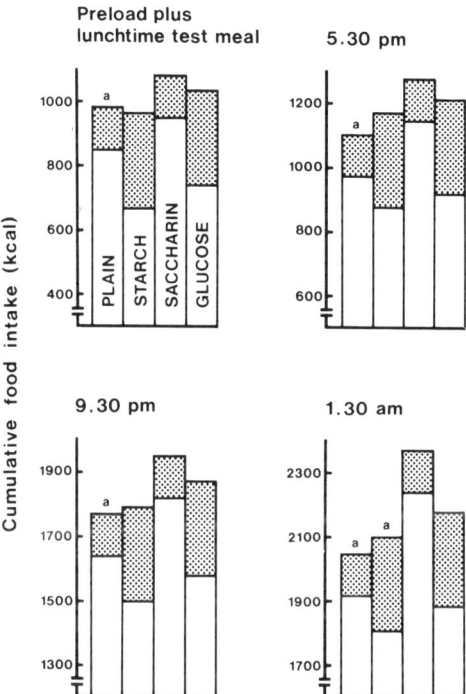

FIG. 3.    Food intake at intervals following the consumption of a low-fat, non-sweet yoghurt preload (plain, energy value = 131 kcal, cell C in Fig. 2) and the same yoghurt supplemented with 50 g starch (295 kcal, D), 163 mg saccharin (131 kcal, B) and 50 g glucose (295 kcal, A). The volume of each of the four preloads was 215 ml, and they were all equally liked. Preloads A and B and preloads C and D were equi-sweet. The study was conducted according to a 'within subjects' design ($N = 21$), and voluntary food intake was measured in a test meal begun at 1·00 pm (and ending at approximately 1·30 pm), 1 hour after the preload was consumed. Thereafter subjects completed a food intake diary until bedtime that night. These cumulative intakes *include* the preloads (indicated by the shaded areas). 'a' denotes significantly different from saccharin (*t*-test, $p < 0.05$, 2-tail). The difference between food intakes following the plain and saccharin preloads increased significantly over time ($F[3, 60] = 2·81$, $p < 0·05$). A preliminary report of this study was presented at the Royal Society of Medicine, London, for the June 1988 meeting of the Association for the Study of Obesity.[20]

calories overall (preload plus test meal) than in either of the two low-calorie conditions.

In other preloading studies a vehicle control has been absent. Rolls and her colleagues[26-28] report two studies in which subjects were fed a dessert (e.g. 'jello' or an instant chocolate pudding) containing either sucrose or aspartame. The size of this preload was not fixed, although on average similar amounts (weight) of the low and high calorie versions were consumed. Despite the resulting differences in calorie intake (up to 200 kcal), no reliable differential effects on measures of hunger, satiation and subsequent food intake were detected. Food intake, nonetheless, tended to be higher following the low calorie desserts, the degree of caloric compensation varying between 26% and 81% of the differences in preload intakes. At the same time there were no significant differences in total (preload plus subsequent meal) food intake. In other words, caloric compensation was incomplete, at least within the time in which intake was measured (up to 2 h after preloading). Finally, in a similar two-condition study in children, no consistent differences were found between the effects of drinks sweetened with sucrose or aspartame, either on subjective ratings of hunger or on food intake.[29] Also, no differences were found between the effects of aspartame and sodium cyclamate. However, few details are available in this preliminary report.

## 7.2 Longer-term Studies

In a series of three studies, Porikos and her colleagues[30-34] measured the daily food intakes of a total of 24 individuals (13 obese and 11 lean, men and women) living continuously in a metabolic ward for between 15 and 30 days. A varied menu of foods and drinks was available throughout. After a baseline period many of the sucrose-containing items were replaced covertly by equivalent aspartame-sweetened products for 6–12 days, followed by a further baseline (sucrose) period. During aspartame substitution the energy intakes of both lean and obese subjects fell considerably before increasing to stabilise at about 85% of the baseline level (3400–4100 kcal/day). There was partial compensation for the change in energy density of the diet, amounting to about 40% of the energy 'saving' due to the consumption of aspartame-sweetened products, and significant increases in protein and fat intake after 3 days on the aspartame-containing diet.[32]

Two features of these influential studies are particularly noteworthy. First, the baseline intakes were very large (highly varied and palatable food was available free of charge to subjects throughout the day) and significant weight gains were recorded during the baseline periods, indicating energy intakes above the habitual levels of the subjects. Second, the mandatory consumption of a minimum of two soft drinks per day considerably increased the energy dilution (i.e. reduction in the energy density of the diet consumed) achieved by the substitution of aspartame for sucrose. As a result, complete caloric compensation during aspartame substitution would have required the consumption of very large and probably unacceptable amounts of food.[35] Indeed, as Booth[36] has pointed out, the degree of caloric compensation observed may represent complete compensation in relation to the habitual (pre-study) level of intake.

This conclusion receives support from a further study on caloric dilution, also carried out on subjects living continuously in a residential laboratory.[37] The covert substitution of reduced-calorie foods and drinks (products with reduced fat and aspartame substituted for sucrose) resulted in a reduction of energy intake in this group of items of 500 kcal/day. There was, however, rapid (within 1–3 days) and complete compensation for this deficit arising from an increase in the amounts of non-calorie-manipulated items consumed. Subsequently, when only the regular items were again available during a final 3-day period, the subjects failed to compensate for the increase in caloric density and *overate* significantly. Baseline intake was moderate (about 2500 kcal/day for adult male subjects) and there were no significant changes in body weight over the 14 days of the experiment.

Other studies examining the effects of consuming aspartame-containing foods and drinks have been conducted on free-living volunteers. Kanders *et al.*[38] studied obese men and women who had entered a multidisciplinary weight loss program. After 3 weeks on a 'balanced deficit diet' one group ('experimental' group) of participants was instructed to supplement their diets with low-calorie foods and drinks sweetened with aspartame, while a second ('control') group was told to avoid the use of all aspartame- and saccharin-sweetened products. The data showed a slightly greater weight loss among the women (but not the men) from the experimental group, and the authors conclude that 'adding aspartame may facilitate compliance to a hypocaloric diet'. However, it should be noted that they regard this as only a pilot study (no statistical analyses are presented). Moreover, it confounds effects arising from the 'addition to' and 'withdrawal from' the diet of items containing intense sweeteners; the decrease in daily aspartame intake in the control group was in fact rather greater than the increase in aspartame consumption displayed by the experimental group.

Two studies based on dietary surveys found that the use of saccharin[39] and aspartame[40] was associated with lower calorie intake. An earlier study, however, found no association between intense sweetener use and long-term weight loss in a group of obese individuals participating in a weight-reducing programme.[41] More recently, data on 78 000 women aged 50–69 years from a retrospective survey carried out by the American Cancer Society showed that 'the rate of weight gain of AS (artificial sweetener) users was significantly greater than in non-users irrespective of initial relative weight'.[42] In addition, 'the proportion of AS users who gained 10 pounds or more was significantly greater than the proportion of non-users who gained 10 pounds or more at each weight level', although in one weight category, the most obese, more intense sweetener users than non-users lost 10 pounds or more. These findings relate to weight changes over a 1-year period when saccharin was the predominant intense sweetener available.

Unfortunately it is not possible to conclude that the differences in food intake and body weight reported in these surveys were due to the use (or non-use) of intense sweeteners *per se*, because users and non-users may differ in other respects. In particular, users of intense sweeteners may attempt to exert a greater degree of restraint over their eating, resulting in reduced overall intake or perhaps in some circumstances a breakdown in the control of intake.[43] Indeed, in Parham and Parham's[39] survey the substitution of saccharin for sugar did not account completely for the lower energy intake of saccharin users, and their intake of certain important non-sugar-containing foods (e.g. bread) was also significantly lower.

Recently, Addington and Grunewald[44] have reported a study in which young college women consumed an aspartame- or sucrose-sweetened drink every day in addition to their usual diet. Importantly, the subjects were not aware whether the drink contained sugar or an intense sweetener. Although weight gains were similar over the 1-month period of the study, the women receiving the aspartame-sweetened drink reported stronger appetite and desire for high carbohydrate foods 2 h after receiving the drinks and reported consuming more sweet foods during the month than the women receiving the sucrose-sweetened drink.

### 7.3 Post-ingestive Effects of Intense Sweeteners

Ryan-Harshman et al.[45] examined the effects of large doses of up to 10·08 g of aspartame and one of its two constituent amino acids, phenylalanine, ingested in capsules (as many as 24 capsules in some cases). The results apparently failed to support any action of aspartame or phenylalanine on hunger or food intake measured in a test meal, although there were significant effects on plasma amino acid concentrations and ratios. However, no data are shown for the effects of the different treatments on hunger, and few statistical details are presented (e.g. although food intakes were analysed using ANOVA, no $F$-values are given). It is worth noting, therefore, that food intake was 13% lower than control (placebo) level after the highest dose (10·08 g) of phenylalanine and on average 9% lower after aspartame. This is equivalent to the degree of suppression of intake in a test meal brought about by 50 g of oral glucose[22] or by a potent anorexic drug.[46,47] In addition, other results indicate an inhibitory effect of phenylalanine on food intake,[19] and in a recent study[48] 234 mg of aspartame administered in two capsules with water brought about a significant reduction (12% on average) in test meal food intake compared with both the effect of drinking a solution of the same amount of aspartame and the effect of drinking plain water (both control conditions involved the ingestion of placebo capsules). Finally, a possible post-ingestive action of saccharin is indicated by its longer-term stimulatory effect on food intake evident in Fig. 3 (see below).

# 8 INTERPRETATIONS, IMPLICATIONS AND CONCLUSIONS

## 8.1 Additive and Substitutive Effects

The interpretation of experiments involving intense sweeteners depends crucially upon whether the experimenters have used a design that permits the identification of 'additive' or 'substitutive' effects. For example, referring to Fig. 2, the effect of an addition could be deduced by comparing conditions B and C, since B is derived from C by maintaining a constant caloric value but adding sweetness. This comparison is equivalent to asking what the effect is of making something sweet that formerly was bland. Alternatively, the effect of substitution can be deduced by comparing conditions A and B, since B is formed from A by removing calories (which are sweet) and replacing the sweetness. This is equivalent to asking what the effect is of altering caloric content while preserving the same level of sweetness. It follows from this that studies that have used only a limited design (two cells in Fig. 2) cannot, by definition, examine both substitutive and additive effects. Significantly, studies in which a 2-cell design[28] is of the substitutive type have nothing to say about the possible additive effects of sweetness. It is therefore important that all the studies permitting the disclosure of *additive* effects[20–23,25] have demonstrated some stimulatory effect of sweetness on either hunger or food intake, or both. For example, overall intake was found to be higher following the consumption of a sweetened compared with an unsweetened preload[23] (Fig. 3), and Brala and Hagen[25] found that test meal intake was significantly reduced when the perceived sweetness of milkshake preloads was suppressed by gymnemic acid (note that sweetness perception returned to normal at least 30 min before the test meal was eaten).

One way in which sweet tastes could be expected to promote increases in appetite is through their capacity to trigger pre-absorptive digestive responses[16] (see above). Supporting this interpretation is evidence that the taste of saccharin stimulates insulin release[17,49,50] and a fall in blood glucose levels.[51]

What can be concluded about the effects of the *substitution* in foods and drinks of intense sweeteners for sucrose and other sugars? One general finding is that items containing intense sweeteners have a lower satiating power than equivalent (higher calorie) items containing sugar. In all acute and longer-term studies, some degree of caloric compensation took place in response to the energy dilution achieved by the use of intense sweeteners. The extent of this compensation was variable, perhaps partly due to differences in baseline levels of food intake. Under certain circumstances complete caloric compensation was observed[37] (compare the intakes after saccharin and glucose shown in the first panel of Fig. 3). Since these studies involve the covert manipulation of caloric density, the results, irrespective of the accuracy of caloric compensation, provide evidence for the physiological regulation of short-term energy intake. This is an important conclusion; however, it need not necessarily undermine the intended action of intense sweeteners in, for example, promoting weight loss or dietary compliance. This is because under 'natural' (uncontrolled) circumstances many factors probably interact to determine the effects of intense sweeteners (see below).

Findings on the post-ingestive actions of intense sweeteners are relevant to both their addition and substitution in foods and drinks. One effect of aspartame may be an inhibition of appetite, perhaps because of the release of cholecystokinin promoted by phenylalanine, one of its breakdown products.[19] On the other hand, the relatively long-lasting rise in food intake brought about by saccharin (Fig. 3) suggests an (as yet unknown) physiological effect of this substance. Although it is not metabolised, saccharin may trigger receptors within the alimentary canal,[11] or act following its slow absorption from the gut.[52,53] Saccharin injected parenterally, for example, has been found to potentiate insulin-induced hypoglycaemia strongly,[54] and (along with cyclamate) it may also influence glucose metabolism through an ability to inhibit the activity of glucose-6-

phosphatase.[55] Certainly it is unlikely that the taste of saccharin could have produced increases in intake beyond the test meal in the above experiment.

## 8.2 Future Studies

While laboratory studies have revealed important information about the role of intense sweeteners in appetite control, further work is required to compare the effects of the different intense sweeteners and to examine the relationship between their possible physiological modes of action (e.g. stimulation of cephalic phase insulin secretion and cholecystokinin release) and influences on hunger and food intake. There are now good reasons to suppose that not all the intense sweeteners will produce identical response profiles. There are also a number of issues relevant to the potential efficacy of intense sweeteners used as an adjunct to dietary control which require systematic investigation. For example, under natural circumstances the choice and consumption of a product containing an intense sweetener will be guided by the knowledge that it has relatively low calorie content. The extent to which such knowledge might interact with sensory and metabolic influences to modify appetite and eating behaviour is largely unknown. Rolls *et al.*[28] found no difference in the response of subjects who were either informed or not informed that they were being given either high- or low-energy versions of the same preload. However, in another preloading study examining the covert manipulation of energy density, children displayed more accurate caloric compensation than adults,[56] indicating that in adults cognitive factors may overcome or interfere with physiological signals. (A further possibility is that there are age-related changes in the physiological system regulating appetite.) This is also important because dieters exert a high degree of cognitive control over eating,[43] and they are probably among the most frequent users of intense sweeteners. Studies examining the relationship between dietary restraint and the effects of intense sweeteners would be particularly valuable. Finally, it should be pointed out that the preloading studies reviewed here are relevant to the unconditioned effects of covertly altering caloric density. Accordingly they probably set a lower limit on the accuracy of caloric compensation. With repeated experience more accurate adjustments in voluntary intake might be expected.[12] Indeed, it has recently been reported that precise caloric compensation can come into play after as few as five exposures to a novel calorie-reduced food.[57]

## 8.3 Implications

It is worth pointing out that this review has been primarily directed to an evaluation of two scientific issues: (a) the effect of sweet taste on appetite and food intake, and (b) the capacity of the human physiological system to detect the calorie content of ingested fluids and foods (sweet and bland). These are quite separate phenomena. However, a good deal of confusion surrounds the interpretation of certain experiments because the modes of operation of these two phenomena are confounded. (Are studies designed to examine the effect of sweetness, caloric compensation or a mixture of both?) Because these two phenomena coexist for much of the time, unless studies are specifically designed to identify their separate effects, then the outcome will remain ambiguous. It is clear that the use of intense sweeteners is implicated in both the effect of sweetness itself and the adjustment of the calorie value of foods and drinks. Consequently, in the short term it is important to evaluate the effects of intense sweeteners with respect to these issues. As discussed above, studies indicate in general that sweetness tends to exert a stimulatory effect on intake, and the physiological system shows evidence of a capacity to monitor caloric value. Examination of the action of intense sweeteners has an important theoretical function in throwing light upon the mechanisms regulating short-term appetite and food intake.

There is, however, clearly a need for some debate about the extent to which the effects of sweeteners observed in short-term studies in the laboratory may be seen when people use intense sweeteners as part of a habitual dietary regimen. Some would argue that short-term effects are irrelevant and quickly get submerged by a variety of other influences; others would argue that the short-term effects continue to be expressed in an insidious way but are simply more difficult to detect because of the operation of many naturally fluctuating factors. One major logical difficulty is that the short-term studies normally use subjects selected (randomly or systematically) to preserve uniformity. On the other hand, under natural circumstances, individuals choosing to use sweeteners may possess particular attitudes, motivation, life-styles and physiological characteristics that make them unlike subjects evaluated in the laboratory. For the moment there is no ready resolution to this issue, but it seems clear that there is a need to integrate the approaches represented by short-term laboratory studies using sweeteners in controlled preloads and meals, medium-term studies on the effect of sweeteners incorporated into the diet, and long-term studies on the effects of sweetener use upon body weight. This will strengthen our understanding of appetite by identifying the relationship between physiological and socio-psychological factors operating to control patterns of food consumption.

## 9 POSTSCRIPT: INTENSE SWEETENERS AND THE LOW-CALORIE FOOD INDUSTRY

What are the implications of current findings on sweetness and caloric compensation for the industrial sector developing low-calorie foods, many of which make use of intense sweeteners? It is our view that manufacturers can do no more than place on the market products that give people the opportunity to construct an appropriate diet. It is not possible to guarantee that consumers will use high-sweetness/low-calorie products judiciously or rationally. Products may be abused or misused. People may eat two or three highly palatable, low-calorie items and thereby exceed the calorie intake from a single high-calorie product, or they may eat a low-calorie item in addition to normal meals.

The presence of sweet, low-calorie items in the marketplace does not necessarily make it any easier to undereat or to lose weight in the long term. This will always be difficult; if it were not so, millions of people would by now be freed from the burden of obesity. People will always have to find ways of managing the pleasing sensations of the sweet taste and the tendency of low-calorie foods to suppress appetite to a smaller extent than conventional foods. Undereating in our society is not easy. In order for low-calorie products (or any dietary regime) to lead to significant weight loss or weight control it will be necessary for people to act with self-restraint, self-control and self-discipline, and to remain vigilant. How strong is the evidence that people can do this? Unfortunately there appears to be a good deal of evidence to the contrary.

The effectiveness of low-calorie products in the control of body weight depends on informed decision, strong motivation and the ability to comply with recommendations to adjust life-style (calorie intake, eating profile, physical exercise, etc.). Low-calorie products provide people with the opportunity to incorporate one type of dietary adjustment into their life-style. The existence of low-calorie products *per se* is not a sufficient condition for weight loss or weight maintenance. A comparison may be made with a drug treatment. If the drug is taken only once a week instead of twice per day it will not be effective. Consumers must comply with instructions in order to make the treatment work. However, a low-calorie food is not a drug. Whereas taking a drug as prescribed may be enough to ensure effective treatment, simply taking a low-calorie product is not sufficient; control must be exerted continuously over all aspects of dietary intake and energy output. It may be debated who is competent to take advantage of the presence of low-calorie products in the food supply. If consumers are to benefit fully from the use of low-calorie foods and beverages they should understand how such products may best be incorporated into their diet. They should also be aware that a diet containing low-calorie products will not automatically aid weight control. In its promotion of low-calorie products, the food industry can help to educate the consumer in these facts. It is certainly known that many factors will conspire against the intention to lose weight even on the most potent and effective regime. Two factors which should not be underestimated comprise part of the psychobiological system controlling appetite; these are the stimulatory action of sweetness and the caloric suppression of hunger.

## ACKNOWLEDGEMENT

P.J.R. is supported by the Agricultural and Food Research Council (AFRC).

## REFERENCES

1. Blundell, J. E., Rogers, P. J. & Hill, A. J., Evaluating the satiating power of foods; implications for acceptance and consumption. In *Food Acceptance and Nutrition*, ed. J. Solms, D. A. Booth, R. M. Pangborn & O. Raunhardt. Academic Press, London, 1987, pp. 205–19.
2. Kissileff, H. R., Satiating efficiency and a strategy for conducting food loading experiments. *Neurosci. Biobehav. Rev.*, **8** (1984) 129–35.
3. Kissileff, H. R., Gruss, L. P., Thornton, J. & Jordan, H. A., The satiating efficiency of foods. *Physiol. Behav.*, **32** (1984) 319–22.
4. Naim, M. & Kare, M. R., Nutritional significance of sweetness. In *Nutritive Sweeteners*, ed. G. G. Birch & K. J. Parker. Applied Science, London, 1982, pp. 171–93.
5. Van Itallie, T. B. & Abraham, S., Some hazards of obesity and its treatment. In *Recent Advances in Obesity Research*, IV, ed. J. Hirsch & T. B. Van Itallie. Libby, London, 1985, pp. 1–19.
6. Garrow, J., Does plumpness matter? *Nutr. Bull.*, **7** (1982) 49–53.
7. Stunkard, A. J. & Stellar, E., *Eating and Its Disorders*. Raven Press, New York, 1984.
8. Pilgrim, F. J. & Kamen, J. M., Predictors of human food consumption. *Science*, **139** (1963) 501–2.
9. Le Magnen, J., Effets d'une pluralite de stimuli alimentaires sur le déterminisme quantitif de l'ingestion chez le rat blanc. *Arch. Sci. Physiol.*, **14** (1960) 411–19.
10. Wooley, S. C., Physiologic versus cognitive factors in short-term food regulation in the obese and non-obese. *Psychosom. Med.*, **34** (1972) 62–8.
11. Mei, N., Intestinal chemosensitivity. *Physiol. Rev.*, **65** (1985) 211–37.
12. Booth, D. A., Satiety and appetite are conditioned reactions. *Psychosom. Med.*, **39** (1977) 76–81.

13. Le Magnen, J., *Hunger.* Cambridge University Press, Cambridge, 1985.
14. Sjostrom, L., Garellick, G., Krotkiewski, M. & Luyckx, A., Peripheral insulin in response to the sight and smell of food. *Metabolism,* **29** (1980) 901–9.
15. Simon, C., Schlienger, J. L., Sapin, R. & Imler, M., Cephalic phase insulin secretion in relation to food presentation in normal and overweight subjects. *Physiol. Behav.,* **36** (1986) 465–9.
16. Geiselman, P. J. & Novin, D., The role of carbohydrates in appetite, hunger and obesity. *Appetite,* **3** (1982) 203–23.
17. Powley, T. L. & Berthoud, H.-R., Diet and cephalic phase insulin responses. *Amer. J. Clin. Nutr.,* **42** (1985) 991–1002.
18. Blundell, J. E. & Hill, A. J., Artificial sweeteners and the control of appetite; implications for eating disorders. In *The Future of Predictive Safety Evaluation,* ed. A. N. Worden, D. V. Parke & J. Marks. MTP Press, Lancaster, 1986, pp. 263–81.
19. Blundell, J. E., Rogers, P. J. & Hill, A. J., Uncoupling sweetness and calories; methodological aspects of laboratory studies on appetite control. *Appetite,* **11** (1988) Suppl., 54–61.
20. Rogers, P. J. & Blundell, J. E., Separating the actions of sweetness and calories: effects of saccharin and carbohydrates on hunger and food intake in human subjects. *Physiol. Behav.,* **45** (1989) in press.
21. Blundell, J. E. & Hill, A. J., Paradoxical effects of an intense sweetener (aspartame) on appetite. *Lancet,* **i** (1986) 1092–3.
22. Rogers, P. J., Carlyle, J.-A., Hill, A. J. & Blundell, J. E., Uncoupling sweet taste and calories: a comparison of the effects of glucose and three intense sweeteners on hunger and food intake. *Physiol. Behav.,* **43** (1988) 547–52.
23. Guss, J. L., Kissileff, H. R. & Pi-Sunyer, F. X., The effects of fructose and glucose preloads on food intake. Paper presented to the Eastern Psychological Association, April 1988.
24. Rogers, P. J. & Blundell, J. E., Comparison of the effects of aspartame, acesulfame-K and glucose sweetened preloads on hunger and food intake. In preparation.
25. Brala, P. M. & Hagen, R. L., Effects of sweetness perception and caloric value of a preload on short-term intake. *Physiol. Behav.,* **30** (1983) 1–9.
26. Rolls, B. J., Hetherington, M., Burley, V. J. & van Duijvenvoorde, P. M., Changing hedonic responses to foods during and after a meal. In *Interaction of the Chemical Senses with Nutrition,* ed. M. R. Kare & J. G. Brand. Academic Press, New York, 1986, pp. 247–68.
27. Rolls, B. J., Sweetness and satiety. In *Sweetness,* ed. J. Dobbing. Springer-Verlag, Berlin, 1986, pp. 161–73.
28. Rolls, B. J., Hetherington, M. & Laster, L. J., Comparison of the effects of aspartame and sucrose on appetite and food intake. *Appetite,* **11** (1988) Suppl., 62–7.
29. Anderson, G. H., Saravis, S., Schacher, R., Leiter, L. & Zlotkin, S., Aspartame: effects on appetite and food intake of children. *FASEB J.,* **2** (1988) A1197.
30. Porikos, K. P., Booth, G. & Van Itallie, T. B., Effect of covert nutritive dilution on the spontaneous food intake of obese individuals: a pilot study. *Amer. J. Clin. Nutr.,* **30** (1977) 1638–44.
31. Porikos, K. P., Control of food intake in man: response to covert caloric

dilution of a conventional and palatable diet. In *The Body Weight Regulatory System: Normal and Disturbed Mechanisms*, ed. L. A. Croffi, W. P. T. James & T. B. Van Itallie. Raven Press, New York, 1981, pp. 83–7.

32. Porikos, K. P., Hesser, M. F. & Van Itallie, T. B., Caloric regulation in normal-weight men maintained on a palatable diet of conventional foods. *Physiol. Behav.*, **29** (1982) 293–300.

33. Porikos, K. P. & Pi-Sunyer, F. X., Regulation of food intake in human obesity: studies with caloric dilution and exercise. *Clin. Endocrinol. Metab.*, **13** (1984) 547–61.

34. Porikos, K. P. & Van Itallie, T. B., Efficacy of low-calorie sweeteners in reducing food intake: studies with aspartame. In *Aspartame: Physiology and Biochemistry*, ed. D. Lewis, L. J. Stegink & J. Filer. Marcel Dekker, New York, 1984, pp. 273–86.

35. Blundell, J. E., Hill, A. J. & Rogers, P. J., Effects of aspartame on appetite and food intake. In *Dietary Phenylalanine and Brain Function*, ed. R. J. Wurtman. Birkhauser, Boston, 1988.

36. Booth, D. A., Evaluation of the usefulness of low-calorie sweeteners in weight control. In *Developments in Sweeteners*, Vol. 3, ed. T. H. Grenby. Elsevier Applied Science, London, 1987, pp. 275–95.

37. Foltin, R. W., Fischman, M. W., Emurian, C. S. & Rachlinski, J. J., Compensation for caloric dilution in humans given unrestricted access to food in a residential laboratory. *Appetite*, **10** (1988) 13–24.

38. Kanders, B. S., Lavin, P. T., Kowalchuk, M. B., Greenberg, I. & Blackburn, G. L., An evaluation of the effect of aspartame on weight loss. *Appetite*, **11** (1988) Suppl., 73–84.

39. Parham, E. S. & Parham, A. R., Saccharin use and sugar intake by college students. *J. Amer. Dietet. Assoc.*, **76** (1980) 560–3.

40. Smith, J. L. & Heybach, J. P., Evidence for the lower intake of calories and carbohydrate by 19–50 year old female aspartame users from the continuing survey of food intakes by individuals (CSFII85). *FASEB J.*, **2** (1988) A1197.

41. McCann, M. B., Trulson, M. F. & Stulb, S. C., Non-caloric sweeteners and weight reduction. *J. Amer. Dietet. Assoc.*, **32** (1956) 327–30.

42. Stellman, S. D. & Garfinkel, L., Artificial sweetener use and one-year weight change among women. *Preventive Med.*, **15** (1986) 195–202.

43. Polivy, J. & Herman, C. P., Dieting and binging: a causal analysis. *Amer. Psychol.*, **40** (1985) 193–201.

44. Addington, E. E. & Grunewald, K. K., Aspartame- or sugar-sweetened beverages: effects on weight gain, appetite and food intake in young women. *FASEB J.*, **2** (1988) A1197.

45. Ryan-Harshman, M., Leiter, L. A. & Anderson, G. A., Phenylalanine and aspartame fail to alter feeding behaviour, mood and arousal in men. *Physiol. Behav.*, **39** (1987) 247–53.

46. Rogers, P. J. & Blundell, J. E., Effect of anorexic drugs on food intake and the microstructure of eating in human subjects. *Psychopharmacology*, **66** (1979) 156–65.

47. Hill, A. J. & Blundell, J. E., Model system for investigating the actions of anorectic drugs: effect of d-fenfluramine on food intake, nutrient selection, food preferences, meal patterns, hunger and satiety in healthy human subjects. In

*Advances in BioSciences*, Vol. 60, ed. E. Ferrari & F. Brabilla. Pergamon Press, Oxford, 1986, pp. 377–89.

48. Rogers, P. J., Pleming, H. C. & Blundell, J. E., Postingestive effect of an intense sweetener: aspartame consumed without tasting inhibits hunger and food intake. *Physiol. Behav.* (submitted).

49. Berthoud, H. R., Bereiter, D. A., Trimble, E. R., Siegel, E. G. & Jeanrenaud, B., Cephalic phase, reflex insulin secretion. *Diabetologia*, **20** (1981) 393–401.

50. Halter, J., Kulkosky, P., Wood, S., Makous, W., Chen, M. & Porte, D., Afferent receptors, taste perception and pancreatic endocrine function in man. *Diabetes*, **24** (1975) 414.

51. Jorgensen, H., The influence of saccharin on blood sugar. *Acta Physiol. Scand.*, **20** (1950) 33–7.

52. Sweatman, T. W. & Renwick, A. G., The tissue distribution and pharmacokinetics of saccharin in the rat. *Toxic. Appl. Pharm.*, **55** (1980) 18–31.

53. Renwick, G. A., The disposition of saccharin in animals and man—a review. *Food Chem. Toxic.*, **33** (1985) 429–35.

54. Macallum, B., The potentiation of insulin by sulphones. *Canad. J. Res.*, **26E** (1948) 232–8.

55. Lygre, D. D., The inhibition by saccharin and cyclamate of phosphotransferase and phosphohydrolase activities of glucose-6-phosphatase. *Biochim. Biophys. Acta*, **341** (1974) 291–7.

56. Birch, L. L. & Deysher, M., Caloric compensation and sensory specific satiety: evidence for self-regulation of food intake by young children. *Appetite*, **7** (1986) 323–31.

57. Chabert, M. & Louis-Sylvestre, J., Caloric compensation in adolescent males. Paper presented at the First European Congress on Obesity. Stockholm, June 1988.

*Chapter* 11

# PERCEPTUAL CHARACTERISTICS OF SWEETENERS

ANNE TUNALEY

*Department of Food Science and Technology,*
*University of California, Davis, USA*

## SUMMARY

*Perceptual characteristics of sweeteners are of psychological origin, resulting from the interactions between a living organism and a sweetener. They play a major role in food acceptance, selection and intake, and are therefore important to the individual and to industry. Perceptual characteristics help determine which sweetener is most appropriate for a specific purpose and application, and whether it should be used alone or in combination with other sweeteners. A review of some previous publications in this area reveals that the majority concentrate on sweetness. In contrast, this chapter encompasses other sensory characteristics, and discusses which considerations are important in investigating the perceptual characteristics of sweeteners. These considerations include various influences on perception, selecting a sensory technique and data analysis. In concluding, suggestions are made on aspects that need further investigation, which could result in greater and more rapid advances in this area of the perceptual characteristics of sweeteners.*

## 1 PERCEPTUAL CHARACTERISTICS

### 1.1 What are They?

Perceptual characteristics are psychological, resulting from the interactions between a living organism and a stimulus, such as a sweetener. As the name suggests, they involve perceptions and the senses. Thus, unlike degrees Brix or pH, they are not physicochemical properties, so the phrases 'perceptual properties' and 'sensory properties' are misleading and incorrect;

'perceptual characteristics' and 'sensory characteristics' are more appropriate descriptions.

In this chapter the term perceptual characteristics will be used in its
general sense, to include sensory and hedonic aspects of sweeteners. It is
extremely important to separate these two areas and not to combine
and confuse them.[1-3] Reliable and valid qualitative and quantitative
information on sweetness, bitterness, viscosity and other sensory
characteristics can be obtained only if the assessments are not influenced by
hedonic related factors.

### 1.2 Why are They so Important?

Perceptual characteristics can play a major role in food selection and
intake. They are very important for sweeteners, as it has long been
recognised that other sensory characteristics in addition to sweetness are
also perceived in them,[4] and it is often these additional characteristics that
help determine acceptance.

Hence, what we perceive affects preference and acceptance, and these
influence choice of purchase and profitability. Therefore perceptual
characteristics are important not only to the individual but also to industry.

## 2 SWEETENERS

### 2.1 The Range

A very large number of compounds, either found in nature or synthesised,
are perceived by humans to be sweet.[5] They come from many sources and
have a wide variety of chemical structures,[6,7] and include acesulfames,
carbohydrates, dihydrochalcones, dipeptides, nitroanilines, oximes,
saccharins and urea derivatives.[8]

In recent years the search for sweeteners has intensified. However, every
major class of synthetic sweeteners was discovered by accident, through
projects that had aims other than producing sweet-tasting compounds.[8]
Saccharin was the first major synthetic sweetener, discovered in 1879.
Others which followed include dulcin (discovered in 1883), cyclamate
(1937), aspartame (1965) and acesulfame-K (1967).

In general, this chapter will concentrate on sweeteners that are of
particular relevance to the food industry.

### 2.2 Factors that Determine Suitability

Sweeteners are important ingredients in numerous foods, drinks and
pharmaceutical products, and they appear in more articles of the diet than

any other food item, except table salt.[9] Hence, as there are a number of sweeteners from which to choose, and each is unique, a number of factors help to decide which one is most suitable for a specific purpose and application. These factors include cost, availability, legislation, processing requirements, energy value and perceptual characteristics.[10]

When substituting one sweetener with another, it is often desirable that there are no significant changes in the perceptual characteristics of the product. Accurate and detailed information on these is therefore essential to aid the selection of a suitable sweetener, whether it is used alone or in combination with others.[11]

## 3 LITERATURE ON PERCEPTUAL CHARACTERISTICS OF SWEETENERS

### 3.1 Brief Review

A number of chapters have been written which contain small sections on perceptual characteristics of sweeteners, but few are dedicated solely to this important area. A published review of these does not appear to exist, so this chapter attempts to fill the gap. It is not intended to be a detailed and comprehensive review, but it includes relevant material written over the past two decades and highlights the salient points.

The majority of previous chapters concentrated on sweetness and were from the proceedings of symposia. A large number referred to hedonic related aspects in their title or contents, or were presented at conferences on food acceptance. The importance of distinguishing between sensory and hedonic assessments was stressed by Pangborn.[1-3] Although the first publication originated from a symposium dedicated to sugar,[1] other sweeteners were mentioned in the results of experiments that were presented as examples of different measurement techniques, which were categorised under four headings: sensitivity or discrimination, perceived strength or intensity, duration or persistence, and affective or hedonic.

Pangborn returned to this topic a year later, and examined the similarities and differences related to sweetness responses. The latter were classified into the two distinct types 'analytical' and 'affective', and techniques used to measure them were discussed.[2]

Although affectivity was not covered in detail in Pangborn's 1987 publication, it was included as the final category in a table of different techniques '... to emphasise its distinction from the previous measures'.[3] The chapter reviewed the effects of additional tastes and other sensory attributes, such as colour, aroma, viscosity and solution temperature, on sweetness. The observation was made that most of the research in this area had been empirical, on aqueous solutions by a variety of techniques. A systematic summary of early literature on interactions of sweetness in aqueous solutions was followed by an examination of more recent research, including that on foods and beverages.

In the proceedings of a symposium on 'Criteria of food acceptance—how man chooses what he eats', Birch discussed the basic tastes of sugar molecules.[12] Interest was expressed in time–intensity studies, as information on temporal aspects can help explain the influence of molecular structure on taste and hence acceptance. It is encouraging to read a chemist acknowledging the relevance of sensory research and the existence in sugars of sensory characteristics in addition to sweetness.

Four chapters directly connected to hedonics of sweetness and sweeteners were published in 1987. Frijters presented the psychophysical approach to the effects of sweetness on human behaviour, and discussed perceived intensity, pleasantness and attitudes.[13] Booth et al. took another approach to a similar topic at the same symposium, and examined the complex roles of sweeteners and sweetness in the perception, acceptance and selection of food.[14]

Frijters then turned to a different topic from his previous sweetness chapter, outlining six determinations to conduct when investigating partial and complete substitution of a sugar by another sweetener.[15] These ranged from the sweetness preference function to a taste profile. Approximately half of the paper was devoted to interactions in mixtures of sweeteners.

At the same symposium, Lindley examined factors, such as intensity, safety, functionality and sweetness quality, which can influence the acceptability of sugars and intense sweeteners.[16] Some of the sensory differences between sweeteners were considered, and the comment was made that most sugars and sugar syrups have very similar 'sensory and functional properties'. The view was expressed that the greater the similarity between products containing sugars or their syrups and those that are sweetened by alternative agents, the higher the probability of consumer acceptance. A very important conclusion was that all sweeteners have features that food manufacturers can exploit.

In 1987 five chapters—more than in any other year—were published on topics related to perceptual characteristics of sweeteners. The one by Pangborn, entitled 'Selected factors influencing sensory perception of sweetness', was reviewed earlier in this section; it was her fourth in the area.[3] The first, published 13 years earlier, had the similar title 'Sensory perception of sweetness',[9] but each approached the subject from a new perspective and presented different experimental data. In the 1974 chapter, a brief history of sweetness and sweeteners was followed by research examples to highlight the three major experimental variables that influence sweetness. These were classified as variations (a) within the subject, (b) inherent in the physical or chemical stimulus and (c) from the test procedures and external environment. The topics covered ranged from sweet quality to pleasantness, and from taste mixtures to sex and age differences in sweetness responses.

The same proceedings also contained a contribution on 'Sweetness and sensory properties of dextrose–levulose syrup'.[17] Results were presented from experiments to determine the suitability of this syrup for soft drinks, but unfortunately only sweetness was considered in any detail.

Other chapters which concentrate on perception of sweetness include one on the psychology of sweetness by Moskowitz.[18] A wide spectrum of aspects was reviewed to illustrate the multidisciplinary nature of this topic. The suggestion was made that, to unify the area, general organising principles must be found from within relevant publications, and then a predictive model developed to help explain the diverse findings.

Spencer[19] referred to the diversity of taste panels when citing experiments on sweetness that involved trained and untrained panels of various sizes. Results were quoted to support the view that, for the valid assessment of sweetness, selected and trained judges in controlled laboratory conditions are necessary. While concentrating on the sweet taste, the relevance of texture and olfaction in this area was also mentioned.

Three authors from a laboratory of human behaviour and metabolism examined the sweet taste in connection with obesity.[20] After discussing changes in the American diet in this century, and the possible associated health risks, they presented data from experiments investigating the perceptions and dietary habits of obese people. Detection thresholds and preferences were the same or lower in obese compared with normal-weight subjects, and all their results refuted the belief that the obese have a heightened craving for the sweet taste. The implications was that obesity is not caused by a strong desire to consume sweet foods, and the finger of suspicion was pointed at fats.

### 3.2 The Present Review

Many articles have been dedicated to diverse aspects of sweeteners, including their chemistry,[21] physiology,[22] nutrition,[7] sociology[23] and legislation.[24] This chapter provides an opportunity to discuss the sometimes ignored but extremely important topic of perceptual characteristics of sweeteners. In contrast to the earlier ones, this is not from the proceedings of a symposium, and will discuss a number of sensory characteristics, not just sweetness. Other unique features are that it looks back to assess past chapters in the area, then looks forward and suggests aspects that need further investigation and improvement in the future.

## 4 INVESTIGATING PERCEPTUAL CHARACTERISTICS OF SWEETENERS

### 4.1 Factors that Influence Perceptual Characteristics

For sweeteners, these factors include concentration, temperature, pH and the form in which the sweetener is presented. An example of the latter is when crystalline forms of sugar alcohols are assessed, especially xylitol, a negative specific heat of solution results in a cooling sensation, similar to that of menthol.[25] When the polyol is presented in solution this effect is absent.

All these and other variables, such as the use of different assessors and techniques, can affect the data, and must be taken into account during interpretation and evaluation. For these reasons, direct comparison of results and conclusions from different experiments is usually difficult.

When conducting studies on hedonics, many factors must be considered, expecially that responses obtained are context-dependent and dynamic, because they may alter when any of the factors that affect them changes.[26]

### 4.2 Selecting a Sensory Technique

A variety of different techniques are available to investigate perceptual characteristics.[3,27] An expert in the field is able to assess the aims and objectives of a specific experiment and select the most appropriate method. It is unfortunate that the literature contains examples of incorrect and inappropriate use of methods and scales. Techniques can be classified as sensitivity, discrimination, quantitative, qualitative and affective measurements.[3] Some are combinations; for example, profiling involves qualitative and quantitative assessments of all the sensory characteristics perceived in samples.

As previous chapters have examined different sensory techniques in detail (see Section 3.1), it would be uneconomical of space to cover them again. However, one under-utilised method which has not been discussed extensively is that of time–intensity.

This technique characterises temporal perceptions which can be the distinguishing feature of some sweeteners, where assessors perceive the same total intensity but very different temporal sensations. With most other sensory techniques, assessors 'average' their responses and give a single intensity. This results in a loss of important information, which time–intensity evaluation records. Over recent years this technique has been improved and has progressed from assessors making measurements at set time intervals, at which they wrote down a number that represented the perceived intensity,[28] to the computerisation of the different steps.[29] As is true with all methods, time–intensity data are valuable alone but become even more meaningful when used in conjunction with additional information from other techniques.

**4.3 Data Analysis**

There are many different statistical techniques, and it is very important to select the most appropriate for a given data-set and experimental objectives. Certain univariate and multivariate techniques can be applied to data on perceptual characteristics,[30] and by using both types of analysis, the maximum amount of information can be obtained.

Multivariate methods have been applied to this area only in recent years, and they need to be investigated and utilised more extensively, as they enable multivariate data to be understood more easily. This is achieved by assisting interpretation through classification and indicating relationships among variables, reducing fluctuation and visualising the structure of complex data.[31]

A few multivariate methods will illustrate some of the advantages of applying these techniques to analyse data on perceptual characteristics.

An empirical method that can be applied to both traditional and free-choice profiling data is generalised Procrustes analysis (GPA).[32] Very few analytical techniques take account of the fact that humans are individuals with differences in sensitivity and perception and that, despite training, their responses vary. However, GPA is one method that does consider idiosyncrasies in the use of descriptors and the measurement scale. It provides individual and group data and retains the subtleties of the results. In addition, interpretation of multidimensional characteristics is made less subjective or biased and more reliable and formal by providing weightings or loading coefficients associated with each characteristic for every assessor.[32]

Two multivariate techniques that can be applied to preference data are internal (MDPREF) and external (PREFMAP) preference mapping.[33,34] Both provide visual representation of samples and preferences. MDPREF resolves subject scores into a set of preference dimensions that represent differences among samples. There is a vector for each individual, and the direction of increasing preference is shown. Correlations of sensory characteristics obtained from laboratory panels can be superimposed.[35] PREFMAP also retains individual differences but uses linear or quadratic regressions to express preference scores in terms of the sensory dimensions of a multidimensional space, as defined by a laboratory panel.[34]

One of the disadvantages with many statistical techniques, especially multivariate ones, is that they are time-consuming. However, an advantage with automated sensory data collection systems is the potential to make the application of analytical techniques easier and faster.[36]

### 4.4 The Nature of Perceptual Characteristics

*4.4.1 Sensory Characteristics*

These can be classified as odours, flavours, mouthfeel and aftertastes. It is often assumed that odours are not detectable in sweeteners, but at certain concentrations, for some sweeteners, subjects and systems, they are perceived. For example, the descriptive term 'liquorice' is often associated with glycyrrhizin, and this can be undesirable in certain products.[37] In contrast, sucrose is usually considered to have no appreciable odour.[38]

The one common feature of all sweeteners is that they taste sweet, and a great deal of time and effort has been devoted to the study of this one sensory characteristic. This is reflected in a number of publications in the field (see Section 3.1), and for that reason sweetness will not be discussed in detail in this chapter.

It is important to determine differences in sweetness between equi-weight concentrations of sweeteners. Such information is needed before sensory characteristics other than sweetness can be investigated qualitatively and quantitatively or preference can be assessed.

Sucrose is often used as the sweetness and sensory standard and hedonic ideal, against which other sweeteners are compared.[16,39] A sweetness value for sucrose of 1 or 100 is often arbitrarily set for comparison with the equivalent sweetening power of other compounds. This method of expressing and presenting equi-sweetness is not ideal, and it is easy to lose sight of what the figures represent. Equi-sweet values in the literature are often very variable, as Table 1 demonstrates. There are a number of possible reasons for the discrepancies;[44] these include different assessors,

TABLE 1

PUBLISHED RELATIVE SWEETNESS RANGES

| Sweetener | Range[a] | References |
|-----------|----------|------------|
| Acesulfame-K | 130–200 | 10, 40 |
| Aspartame | 128–200 | 10, 41 |
| Fructose | 1·0–1·8 | 42, 43 |
| Glucose | 0·50–0·80 | 44, 45 |
| Glycerol | 0·48–1·1 | 42, 46 |
| Lactitol | 0·23–0·35 | 10, 47 |
| Lactose | 0·15–0·60 | 46, 48 |
| Saccharin | 300–500 | 19, 42 |
| Sorbitol | 0·40–0·57 | 10, 49 |
| Stevioside | 110–300 | 50, 51 |

[a] Sweetness intensity of sucrose = 1.

testing media, pH values, temperatures, sensory methods, standards and their concentrations.

The number and nature of sensory characteristics besides sweetness are important factors when determining equi-sweet concentrations of sweeteners. Redlinger and Setser[52] experienced difficulty when determining concentrations of acesulfame-K, saccharin and cyclamate that were of equivalent sweetness to 5% sucrose, in aqueous and lipid model systems. A persistent bitterness was perceived in all these samples, but, while saccharin had a sharp immediate sweetness, this was of slow onset with acesulfame-K. These findings support those of other workers who had problems when obtaining equi-sweet concentrations for saccharin and cyclamate in flavoured beverages and gelatins,[53,54] and for saccharin and acesulfame-K in aqueous solutions.[10] Experiments on saccharin have shown that an accentuating factor is that, with increasing concentration, bitterness increases faster than sweetness.[55]

Intensity and quality of sweetness were investigated by Moskowitz,[55] who reported that 'artificial' sweeteners differed greatly from sugars. A number of studies have found that no sweetener was perceived to be identical to sucrose.[52–56] Even glucose can produce a moderate burning and bitter side-taste at the back of the mouth.[57] In contrast, sucrose has few perceived secondary attributes, and is often described as having a pure, clean sweetness.[58]

In an investigation of flavoured beverages and gelatins,[53] drinks sweetened with sucrose or aspartame were characterised by a sweet-clean taste and aftertaste, and those with calcium cyclamate or sodium saccharin as sweet-chemical and bitter. Samples containing saccharin were considered significantly more astringent and sour than the others, and those made with cyclamate had the highest intensity of medicinal, cloying and sticky-sweet aftertaste. For the majority of characteristics, saccharin differed the most from sucrose, and aspartame the least.

Similar results were obtained with the gelatins, and approximately the same conclusions were drawn in a study of sweetened aqueous and lipid model systems,[52] in which aspartame had a low non-sweet aftertaste and a similar sweetness profile to sucrose. This all links with previous work that described the sweetness of aspartame as 'sucrose-like',[59,60] and it could help to explain why aspartame has been the most commercially successful new sweetener to date.[60] However, one study found that the main differences between sucrose and aspartame were that the latter had a high sweetness that lingered, and a bitter aftertaste.[52] This is very similar to the results in a paper by Samundsen,[61] where the aftertaste of aspartame was described as a lingering sweetness, bitter-sweet and powdery.

When aspartame was assessed in shortbread-type cookies, the non-sweet aftertaste was much higher in the baked than unbaked product,[56] and assessors characterised it as bitter, medicinal and metallic. Interactions with other ingredients in the food system, plus the loss of sweetness during thermal processing,[41] were possible contributory factors. This study also obtained descriptors of bitter and medicinal for acesulfame-K, saccharin and cyclamate, in both raw and baked cookies.[56] When the same sweeteners were assessed in aqueous and lipid model systems, bitterness, metallic flavour and dryness were perceived in all samples,[52] and their aftertastes in solution were not significantly different. This agrees with other experiments that found acesulfame-K and saccharin to be perceptually similar.[56,59,62]

There appear to be disagreements between different studies. In contrast to the above data, acesulfame-K has been reported to have little or no aftertaste.[39,40,63] Redlinger and Setser[52] found that in cream systems the aftertaste of cyclamate was very high, while a previous study found it was almost free of aftertaste.[63] The sweetener concentration, the system in which it is presented and differences between assessors and the affiliations of authors are possible contributory factors in producing such discrepancies.

Whenever sweeteners are assessed, perceptual characteristics change with time, depending on the nature and intensity of sensory characteristics and also on hedonic related factors. Time–intensity techniques measure these changes, and a number of studies have investigated temporal perceptions of natural and synthetic sweeteners, finding that each temporal pattern is highly substance-specific.[53,64,65]

When the technique was applied to orange gelatins containing equi-sweet concentrations of one of four sweeteners,[53] bitterness was the main distinguishing characteristic. The saccharin sample was the most bitter, reaching its maximum intensity after 20 s of oral manipulation. The amount of bitterness in the other samples decreased in the order cyclamate > aspartame > sucrose.

The time–intensity technique has also been applied to less common sweeteners. Some of the results include those on $\beta$-neohesperidin dihydrochalcone, which had a slight delay in taste onset and a lingering aftertaste.[66] Similar curves were obtained from ammoniated glycyr-rhizinate,[67] but the extent of delay and persistence was much greater.[65] These contrast sharply with the sweetness profile of sucrose, which is perceived very quickly and has a sharp cut-off.[68]

Birch and Ogunmoyela[69] studied temporal perception to investigate molecular structure and taste. They found that, in a sucrose-sweetened chocolate drink, sweetness persistence and intensity rose with increasing concentrations of glycerol monostearate or lecithin. They hypothesised that this was probably due to the surfactants lowering viscosity, which enabled sucrose molecules to diffuse more easily to receptors and therefore taste sweeter longer.

Time–intensity studies have assessed synergism, which results when the sweetness of a mixture is greater than the sum of that perceived in its components.[70] Synergistic sweetness effects were found in lactose–xylitol mixtures, but the reverse (suppression) in lactose–galactose mixtures. If the sweetness of lactose can be enhanced by mixing it with other sweeteners, the number of applications for which it is suitable increases.[71]

One group of sensory characteristics that are important for many products are those related to mouthfeel. This is especially true for beverages. At the concentrations used in most soft drinks, sugars are important for viscosity or 'body'. In a study of sugars for this purpose, Moskowitz found that 'viscosity–fluidity' was the most important discriminatory characteristic.[72] As polyols are less sweet than sugars at equi-weight concentrations, greater quantities are needed, and hence the products are more viscous. In contrast, intense or potent sweeteners are present in such small amounts that 'body' has to be provided from other sources such as gels, thickeners and gums.

The overall conclusions from all this sensory information are that sweeteners are not just sweet, but additional characteristics are perceived in all sweeteners, and can include odours, flavours, mouthfeel and aftertastes. Further, no two sweeteners are identical, as they vary in the number, nature and intensity of sensory characteristics. Hence, for each sweetener there are likely to be applications for which it is best suited, whether used alone or in mixtures.

### 4.4.2 Hedonics

Sweeteners probably provide more pleasure to mankind than any other edible substance. Sweetness is considered to be pleasant in a wide variety of contexts,[73] and the ability to perceive and like sweetness is inborn. Facial expressions of neonates after oral administration of sweet solutions have been studied during the first hours after birth.[74] Despite no previous extra-uterine feeding experiences, they show expressions that are interpreted as displaying enjoyment, liking and appreciation.

Sweetness is associated with emotive value judgements that are considered to be psychologically, socially and materially desirable. Sweet foods are attractive far beyond their value in relieving hunger, while sweet drinks are consumed in excess of their value in relieving thirst.

Sucrose has positive hedonic qualities which increase with increasing concentration, up to an optimum. In contrast, glycerol exhibits a non-monotonic function, with pleasantness decreasing and then increasing as a function of concentration.[38] This effect depends on bitterness dissipating at higher concentrations, and sweetness dominating. As with many sweeteners, bitterness can contribute more to the overall dislike/like of samples than the sweetness.[38]

Much of the published data on preference of sweeteners is on specific food systems.[75,76] One recent study on aqueous solutions of nine sweeteners[77] found that with 17 assessors sucrose was the most liked, and a mixed extract of the leaves of *Stevia rebaudiana* Bertoni (SrB) was the least liked. The sugars and sugar alcohols sucrose, fructose, lactitol and sorbitol all had higher mean preference scores than the non-nutritive sweeteners aspartame, saccharin, acesulfame-K and SrB. Results for sensory characteristics obtained from conventional profiling were related to preference assessments from the same judges. Significant negative correlations were observed for bitter and persistent aftertastes and bitter flavour, while clean flavour was positively correlated with preference. These findings support the mean preference results, as SrB, acesulfame-K and saccharin were perceived as bitter and persistent, and had the lowest preference scores, while the sugars and polyols had the highest scores and were perceived as 'clean'. As will all preference results, it must be remembered that they are context-dependent and represent the preferences of individual assessors under a given set of conditions at a particular time.

Preference and acceptability should not be confused; a sample may be scored as 'most preferred' within a set of samples, yet none may be acceptable. Also, acceptability is different from utilisation, as consumers may find a food acceptable on sensory grounds but never purchase or consume it because of other considerations.

The decision to consume a food is usually complex and is based on social, health, sensory and hedonic factors. The situation is dynamic and can change over time and on exposure. For example, in the past few years there has been a great increase in the consumption of carbonated soft drinks containing non-nutritive sweeteners.[78]

A recent study investigated the consumption of regular and diet sodas, plus the beliefs and concerns of the consumers.[79] Of 100 young American females, 30% drank regular and 44% diet sodas. Each group reported loyalty to the type of beverage they usually consumed, plus dislike of the other type, while non-users did not like either type of beverage. Examining consumer beliefs can provide insight into reasons for preference of a specific soda. Each group perceived their 'own' soda superior in taste, more efficient in quenching thirst and more compatible with other menu items. Selection of sodas may also reflect overall consumption patterns, and attitudes to one's body. Users of regular sodas were less concerned about weight-related issues and reported consumption of several foods high in sugar, fat and/or sodium, at significantly higher frequencies than drinkers of diet sodas and non-users.

All this detailed information on the perceptual characteristics of sweeteners can be utilised in a number of areas, but especially in process and product development, quality assurance, correlation of sensory with instrumental data and market research.[80] It is hoped that such knowledge can help manufacturers improve the quality and hence preference and acceptance of sweetened products.

Despite this importance, there are not many publications of detailed investigations on the perceptual characteristics of sweeteners. However, this does not mean that the work has not been performed. Industry has directed a great deal of time, effort and money to this area, but most of the research remains unpublished and therefore unavailable to the general scientific population.

## 5 SUGGESTIONS FOR THE FUTURE

There are a number of areas related to perceptual characteristics of sweeteners that require further investigation. This section refers to those that are likely to be most beneficial in advancing knowledge and understanding, and could result in greater and more rapid advances in this field.

Much of the research on sweeteners has been conducted on model systems, the results from which are sometimes extrapolated to foods. However, this can give spurious information, especially with preference results.[75] Thus the lack of investigations conducted directly on food systems should be rectified.

The search continues for a sweetener that is perceptually and functionally similar to sucrose, non-toxic, with low or no energy value, easily available, cost-effective, physiologically inert and non-cariogenic.[11,81] At present no one sweetener is ideal and fulfils all these requirements, so the use of mixtures appears to be the most obvious solution. The multiple sweetener approach has already stimulated some studies, but additional and more detailed investigations on perceptual characteristics are essential. Information on various combinations and concentrations of sweeteners in different systems could help to minimise the limitations of single sweeteners and maximise the advantages of mixtures.

Investigations are needed to assess the effects on perceptual characteristics of prolonged exposure to certain sweeteners. Presentation of physically identical stimuli on a number of different occasions can evoke perceptions and responses that vary.[82] This applies to both laboratory panels and consumers.

An important consideration is that perception may be affected by the trend towards sweetener substitution, and the increased consumption of reformulated products.[78] The reasons for any effects would probably be complex, but could include changed points of reference due to repeated exposure to alternative sweeteners. One possibility is that sucrose might stop being the sensory standard and hedonic ideal[16,39] for certain individuals. The results of projects to investigate such effects would be extremely interesting and valuable.

Studies should also be conducted to compare laboratory panel and consumer responses related to perceptual characteristics of the same sweet samples. A number of experiments used laboratory panels,[54,83] and others have employed consumers.[84,85] However, investigations on sweeteners that used both types of subjects do not exist, and studies conducted on other products concentrated on preference.[86,87] One suggestion for future research is to use free-choice profiling[88] to obtain detailed information on perceptual characteristics of sweeteners from a trained laboratory panel and also from consumers.

Once data have been collected, it is vital to select the most appropriate statistical technique to obtain the maximum amount of reliable information (see Section 4.3). More investigations of the application of different analytical techniques, especially multivariate ones, to data on perceptual characteristics of sweeteners would be very beneficial.

Finally, it is highly desirable that much more of the perceptual research on sweeteners performed by and for industry is made public. The effects of this release and dissemination of information would not only improve our knowledge and understanding but could facilitate greater and more rapid advances in this important area.

## REFERENCES

1. Pangborn, R. M., Sensory discrimination, perceived intensity, and hedonic responses to sweetness in beverages. In *Sugar: Science and Technology*, ed. G. G. Birch & K. J. Parker. Applied Science Publishers, London, 1979, pp. 383–401.
2. Pangborn, R. M., A critical analysis of sensory responses to sweetness. In *Carbohydrate Sweeteners in Foods and Nutrition*, ed. P. Koivistoinen & L. Hyvonen. Academic Press, New York, 1980, pp. 87–110.
3. Pangborn, R. M., Selected factors influencing sensory perception of sweetness. In *Sweetness*, ed. J. Dobbing. Springer-Verlag, London, 1987, pp. 49–66.
4. Moncrieff, R. W., *The Chemical Senses*. Wiley, New York, 1946.
5. O'Brien Nabors, L. & Gelardi, R. C., *Alternative Sweeteners*. Marcel Dekker, New York, 1986.
6. Birch, G. G. & Lee, C. K., The theory of sweetness. In *Developments in Sweeteners*, Vol. 1, ed. C. A. M. Hough, K. J. Parker & A. J. Vlitos. Applied Science Publishers, London, 1979, pp. 165–86.
7. Bruce, A., Sweeteners in nutrition policy. In *Carbohydrate Sweeteners in Foods and Nutrition*, ed. P. Koivistoinen & L. Hyvonen. Academic Press, New York, 1980, pp. 45–59.
8. van der Wel, H., van der Heijden, A. & Peer, H. G., Sweeteners. *Food Reviews International*, **3**(3) (1987) 193–268.
9. Pangborn, R. M., Sensory perception of sweetness. In *Symposium: Sweeteners*, ed. G. E. Inglett. AVI, Westport, CT, 1974, pp. 23–44.
10. Tunaley, A., Thomson, D. M. H. & McEwan, J. A., Determination of equi-sweet concentrations of nine sweeteners using a relative rating technique. *International Journal of Food Science and Technology*, **22** (1987) 627–35.
11. Gelardi, R. C., The multiple sweetener approach and new sweeteners on the horizon. *Food Technology*, **41**(1) (1987) 123–4.

12. Birch, G. G., Basic tastes of sugar molecules. In *Criteria of Food Acceptance: How Man Chooses What He Eats*, ed. J. Solms & R. L. Hall. Forster Verlag, Zürich, 1981, pp. 282–91.
13. Frijters, J. E. R., Sensory sweetness perception, its pleasantness, and attitudes to sweet foods. In *Sweetness*, ed. J. Dobbing. Springer-Verlag, London, 1987, pp. 67–80.
14. Booth, D. A., Conner, M. T. & Marie, S., Sweetness and food selection: measurement of sweeteners' effects on acceptance. In *Sweetness*, ed. J. Dobbing. Springer-Verlag, London, 1987, pp. 143–60.
15. Frijters, J. E. R., Aspects of sugar substitution in sweet foods and drinks. In *Food Acceptance and Nutrition*, ed. J. Solmes, D. A. Booth, R. M. Pangborn & O. Raunhardt. Academic Press, London, 1987, pp. 115–27.
16. Lindley, M. G., Acceptance effects of sugars and intense sweeteners. In *Food Acceptance and Nutrition*, ed. J. Solmes, D. A. Booth, R. M. Pangborn & O. Raunhardt. Academic Press, London, 1987, pp. 99–114.
17. Brooks, G. A., Warnecke, M. O. & Long, J. E., Sweetness and sensory properties of dextrose–levulose syrup. In *Symposium: Sweeteners*, ed. G. E. Inglett. AVI, Westport, CT, 1974, pp. 97–110.
18. Moskowitz, H. R., The psychology of sweetness: historical trends and current research. In *Sweeteners and Dental Caries*, ed. J. H. Shaw & G. G. Roussos. Information Retrieval Inc., Washington, 1978, pp. 41–74.
19. Spencer, H. W., Taste panels and the measurement of sweetness. In *Sweetness and Sweeteners*, ed. G. G. Birch, L. F. Green & C. B. Coulson. Applied Science Publishers, London, 1971, pp. 112–29.
20. Drewnowski, A., Gruen, R. K. & Grinker, J., Carbohydrates, sweet taste, and obesity: changing consumption patterns and health implications. In *Food Carbohydrates*, ed. D. R. Lineback & G. E. Inglett. AVI, Westport, CT, 1982, pp. 153–69.
21. Birch, G. G., Chemical aspects of sweetness. In *Sweetness*, ed. J. Dobbing. Springer-Verlag, London, 1987, pp. 3–13.
22. Ylikahri, R. H. & Pelkonen, R., Carbohydrate sweeteners in metabolism and diseases. In *Carbohydrate Sweeteners in Foods and Nutrition*, ed. P. Koivistoinen & L. Hyvonen. Academic Press, New York, 1980, pp. 15–35.
23. Fischler, C., Attitudes towards sugar and sweetness in historical and social perspective. In *Sweetness*, ed. J. Dobbing. Springer-Verlag, London, 1987, pp. 83–98.
24. Elias, P. S., Legislative aspects of artificial sweeteners and other food additives. In *Sweetness and Sweeteners*, ed. G. G. Birch, L. F. Green & C. B. Coulson. Applied Science Publishers, London, 1971, pp. 139–59.
25. Emodi, A., Xylitol: its properties and food applications. *Food Technology*, **32**(1) (1978) 28–32.
26. Frijters, J. E. R., Sensory evaluation: a link between food and food acceptance research. *Lebensmittel Wissenschaft und Technologie*, **8** (1975) 294–7.
27. Stone, H. & Sidel, J. L., *Sensory Evaluation Practices*. Academic Press, London, 1985.
28. Neilson, A., Time–intensity studies. In *Flavor Research and Food Acceptance*, Arthur D. Little Inc. (Sponsor). Reinhold, New York, 1957, Ch. 7.
29. Lee, W. E., III & Pangborn, R. M., Time–intensity: the temporal aspects of sensory perception. *Food Technology*, **40**(11) (1986) 71–8.

30. Piggott, J. R. (ed.), *Statistical Procedures in Food Research*. Elsevier Applied Science, London, 1986.
31. Piggott, J. R. & Sharman, K., Methods to aid interpretation of multidimensional data. In *Statistical Procedures in Food Research*, ed. J. R. Piggott. Elsevier Applied Science, London, 1986, pp. 181–232.
32. Arnold, G. M. & Williams, A. A., The use of generalised Procrustes techniques in sensory analysis. In *Statistical Procedures in Food Research*, ed. J. R. Piggott. Elsevier Applied Science, London, 1986, pp. 233–53.
33. Carroll, J. D. & Chang, J. J., Analysis of individual differences in multidimensional scaling via an N-way generalization of 'Eckart–Young' decomposition. *Psychometrika*, **35** (1970) 283–320.
34. Schiffman, S. S., Reynolds, M. L. & Young, F. W., *Introduction to Multidimensional Scaling: Theory, Methods and Applications*. Academic Press, New York, 1981.
35. Macfie, H. J. H., Data analysis in flavour research: achievements, needs and perspectives. In *Flavour Science and Technology*, ed. M. Martens, G. A. Dalen & H. Russwurm, Jr. Wiley, Chichester, 1987, pp. 423–37.
36. Winn, R. L., Touch screen systems for sensory evaluation. *Food Technol.*, **42**(11) (1988) 68–70.
37. Crammer, B. & Ikan, R., Glycyrrhizin. *Chem. Soc. Rev.*, **6** (1977) 431.
38. Cardello, A. V., Ball, D. H., Alabran, D. M., Morrill, A. & Powell, G. G., The taste, odor and hedonic quality of polyglycerols. *Chemical Senses*, **9**(3) (1984) 285–301.
39. Nicol, W. M., Sucrose—the optimum sweetener. In *Nutritive Sweeteners*, ed. G. G. Birch & K. J. Parker. Applied Science Publishers, London, 1982, pp. 17–36.
40. Von Rymon Lipinski, G. W. & Huddart, B. E., Acesulfame-K. *Chemistry and Industry*, **11** (1983) 427–32.
41. Beck, I. C., Sweetness, character, and applications of aspartic acid-based sweeteners. In *Symposium: Sweeteners*, ed. G. E. Inglett. AVI, Westport, CT, 1974, pp. 164–81.
42. Nieman, C., Relative susskraft von zuckerarten. *Zucker und Susswaren Wirtschaft*, **11**(9) (1958) 465–7.
43. McBride, R. L., Category scales of sweetness are consistent with sweetness-matching data. *Perception and Psychophysics*, **34** (1983) 175–9.
44. Pangborn, R. M., Relative taste intensities of selected sugars and organic acids. *J. Food Sci.*, **28** (1963) 726–33.
45. Nieman, C., Sweetness of glucose, dextrose and sucrose. *Wissenschaft und Forschung*, **13** (1960) 706–7.
46. Dermer, O., The science of taste. *Proc. Okla. Acad. Sci.*, **27** (1946) 9–20.
47. van Velthuijsen, J. A., Food additives derived from lactose: lactitol and lactitol palmitate. *Journal of Agricultural and Food Chemistry*, **27** (1979) 680–6.
48. Nickerson, T. A., Lactose. In *Fundamentals of Dairy Chemistry*, ed. B. H. Webb, J. A. Johnson & J. A. Alford. AVI, Westport, CT, 1974, p. 224.
49. Hough, C. A. M., Sweet polyhydric alcohols. In *Developments in Sweeteners*, Vol. 1, ed. C. A. M. Hough, K. J. Parker & A. J. Vlitos. Applied Science Publishers, London, 1979, pp. 69–85.
50. Isima, N. & Kakayama, O., Sensory evaluation of stevioside as a sweetener. *Reports of the National Food Research Institute* (Tokyo), **31** (1976) 80–5.

51. Crammer, B. & Ikan, R., Progress in the chemistry and properties of rebaudiosides. In *Developments in Sweeteners*, Vol. 3, ed. T. H. Grenby. Elsevier Applied Science, London, 1988, pp. 45–64.

52. Redlinger, P. A. & Setser, C. S., Sensory quality of selected sweeteners: aqueous and lipid model systems. *J. Food Sci.*, **52**(2) (1987) 451–4.

53. Larson-Powers, N. & Pangborn, R. M., Paired comparison and time–intensity measurements of the sensory properties of beverages and gelatins containing sucrose or synthetic sweeteners. *J. Food Sci.*, **43**(1) (1978) 41–6.

54. Larson-Powers, N. & Pangborn, R. M., Descriptive analysis of the sensory properties of beverages and gelatins containing sucrose or synthetic sweeteners. *J. Food Sci.*, **43**(1) (1978) 47–51.

55. Moskowitz, H. R., Sweetness and intensity of artificial sweeteners. *Perception and Psychophysics*, **8**(1) (1970) 40.

56. Redlinger, P. A. & Setser, C. S., Sensory quality of selected sweeteners: unbaked and baked flour doughs. *J. Food Sci.*, **52**(5) (1987) 1391–3.

57. Cameron, A. T., The taste sense and the relative sweetness of sugars and other sweet substances. *Sugar Research Foundation Scientific Report*, **9** (1947) 1–72.

58. Godshall, M. A., The role of carbohydrates in flavor development. *Food Technology*, **42**(11) (1988) 71–8.

59. Inglett, G. E., Sweeteners—a review. *Food Technology*, **35** (1981) 37.

60. Homler, B., Properties and stability of aspartame. *Food Technology*, **38**(7) (1984) 50.

61. Samundsen, J. A., Has aspartame an aftertaste? *J. Food Sci.*, **50** (1985) 1510.

62. Thomson, D. M. H., Tunaley, A. & van Trijp, H. C. M., A reappraisal of the use of multidimensional scaling to investigate the sensory characteristics of sweeteners. *J. Sensory Studies*, **2** (1987) 215–30.

63. O'Brien, L. & Gelardi, R., Alternative sweeteners. *Chem. Technol.*, **11** (1981) 274.

64. DuBois, G., Crosby, G., Stephenson, R. & Wingard, R., Jr, Dihydrochalcone sweeteners: synthesis and sensory evaluation of sulfonate derivatives. *J. Agric. Food Chem.*, **25**(4) (1977) 763.

65. Swartz, M., Sensory screening of synthetic sweeteners using time–intensity evaluations. *J. Food Sci.*, **45**(3) (1980) 577–81.

66. Crosby, G., New sweeteners. *CRC Crit. Rev. Food Sci. Nutr.*, **7** (1976) 297.

67. Cook, M. & Gominger, B., Glycyrrhizin. In *Symposium: Sweeteners*, ed. G. E. Inglett. AVI, Westport, CT, 1974, pp. 211–15.

68. Crammer, B. & Ikan, R., Sweet glycosides from the stevia plant. *Chem. in Britain*, **22** (1986) 915.

69. Birch, G. G. & Ogunmoyela, G., Effect of surfactants on the taste and flavor of drinking chocolate. *J. Food Sci.*, **45**(4) (1980) 981–4.

70. Moskowitz, H. R., Models of additivity for sugar sweetness. In *Sensation and Measurements*, ed. H. R. Moskowitz, B. Scharf & J. C. Stevens. Reidel, The Netherlands, 1974, pp. 195–226.

71. Harrison, S. K. & Bernard, R. A., Time–intensity sensory characteristics of saccharin, xylitol and galactose and their effect on the sweetness of lactose. *J. Food Sci.*, **49**(3) (1984) 780–6.

72. Moskowitz, H. R., Perceptual attributes of the tastes of sugars. *J. Food Sci.*, **37**(4) (1972) 624–6.

73. Beauchamp, G. K. & Cowart, B. J., Development of sweet taste. In *Sweetness*, ed. J. Dobbing. Springer-Verlag, London, 1987, pp. 127–40.
74. Steiner, J. E., Facial expressions of the neonate infant indicating the hedonics of food-related chemical stimuli. In *Taste and Development: The Genesis of Sweet Preference*, ed. J. M. Weiffenbach. US Department of Health and Welfare, Bethesda, MD, 1977, pp. 173–88.
75. Trant, A. S. & Pangborn, R. M., Discrimination, intensity and hedonic responses to color, aroma, viscosity and sweetness of beverages. *Lebensmittel Wissenschaft und Technologie*, **16** (1983) 147–52.
76. Lucas, F. & Bellisle, F., The measurement of food preferences in humans: do taste-and-spit tests predict consumption? *Physiology and Behavior*, **39**(6) (1987) 739–43.
77. Tunaley, A., Thomson, D. M. H. & McEwan, J. A., An investigation of the relationship between preference and the sensory characteristics of nine sweeteners. In *Food Acceptability*, ed. D. M. H. Thomson. Elsevier Applied Science, London, 1989, pp. 387–400.
78. Peterkin, B. B., Women's diets: 1977 and 1985. *Journal of Nutrition Education*, **18** (1986) 251–7.
79. Tuorila, H., Pangborn, R. M. & Schutz, H. G., Choosing a beverage: comparison of preferences and beliefs related to the consumption of regular vs diet sodas. (1989) (in preparation).
80. Piggott, J. R. & Canaway, P. R., Finding the word for it: methods and uses of descriptive sensory analysis. In *Flavour '81*, ed. P. Schreier. De Gruyter, Berlin, 1981, pp. 33–46.
81. Würsch, P. & Daget, N., Sweetness in product development. In *Sweetness*, ed. J. Dobbing. Springer-Verlag, London, 1987, pp. 247–60.
82. Thurstone, L. L., A law of comparative judgement. *Psychology Review*, **34** (1927) 273–86.
83. Tunaley, A., Development and application of techniques to investigate the perceptual characteristics of sweeteners. PhD thesis, University of Reading, UK, 1988.
84. Prattala, R., Rasanen, L. & Ahlstrom, A., Consumer behaviour in relation to sweetness of food. In *Carbohydrate Sweeteners in Foods and Nutrition*, ed. P. Koivistoinen & L. Hyvonen. Academic Press, New York, 1980, pp. 111–25.
85. Tuorila, H., Hedonic responses and attitudes in the acceptance of sweetness, saltiness and fattiness of foods. In *Food Acceptance and Nutrition*, ed. J. Solmes, D. A. Booth, R. M. Pangborn & O. Raunhardt. Academic Press, London, 1987, pp. 337–51.
86. Calvin, L. D. & Sather, L. A., A comparison of student preference panels with a household consumer panel. *Food Technology*, **13**(8) (1959) 469–72.
87. Shepherd, R. & Griffiths, N. M., Preference for eggs from different production methods assessed by consumers and in-house panels. In *Flavour Science and Technology*, ed. M. Martens, G. A. Dalen & H. Russwurm, Jr. Wiley, Chichester, 1987, pp. 407–12.
88. Williams, A. A. & Langron, S. P., A new approach to sensory profile analysis. In *Flavour of Distilled Beverages: Origin and Development*, ed. J. R. Piggott. Ellis Horwood, Chichester, 1983, pp. 219–24.

Chapter 12

# LYCASIN®* AND THE PREVENTION OF DENTAL CARIES

ANDREW J. RUGG-GUNN

*Departments of Oral Biology and Child Dental Health,
University of Newcastle upon Tyne Dental School, UK*

## SUMMARY

*Lycasin® is a hydrogenated glucose syrup obtained from the enzymic hydrolysis of corn starch. It was originally made in Sweden but since the beginning of the 1980s the only Lycasin available has been Lycasin 80/55 which is made in France. The dental properties of Lycasin 80/55 are superior to those of the original Swedish Lycasin. This paper reviews the evidence on the dental aspects of Lycasin 80/55. The evidence comes from many types of experiment: acid production* in vitro, *plaque pH experiments, animal experiments and enamel slab experiments. From these data it is reasonable to conclude that Lycasin is likely to be non-cariogenic or virtually so. Substitution of dietary sugars by Lycasin in confectionery and syrup medicines is likely to be of significant benefit to dental health.*

## 1 INTRODUCTION

Lycasin was originally manufactured by the Lyckeby Starch Refining Company, Sweden. In the 1970s production was transferred to Roquette Frères, France. The manufacturing process was changed, resulting in a new product, Lycasin 80/55.

Swedish Lycasin was extensively tested in animal experiments,[1] in plaque pH experiments,[2] in incubation experiments[3] and in a clinical trial (the Roslagen study) of Lycasin-containing candies.[4] Frostell and Birkhed[5] and Birkhed and Edwardsson[6] compared the acidogenicity of Swedish Lycasin (candy quality), Lycasin 80/55 (made in France) and dietary sugars, and

* Lycasin is a registered trade mark of Roquette Frères, Lestrem, France.

concluded that both Lycasins were significantly less fermentable than glucose and that acid production from Lycasin 80/55 was less than half that from Swedish Lycasin. Because of the superior properties of Lycasin 80/55 and the fact that Swedish Lycasin is no longer available, only Lycasin 80/55 will be considered further. The purpose of this chapter is to review the properties of Lycasin (80/55) and the benefits to dental health that might result from substitution of Lycasin for dietary sugars. Reviews of the cariogenicity of other non-sugar bulk sweeteners have been published, for example those on xylitol,[7] sorbitol[8] and lactitol.[9]

### 1.1 Composition of Lycasin

Lycasin, a hydrogenated glucose syrup, is a clear, colourless sweet-tasting aqueous solution of sorbitol, hydrogenated oligosaccharides and poly-saccharides prepared by the catalytic hydrogenation of glucose syrup. On a dry-weight basis it has the following approximate composition, free D-sorbitol (7%), 4-O-α-D-glucopyranosyl-D-glucitol (52%), hydrogenated tri-to hepta-saccharides (23%), hydrogenated higher polysaccharides (18%) and hydrogenated polysaccharides with a degree of polymerisation greater than 20 (3% max.). It has been found that, to maintain the non-cariogenic properties of Lycasin, the concentration of hydrogenated polysaccharides with a degree of polymerisation greater than 20 should be kept to a maximum level of 3%. Lycasin has a very agreeable sweet taste; its sweetness is 0·75 times that of sucrose on an equal weight basis. The technological properties of Lycasin, which have been discussed by Sicard and Leroy,[10] are good, so its use in foods, confectionery and pharma-ceutical products is likely to increase.[11]

## 2 METHODS FOR ESTIMATING CARIOGENICITY

The best test of cariogenicity is the clinical trial involving human subjects. However, caries is a chronic disease, so trials have to last a minimum of 2 years, and it is difficult to keep large numbers of human volunteers on the same diet for this length of time. Because of the difficulties of undertaking clinical trials in the field of diet and dental caries, it is necessary to look at alternative ways of assessing the cariogenicity of foods. These can be considered under six headings.

### 2.1 Incubation Experiments

These are the simplest tests and examine the ability of plaque micro-organisms to metabolise the test food to acid. These experiments are done outside the mouth and can be classed as 'test-tube experiments'. Saliva, which contains oral micro-organisms, or pure cultures of oral micro-organisms have been substituted for plaque. This is a very stringent test, since some acid may be formed from some foods but at such a slow rate as to be of little clinical relevance. Conversely, rapid acid production indicates that the food under test is potentially cariogenic.

This type of experiment can be extended by adding dental enamel or hydroxyapatite to the incubation mixture and measuring the amount of calcium and phosphate that dissolves from the mineralised tissue.

## 2.2 Plaque pH Experiments

These can be seen as incubation experiments *in vivo*. The pH in the dental plaque on the teeth is monitored during and after the test food is eaten. Again, it is pH change or the acidogenicity of the food that is being measured, not cariogenicity.

There are three main methods of assessing plaque pH: (1) by micro-electrode probes (antimony or glass) inserted into the plaque, (2) by indwelling micro-electrodes (glass or iridium oxide) built into intra-oral appliances which are worn for several days to allow plaque to accumulate, and (3) by removing small samples of plaque from representative teeth and measuring the pH of these samples outside the mouth. Each method has its advantages but, regardless of the method used, there is close agreement between the methods in the order in which foods are ranked according to acidogenicity, although they differ in the absolute pH values recorded.

The indwelling glass micro-electrode system was developed in Zürich 20 years ago, and since then has been used to identify dentally safe snacks. In 1969 the Swiss Office of Health introduced legislation for the labelling of foods with regard to dental health. The products could be labelled 'zahnschonend' ('ménage les dents', 'safe for teeth') only if they did not depress plaque pH below 5·7 either during consumption or up to 30 min later, as measured by the Zürich micro-electrode system. Many snack foods made in many countries have been tested by this system.[12]

## 2.3 Changes in Oral Flora

A number of studies have suggested that micro-organisms of the *Strep. mutans* group are more capable of causing caries than other organisms. In fact caries has been labelled a 'specific plaque' disease.[13] It has been suggested that any potential preventive effect of foods can be assessed by their ability to alter plaque flora in one direction or another.[14] However, the unique relationship between oral levels of *Strep. mutans* and the development of dental caries is not universally accepted, and for the present this must be considered a weak cariogenicity test.

## 2.4 Animal Experiments

The rat is the most commonly used animal, although hamsters, mice and monkeys have been used. The main advantage of animal experiments is that the dietary regime—the type and amount of food and the frequency of eating—can all be carefully controlled. There are certain drawbacks which have a bearing on the interpretation of animal experiments. First, the animals may not like some foods. Some polyols, for example, may cause severe intestinal disturbances and loss of weight at very high feeding levels in some feeding experiments, but in other studies high levels are well tolerated.[15] In order to standardise experimental conditions, in some tests, the animals are super-infected with specific micro-organisms (e.g. *Strep. mutans*). Comparisons between foods can be affected by this super-infection, giving a false indication of the relative cariogenicity of some foods or some sugars.

## 2.5 Artificial Mouth Experiments

In these experiments, extracted teeth are mounted in a chamber and exposed to artificial saliva and organisms so that a 'plaque' builds up on the tooth surface. Test foods are then introduced into the 'mouth' and the subsequent development of caries is assessed quantitatively after sectioning the teeth.

The conditions in the mouths of living people are constantly changing and it is impossible to reproduce all these changes, many of which may be protective. Because of these limitations it is unlikely that artificial mouths will become a wholly reliable guide to the cariogenicity of foods.

## 2.6 Enamel Slab Experiments

In order to overcome some of the drawbacks of the artificial mouth, intra-oral appliances have been made capable of holding slabs of enamel. Plaque forms on the surface of the slabs, which remain in the mouth for 1–4 weeks. On a number of occasions, throughout the day, the appliance is removed from the mouth and the slabs put into solutions or suspensions of foods in order to simulate the eating of that food. The appliance is then re-inserted into the mouth. The development of caries is measured quantitatively by examining the enamel slabs. Although this system enables different solutions to be compared in the same experiment, a major drawback is that the test foods are not actually eaten, so the stimulatory effect of the food on salivary flow (which varies with different foods) is absent. If the appliance containing the enamel slabs remains in the mouth during eating, the experiment can be conducted as a cross-over trial, where the slabs are changed after each test period.

## 2.7 Summary

There are a number of methods that can be used to assess the potential cariogenicity of foods; each has its advantages and its limitations. While the clinical trial can be regarded as the ultimate test of the caries preventive effect of a food, other tests are also capable of giving an accurate assessment. Because each test has its limitations, it is necessary to combine evidence from as many sources as possible in order to judge accurately the cariogenicity of any food.[16]

## 3 DENTAL ASPECTS OF LYCASIN

### 3.1 Acid Production *in vitro*

The rate of acid production from the metabolism of Lycasin by plaque micro-organisms has been investigated *in vitro* by Frostell and Birkhed.[5] Plaque was removed from 10 subjects and suspended in buffer solution at pH 6·8. An aqueous solution of the test substrate was then added, and the amount of alkali required per unit time to keep the pH constant was recorded by an automatic recording titration system.[17] The rate of acid production from Lycasin was 38% that from glucose.

At about the same time Havenaar *et al.*[18] reported on the ability of various plaque bacteria to produce acid from Lycasin. This was assessed colorimetrically after the substrate and bacterial suspension had been incubated in a broth agar containing phenol red for 72 h. At least 20 strains of each of 8 species were tested. Lycasin was fermented by *S. mutans* and *A. odontoliticus*, but not by *S. sanguis*, *S. mitior* or *S. milleri*. It was fermented slightly by *A. viscosus*, *A. naeslundii* and *Lactobacillus*. In contrast, glucose was metabolised by all these organisms.

The adaptation of *S. mutans* strains to ferment Lycasin was tested by frequent sub-culturing in the same series of experiments.[18] A marked increase in fermentation of Lycasin by *S. mutans* was recorded (Fig. 1a,b), but this property was lost when the adapted strain was sub-cultured once in glucose, indicating that the adaptation to ferment Lycasin was rather unstable (Fig. 1c). Havenaar *et al.*[18] stated that, since sugar substitutes are virtually always consumed in combination with other sugars (and

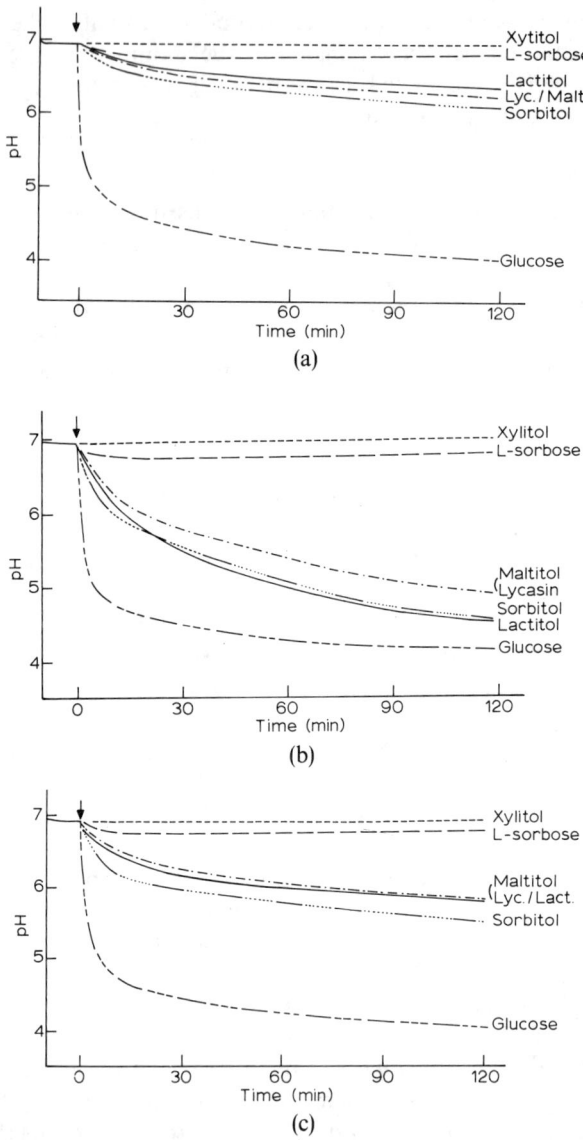

FIG. 1. Fermentation of sweeteners by *Strep. mutans*: (a) using non-adapted strains, (b) using adapted strains obtained by frequent sub-culturing in solutions of the appropriate sweetener, and (c) using dis-adapted strains obtained by sub-culturing once in glucose. From Havenaar *et al.*[18] with permission of S. Karger AG, Basel.

additionally since carbohydrate will always be present in the oral cavity), it seems unlikely that gross adaptation will occur *in vivo*.

Soyer and Frank[19] investigated the growth of one strain of *S. mutans* (ATCC 25175) in the presence of various sugars and sugar substitutes. Two different media were used. In the complex medium (brain heart infusion) Lycasin reduced the bacterial growth (and acid production) only slightly, while in the synthetic medium (which did not contain sugars) the bacteria did not grow or produce acid in the presence of Lycasin. As part of a series of experiments, Grenby and Saldanha[20] also reported that consistently less acid was produced from Lycasin than from sugar-based sweets when these were incubated with oral micro-organisms. They also showed that polysaccharide synthesis, which is considered a potentially harmful property of dental plaque, was much less when oral micro-organisms were incubated with Lycasin than with sucrose. When hydroxyapatite was added to the incubation mixtures, much less mineral dissolved in the Lycasin-containing incubates than in those containing sucrose.

Acid production by dental plaque bacteria from constituents of Lycasin was investigated by Würsch and Koellreutter[21] using the 'pH-stat' described by Birkhed.[17] The slow rate of fermentation of Lycasin was attributed, in part, to the inhibition of α-amylase by maltotriitol which is present in Lycasin at 18% by weight.

Bibby and Fu[22] compared the depression in the pH of dental plaque *in vitro* after exposure to sugars and Lycasin using a newly developed test system. This involved packing freshly collected plaque into an iron gauze basket into which a micro pH electrode was inserted. The electrode and basket assembly were then inserted into 0·1%, 1·0% or 10% solutions of the sweetener, and the pH was recorded after 20 min incubation (Fig. 2). The authors suggest that this test is more severe than those proposed previously because their centrifuged plaque contained a more concentrated cell mass and probably a wide variety of bacterial types with a broader and more powerful enzymic potential. The pH fell when the system was inserted in the solution of Lycasin but the final pH was higher than with a glucose solution. Raising the concentration of Lycasin resulted in a lower final pH.

Soderling *et al.*[23] investigated the effect upon acid production of combining xylitol with various other sweeteners. These combinations were incubated for 4 h with salivary sediment obtained from 10 subjects. Lycasin alone was fermented slowly. The addition of xylitol resulted in a very slightly smaller pH fall, and led the authors to conclude that combinations of xylitol and Lycasin should promote remineralisation of enamel as efficiently as xylitol alone.

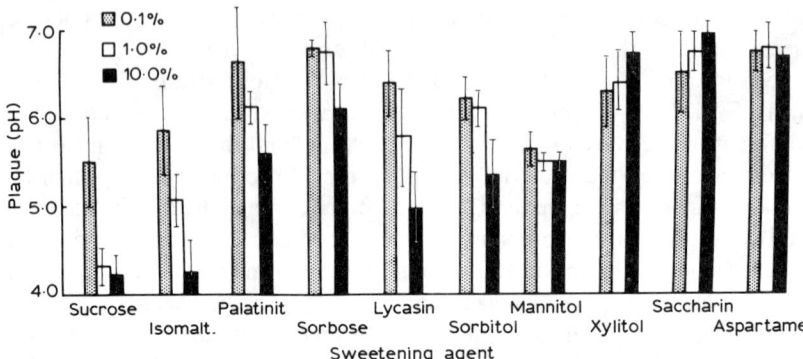

FIG. 2.    The mean ($\pm$ sd) pH of plaque samples after incubation for 20 min in 0·1%, 1·0% and 10·0% solutions of various sweeteners. For saccharin and aspartame sweetness equivalents to sucrose solutions were used. From Bibby and Fu[22] with permission of the editor of *Journal of Dental Research*.

The extent to which Lycasin can be broken down by a specific amyloglucosidase to mono-, di- and possibly oligo-saccharide fragments that could be fermented to acid by oral micro-organisms was recently investigated by Grenby and Bull.[24] They found that the release of D-glucose from Lycasin was small, similar to that from starch and much less than that from glucose syrup.

In summary, incubation experiments *in vitro* have shown that Lycasin can be fermented by plaque organisms but the rate is slow compared with sugars. In combination with xylitol, acid production is very slow. Only some of the organisms capable of fermenting sugars can ferment Lycasin. From the limited evidence available, it would appear that some strains can adapt to ferment Lycasin, but it seems unlikely that gross adaptation will occur *in vivo*.

### 3.2 Plaque pH Experiments
A variety of methods of monitoring plaque pH have been used in a number of European centres of dental research. Experiments conducted in Zürich using the indwelling glass electrode technique showed that it was possible to manufacture confections containing Lycasin which were 'zahnschonend' (safe for teeth).[25] In a subsequent publication,[26] the same team reported that a 10% aqueous solution of Lycasin and a hard candy made from Lycasin did not depress plaque pH significantly and could be classified as 'safe for teeth'. Both experiments showed that the composition of Lycasin was critical, since Lycasins that deviated from the French Lycasin 80/55

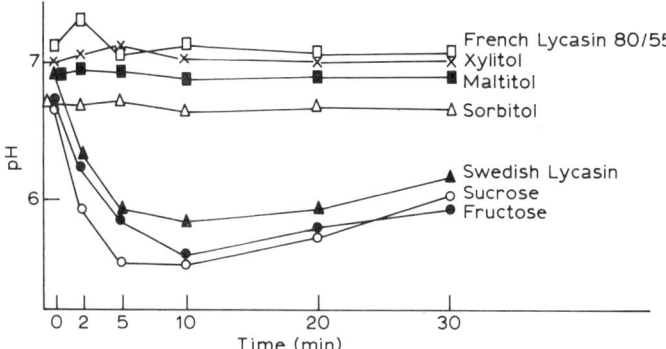

FIG. 3.   Changes in plaque pH after rinsing for 30 s with 50% solutions of various sweeteners. The results are the mean values obtained from 18–20 subjects. From Birkhed and Edwardsson[6] with permission of S. Karger AG, Basel.

formula were fermented more readily and could not be called 'safe for teeth'. In view of this, in the UK the 1983 Sweeteners in Food Regulations laid down strict compositional requirements for Lycasin.[27]

A number of review articles have been published in German, French and English by the Zürich team, all commenting that Lycasin solutions and a number of confections (candies and toffees) sweetened with Lycasin are classed as 'safe for teeth'.[12,28-31]

Frostell and co-workers[5,32] also conducted experiments with various Lycasins. They monitored plaque pH using the plaque sampling method and tested the effect of rinsing with 10 ml of a 50% solution of Lycasin 80/55 in 18 subjects. The pH of the dental plaque did not fall below the resting value of 7·1. In contrast, plaque pH fell to around 5·5 after rinsing for 30 s with a 50% solution of sucrose or fructose.[6] The results of these experiments are shown in Fig. 3, where the negligible effect of Lycasin 80/55, xylitol, maltitol and sorbitol upon plaque pH can be seen. Commenting on these results, Birkhed and Edwardsson[6] classified Lycasin as 'safe for teeth', and they suggested that the 'critical pH' value of 6·5 under their system, using the plaque sampling method, was equivalent to a 'critical pH' of 5·7 under the Zürich system where an indwelling glass electrode is used.

In Germany, Gehring and Hufnagel[33] used a glass pH-sensitive probe to monitor plaque pH, in 6 subjects, before and after they rinsed with a 20% Lycasin solution for 2 min. Plaque pH fell to 5·7 after a Lycasin rinse. After rinsing with a comparable sucrose rinse or a fructose rinse, the pH fell to 4·6 in each case.

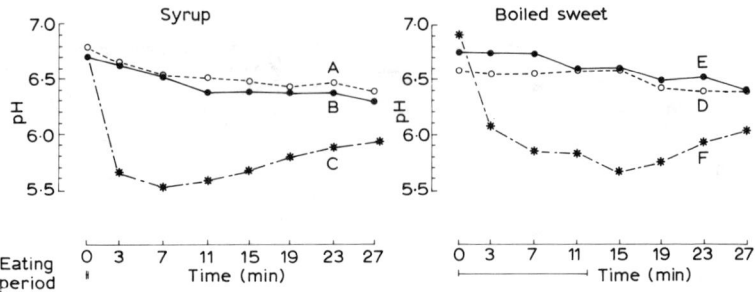

FIG. 4.  Plaque pH curves for Lycasin, sorbitol and sucrose/glucose, when taken as a 70% syrup or as a boiled sweet. Curves are the mean curves for 12 subjects. ○, Lycasin; ●, sorbitol; ✳, sucrose/glucose. From Rugg-Gunn[34] with permission of the editor *Caries Research* (S. Karger AG, Basel).

Rugg-Gunn[34] reported experiments which compared the effect upon plaque pH of Lycasin, sorbitol and sucrose/glucose. These sweeteners were taken as (a) 70% syrups or (b) boiled sweets. The syrup (10 ml) was taken from a spoon and held in the mouth for 10 s before being swallowed or spat out. The boiled sweet, which weighed 5·4 g, was held in the mouth until dissolved. The order in which each of the 12 subjects tested each of the 6 test substances was randomised. The results are shown graphically in Fig. 4. The effects of sorbitol and Lycasin upon plaque pH, when taken as a 70% syrup or as a boiled sweet, were similar, and very slight when compared with sucrose/glucose. The differences between sucrose/glucose and the two non-sugar sweeteners were statistically highly significant ($p < 0.002$). Although the boiled sweet was in the mouth for a very much longer time than the syrup (12 min compared with 10 s), there was little difference in the plaque pH curves produced. The pH minima for Lycasin and sorbitol were at 27 min, the end of the experimental period. It is impossible to say whether this fall in pH would continue, but the data imply that both non-sugar sweeteners were being fermented at a very low level.

A number of reviews of the acidogenicity of sugar substitutes (including Lycasin) have been published recently, apart from those already mentioned. A report by Blot-Calmy[35] in France discussed the intra-oral telemetric technique used in Zürich and, while noting that Lycasin sweets were classed as 'ménage les dents' (safe for teeth), remarked that the technique was a valid and reliable means of predicting non-cariogenicity or low cariogenicity. Hefti[36] from Basel, Switzerland, also supported the validity of classifying foods as 'zahnschonend' and listed Lycasin as one of the sweeteners that passed the Zürich intra-oral plaque pH test. Edgar and Dodds[37] summarised the substantial literature on the acidogenicity of sweeteners but cautioned that the results must be interpreted in conjunction with evidence from other sources. They remarked that Lycasin exhibited a low acidogenicity in plaque pH testing.

### 3.3  Animal Experiments and Alteration in Plaque Flora

It has been mentioned previously that rats do not like eating polyols and, because of this, investigations into the cariogenicity of polyols in these animals are difficult. However, a number of rat experiments have been conducted into the cariogenicity of Lycasin, and the series reported by Havenaar et al.[15] can be considered to be amongst the best ever undertaken in the field of diet and caries in animals. Their main study included a comparison of the cariogenicity of starch, sucrose and Lycasin, and consisted of two experiments. In the first, the experimental food was given 18 times per day while, in the second, the experimental food was given only 14 times per day and was alternated with a sugar diet given 4 times per day. This latter dietary regime was used to obtain similar bacterial colonisation patterns in the rats in the different groups. The proportion of Lycasin in the diets was 20% in experiment I and 25% in experiment II.

The body weight and the weight of body organs were reported. All the rats were in good health, and examination showed that non-specific influences of the dietary regimes were minimal. The incidence of fissure caries is shown in Table 1. Diets containing Lycasin, xylitol or L-sorbose produced similar low levels of caries, much lower than those caused by sucrose or sorbitol. The authors concluded that Lycasin was 'hardly cariogenic or non-cariogenic' in both experiments.

TABLE 1

INCIDENCE OF FISSURE CARIES OCCURRING IN RATS FED DIETS CONTAINING A STANDARD DIET (SSP) PLUS TEST DIETS[15]

| Diet SSP plus: | Experiment I[a] | Experiment II[b] |
|---|---|---|
| Starch | 5·4 | 6·0 |
| Starch and sucrose | 13·9 | 12·8 |
| Starch and Lycasin | 4·3 | 5·7 |
| Starch and xylitol | 3·4 | 3·4 |
| Starch and sorbitol | 10·4 | 11·1 |
| Starch and L-sorbose | 4·4 | 2·8 |

[a] Diets contained 20% of the test sweeteners and 30% starch and were given 18 times per day.
[b] Diets were given 14 times per day and alternated with diet SSP (containing 20% sucrose and 10% glucose) 4 times per day.

TABLE 2

RESULTS FROM THREE EXPERIMENTS WHICH COMPARED THE CARIOGENICITY OF LYCASIN
WITH THAT OF OTHER SWEETENERS[39]

(a)  Diets in powdered form at 16% by weight

|  | Sucrose | Lycasin 80/55 | Sorbitol |
|---|---|---|---|
| Caries score[a] | 7·5 (2·9) | 2·1 (0·5) | 0·8 (0·2) |
| Plaque score[a] | 9·7 (0·5) | 8·4 (1·0) | 8·9 (1·1) |

(b)  Diets in pelleted form at 25% by weight

|  | Sucrose | Lycasin 80/55 | Starch |
|---|---|---|---|
| Caries score | 4·3 (0·9) | 1·4 (0·4) | 1·0 (0·2) |
| Plaque score | 4·4 (0·8) | 2·0 (0·3) | 5·3 (0·8) |

(c)  Diets in pelleted form at 25% by weight

|  | Sucrose | Lycasin 80/55 | Lycasin 05/60 | Starch |
|---|---|---|---|---|
| Caries score | 3·7 (0·5) | 1·1 (0·4) | 1·5 (0·3) | 1·7 (0·3) |
| Plaque score | 5·7 (0·6) | 4·9 (0·6) | 5·3 (0·6) | 6·8 (0·7) |

[a] Scores are given as means ($\pm$ standard errors).

In further studies, Havenaar et al.[38] investigated whether selection or adaptation of oral bacteria occurred when Lycasin was consumed frequently over a long period. Rats were inoculated with S. mutans alone or with S. mutans plus A. viscosus. More rats were then inoculated with plaque from the original animals. This was repeated to produce five consecutive transmissions of plaque flora. No alterations were observed in the numbers or percentages of S. mutans or A. viscosus, or in the fermentation rate of Lycasins by plaque flora in vitro. In agreement with the previous experiments,[15] Lycasin was virtually non-cariogenic compared with sucrose, irrespective of whether the rats were inoculated with S. mutans alone or in combination with A. viscosus, with the original strains or with plaque from the preceding experiments on rats on a Lycasin diet. The authors concluded that their study showed clearly that there was no adaptation or selection of plaque flora towards Lycasin and no alteration in the extremely low cariogenicity of Lycasin.

More recently, Grenby[39] has published a series of studies on laboratory rats. His first studies investigated the tolerance of caries-active Osborne–Mendel rats to diets containing different concentrations of Lycasin taken in either a powder or pellet form. It was found that animals could tolerate up to 15–16% of Lycasin in dry-powder diets and up to 25% in pelleted diets before gastro-intestinal disturbance. The remaining studies investigated the cariogenicity of Lycasin compared with other sweeteners (Table 2). When the diets were given in a powdered form at 16% by weight, animals on the

Lycasin diet developed 72% less caries, and those on the sorbitol diet 89% less caries, compared with the sucrose control. When the diets were given at 25% in a pelleted form, rats on the Lycasin diet had 67% less caries than those on the sucrose diet and similar caries scores to rats on the starch diet. In the final experiment, an experimental grade of hydrogenated glucose syrup (Lycasin 05/60) was tested against Lycasin 80/55, sucrose and starch. All sweeteners were taken at 25% in a pelleted diet. Rats in both Lycasin groups and in the starch group developed similar levels of caries, significantly below the scores of those on the sucrose diet. This was the first report of cariogenicity testing of hydrogenated glucose syrup 05/60 which differed from Lycasin (80/55) in containing much less maltitol and more oligo- and higher saccharides.[39] In this series of experiments, Grenby[39] found no clear relation between caries and plaque scores (Table 2). In only one of the three studies was the plaque level in rats eating Lycasin lower than that in the sucrose-eating rats, a finding which could indicate a limited capacity of plaque organisms to form polysaccharides from Lycasin.

In order to avoid possible systemic side-effects of feeding polyols to rats, Firestone et al.[40] devised a different approach, in which the substances under test were applied topically to the teeth of rats 5 times per day, as a 50% solution. The authors concluded that the topical application of Lycasin was non-cariogenic, although the relevance of these findings is unclear since little difference in caries development was recorded in those animals receiving topical application of sucrose, compared with those receiving a topical application of water.

Leach and co-workers[41,42] have published two abstracts in which they reported that Lycasin was capable of reversing carious lesions in rats that were previously fed sucrose. However, no experimental details have yet been given.

In summary, two series of experiments on the cariogenicity of dietary Lycasin in animals have been fully described.[15,39] Both series were well conducted and reported, and indicated that Lycasin is non-cariogenic, or virtually so, in rats. Adaptation of oral bacteria to Lycasin does not appear to occur to any significant extent. Laboratory animal trials of Lycasins and other sweeteners have been reviewed.[43] The author concluded that 'Under the experimental conditions described, Lycasin 80/55 is virtually non-cariogenic'.

### 3.4 Enamel Slab Experiments in Humans

One report of an investigation into the effect of Lycasin on carious lesions in enamel slabs worn *in vivo* has been published.[44] Five dental students wore intra-oral appliances containing slabs of human enamel for two experimental periods on each of 14 days. Pre-cavitation (white spot) lesions had been created in the enamel beforehand, by immersing the teeth (human premolars extracted for orthodontic reasons) in an acid solution for 7 days. Half of each lesion was worn for the first 14-day test period while the subjects consumed 15 Lycasin candies per day. After a gap of a few days, the second half was worn for 2 weeks while the subjects abstained from eating between meals. Both halves of the lesion were then sectioned and compared microradiographically with a section of the original lesion. The results showed that the lesions worn during the first period had remineralised when Lycasin candies were consumed, and that this remineralisation was greater (51% recovery of mineral) than in lesions worn during the second period when there was no between-meal eating (26% recovery of mineral). The authors suggest that this pilot study demonstrated that eating Lycasin-containing confectionery could aid the remineralisation of pre-cavitation carious lesions.

## 4 SUMMARY

Evidence of the non-cariogenicity of Lycasin is based on five types of investigations: *in vitro* incubation experiments, *in vivo* plaque pH experiments, animal experiments, the effect on oral flora, and one human enamel slab experiment. No clinical trial has been reported. The incubation experiments and plaque pH studies have both been duplicated in independent research centres. The recent report by Grenby[39] of extensive animal experiments fully supports the excellent animal studies of Havenaar et al.[15,38] Despite the lack of data from human clinical trials, it is reasonable to conclude that Lycasin is likely to be non-cariogenic in man. Because Lycasin can be metabolised slowly by oral micro-organisms, it is possible that it could be cariogenic if eaten frequently by very caries-susceptible people. This remote possibility would be of negligible public health importance.

## 5 PLACE OF LYCASIN IN THE DIET

It has been said repeatedly that sugars are the most important dietary item in the development of dental caries. Mainly because of this, many reports[45-47] have urged that sugar consumption should be reduced. One

TABLE 3

MEAN DAILY INTAKE OF ADDED SUGARS FROM VARIOUS SOURCES IN 405
11–13-YEAR-OLD ENGLISH ADOLESCENTS[48]

| Dietary source | Daily intake (g/day) | Added sugars intake (%) |
|---|---|---|
| Confectionery | 23 | 28 |
| Table sugar | 20 | 24 |
| Soft drinks | 14 | 17 |
| Biscuits and cakes | 10 | 12 |
| Sweet puddings | 7 | 9 |
| Syrups and preserves | 3 | 3 |
| Breakfast cereals | 2 | 3 |
| Other | 3 | 3 |
| | 81 | 100 |

way to achieve this reduction is to substitute non-cariogenic sweeteners for sugars in dietary items where sweetness is desired. It is impractical to expect that all sugars now added to our diet can be removed, or that they can be substituted totally by alternative sweeteners. An important step in deciding where the greatest benefit to dental health can be made by sugar substitutes is to examine the sources of dietary sugars in sections of the community at risk of developing dental caries. Caries development is greatest in adolescents, and a recent survey of the diets of 405 English adolescents[48] has revealed that, of the 81 g of added sugars consumed per person per day, 28% (23 g) came from confectionery, 24% from table sugar, 17% from soft drinks and 12% from biscuits and cakes (Table 3). These children consumed, on average, 57 g of confectionery per day.[49] Not only is confectionery a most important source of dietary sugars, but it is marketed for frequent eating and is therefore likely to be particularly harmful to teeth. Because of this, the use of non-cariogenic sweeteners such as Lycasin in confectionery is likely to make a significant impact on dental health. The habit of snacking is likely to remain, and it is important that snack foods are nutritious foods.

Dietary items can be favourable to dental health by provoking a strong salivary flow which will aid the remineralisation of early carious lesions in teeth. An example is chewing-gum made using a non-cariogenic sweetener. When a sorbitol-based gum was chewed, the plaque pH immediately rose and remained at around 7·5 for the 30 min that the gum was chewed.[50] The total amount of calories ingested from chewing-gum is small compared with other types of confectionery.

Another important area is medicines. Many medicines, particularly paediatric medicines, are prescribed in a syrup form to make them more acceptable. In many countries sucrose is the principal syrup used. Some children have to take these medicines once or more per day for months or years, and it is known that this causes a considerable amount of dental decay, often in children in whom it is particularly important to maintain good dental health.[51] Thus substitution of sugar in these syrups with a non-cariogenic sweetener is highly desirable. Because of its properties, Lycasin may be especially suited to this pharmaceutical application.

In summary, the amount of sugar and the frequency of eating sugar should be reduced. Substitution of dietary sugar by non-cariogenic sweeteners such as Lycasin is a means of achieving this and, if implemented, could reduce significantly the amount of dental disease.

## REFERENCES

1. Frostell, G., Keyes, P. H. & Larson, R. H., Effect of various sugars and sugar substitutes on dental caries in hamsters and rats. *J. Nutr.*, **93** (1967) 65–76.
2. Frostell, G., Effect of mouthrinses with sorbitol and Lycasin on the pH of dental plaque. *Odontol. Rev.*, **24** (1973) 217–26.
3. Bramstedt, F. & Trautner, K., Sugar substitutes and biochemistry of dental plaque. *Deutsch. Zahnärztl. Z.*, **26** (1971) 1135–44.
4. Frostell, G., Blomlof, L., Blomqvist, T. *et al.*, Substitution of sucrose by Lycasin in candy. *Acta Odont. Scand.*, **32** (1974) 235–54.
5. Frostell, G. & Birkhed, D., Acid production from Swedish Lycasin (candy quality) and French Lycasin (80/55) in human dental plaques. *Caries Res.*, **12** (1978) 256–63.
6. Birkhed, D. & Edwardsson, S., Acid production from sucrose substitutes in human dental plaque. In *Health and Sugar Substitutes*, ed. B. Guggenheim. Karger, Basel, 1979, pp. 211–17.
7. Makinen, K. K. & Scheinin, A., Xylitol and dental caries. *Ann. Rev. Nutr.*, **2** (1982) 133–50.
8. Birkhed, D., Edwardsson, S., Kalfas, S. & Svensater, G., Cariogenicity of sorbitol. *Swed. Dent. J.*, **8** (1984) 147–54.
9. den Uyl, C. H., Technical and commercial aspects of the use of lactitol in foods as a reduced-calorie bulk sweetener. In *Developments in Sweeteners*, Vol. 3, ed. T. H. Grenby. Elsevier Applied Science, London, 1987, pp. 65–81.
10. Sicard, P. J. & Leroy, P., Mannitol, sorbitol and Lycasin; properties and food applications. In *Developments in Sweeteners*, Vol. 2, ed. T. H. Grenby, K. J. Parker & M. G. Lindley. Applied Science Publishers, London and New York, 1983, pp. 1–25.
11. Whitmore, D. A., Developments in the properties and applications of Lycasin and sorbitol. *Food Chem.*, **16** (1985) 209–29.

12. Imfeld, T. N., *Identification of Low Caries Risk Dietary Components*. Karger, Basel, 1983.
13. Emilson, C.-G. & Krasse, B., Support for and implications of the specific plaque hypothesis. *Scand. J. Dent. Res.*, **93** (1985) 96–104.
14. Loesche, W. J., The rationale for caries prevention through the use of sugar substitutes. *Int. Dent. J.*, **35** (1985) 1–8.
15. Havenaar, R., Drost, J. S., de Stoppelaar, J. D., Huis in't Veld, J. H. J. & Backer Dirks, O., Potential cariogenicity of Lycasin 80/55 in comparison to starch, sucrose, xylitol, sorbitol and L-sorbose in rats. *Caries Res.*, **18** (1984) 375–84.
16. DePaola, D. P., Executive summary; consensus conference on methods for assessment of the cariogenic potential of foods. *J. Dent. Res.*, **65** (1986) (Spec. issue) 1540–3.
17. Birkhed, D., Automatic titration method for determination of acid production from sugars and sugar alcohols in small samples of dental plaque material. *Caries Res.*, **12** (1978) 128–36.
18. Havenaar, R., Huis in't Veld, J. H. J., Backer Dirks, O. & de Stoppelar, J. D., Some bacteriological aspects of sugar substitutes. In *Health and Sugar Substitutes*, ed. B. Guggenheim. Karger, Basel, 1979, pp. 192–8.
19. Soyer, C. & Frank, R. M., Influence du milieu de culture sur la croissance du *Streptococcus mutans* ATCC 25175 en presence de differents glucides et leurs dérivés. *J. Biol. Buccale*, **7** (1979) 295–301.
20. Grenby, T. H. & Saldanha, M. G., Comparison of Lycasin versus sucrose sweets in demineralization studies of human enamel and hydroxyapatite. *Caries Res.*, **22** (1988) 269–75.
21. Würsch, P. & Koellreutter, B., Maltitol and maltotriitol as inhibitors of acid production in human dental plaque. *Caries Res.*, **16** (1982) 90–5.
22. Bibby, B. G. & Fu, J., Changes in plaque pH by sweeteners. *J. Dent. Res.*, **64** (1985) 1130–3.
23. Soderling, E., Talonpoika, J. & Makinen, K. K., Effect of xylitol-containing carbohydrate mixtures on acid and ammonia production in suspensions of salivary sediment. *Scand. J. Dent. Res.*, **95** (1987) 405–10.
24. Grenby, T. H. & Bull, J. M., Amylolytic breakdown of Lycasin compared with other carbohydrate derivatives. *Caries Res.*, **22** (1988) 276–9.
25. Muhlemann, H. R., Report on polysorb (Lycasin) 80/55—candies. Report to Roquette Frères, 1976.
26. Imfeld, T. & Muhlemann, H. R., Addendum to: Acid production from Swedish Lycasin (candy quality) and French Lycasin (80/55) in human dental plaques (G. Frostell & D. Birkhed). *Caries Res.*, **12** (1978) 262–3.
27. Ministry of Agriculture, Fisheries and Food, Food additives and contaminants committee report on the review of sweeteners in food. FAC/REP/34. HMSO, London, 1982.
28. Imfeld, T. & Muhlemann, H. R., Evaluation of sugar substitutes in preventive cariology. *J. Prev. Dent.*, **4** (1977) 8–14.
29. Imfeld, T., Evaluation of the cariology of confectionery by intra-oral wire-telemetry. *Helv. Odont. Acta*, **21** (1977) 1–28.
30. Muhlemann, H. R., Sugar substitutes and plaque pH telemetry in caries prevention. *J. Clin. Periodontol.*, **6** (1979) suppl. 17, 47–52.

31. Imfeld, T. & Duhamel, I., Evaluation de l'alimentation non-cariogene à l'aide de la telemetrie intra-orale du pH de la plaque interdentaire. *Rev. Odont. Stomatol.*, **9** (1980) 27–38.

32. Frostell, G., Report on Lycasin 80/55. Report to Roquette Frères, 1977.

33. Gehring, F. & Hufnagel, H.-D., Intra- und extraorale pH-Messungen an Zahnplaques des Menschen nach Spulungen mit einigen Zucker- und Saccharoseaustauschstoff-Losungen. *Oralprophylaxe*, **5** (1983) 13–19.

34. Rugg-Gunn, A. J., Effect of Lycasin upon plaque pH when taken as a syrup or as a boiled sweet. *Caries Res.*, **22** (1988) 375–6.

35. Blot-Calmy, B., Prevention and intra-oral telemetry of the pH of dental plaque. *Information Dentaire*, **64** (1982) 2371–9.

36. Hefti, A., Zuckerersatzstoffe in der Kariesprophylaxe. *Schweiz. Med. Mschr.*, **110** (1980) 269–73.

37. Edgar, W. M. & Dodds, M. W. J., The effect of sweeteners on acid production in plaque. *Int. Dent. J.*, **35** (1985) 18–22.

38. Havenaar, R., Drost, J. S., Huis in't Veld, J. H. J., Backer Dirks, O. & de Stoppelaar, J. D., Potential cariogenicity of Lycasin 80/55 before and after repeated transmissions of the dental plaque flora in rats. *Arch. Oral Biol.*, **29** (1984) 993–9.

39. Grenby, T. H., Dental effects of Lycasin in the diet of laboratory rats. *Caries Res.*, **22** (1988) 288–96.

40. Firestone, A. R., Schmid, R. & Muhlemann, H. R., The effects of topical applications of sugar substitutes on the incidence of caries and bacterial agglomerate formation in rats. *Caries Res.*, **14** (1980) 324–32.

41. Leach, S. A. & Green, R. M., Reversal of fissure caries in the albino rat by Lycasin. *Caries Res.*, **17** (1983) 157 (Abstr. 5).

42. Leach, S. A., Connell, R., Speechley, J. A. & Green, R. M., Reversal of dental caries by the sugar substitute Lycasin *in vivo. J. Dent. Res.*, **63** (1984) 334 (Abstr. 1466).

43. Havenaar, R., Dental advantages of some bulk sweeteners in laboratory animal trials. In *Developments in Sweeteners*, Vol. 3, ed. T. H. Grenby. Elsevier Applied Science, London, 1987, pp. 189–211.

44. Leach, S. A., Speechley, J. A., White, M. J. & Abbott, J. J., Remineralisation *in vivo* by stimulating salivary flow with Lycasin: a pilot study. In *Factors Relating to Demineralisation and Remineralisation of the Teeth*, ed. S. A. Leach. IRL Press, Oxford, 1986, pp. 69–79.

45. United States Senate, *Dietary Goals of the United States*, 2nd edn. US Government Printing House, Washington, DC, 1977.

46. National Advisory Committee on Nutrition Education, *Proposals for Nutritional Guidelines for Health Education in Britain* (chairman W. P. T. James). Health Education Council, London, 1983.

47. Royal College of Physicians, Obesity (chairman D. Black). *J. Roy. Coll. Physicians*, **17** (1983) 5–65.

48. Rugg-Gunn, A. J., Hackett, A. F., Appleton, D. R. & Moynihan, P. J., The dietary intake of added and natural sugars in 405 English adolescents. *Hum. Nutr. Appl. Nutr.*, **40A** (1986) 115–24.

49. Rugg-Gunn, A. J., Hackett, A. F., Appleton, D. R., Jenkins, G. N. & Eastoe, J. E., Relationship between dietary habits and caries increment assessed over

two years in 405 English adolescent school children. *Arch. Oral Biol.*, **29** (1984) 983–92.

50. Rugg-Gunn, A. J., Edgar, W. M. & Jenkins, G. N., The effect of eating some British snacks upon the pH of human dental plaque. *Brit. Dent. J.*, **145** (1975) 95–100.

51. Hobson, P., Sugar-based medicines and dental disease. *Community Dental Health*, **2** (1985) 57–62.

Chapter 13

# LATEST DENTAL STUDIES ON XYLITOL AND MECHANISM OF ACTION OF XYLITOL IN CARIES LIMITATION

Kauko K. Mäkinen

*University of Michigan School of Dentistry,*
*Ann Arbor, Michigan, USA*

## SUMMARY

*Including the Turku studies in the early 1970s, about ten human clinical studies or field trials have tested the efficacy of xylitol as a caries-curbing agent. The results all point to the consumption of small daily quantities of xylitol decreasing the incidence of dental caries both in children and in adults. The effective daily dose levels of xylitol ranged widely, from about 1 g to 20 g per subject, although clinically significant results have normally been obtained with 7–10 g, usually in the form of chewing-gum. Current information suggests that the caries-preventive effect of xylitol is both dose- and concentration-dependent. Consumption of the recommended quantities of xylitol (7–10 g/day) has been associated with up to 50% reduction in the amount of dental plaque. Plaque grown in the presence of xylitol has been shown to be looser, less acidic and less inflammatory than plaque grown under control conditions, and may contain more calcium to aid the remineralization of incipient caries lesions. Physicochemical studies have shown that xylitol and some other polyols stabilize calcium phosphate solutions, which also contributes to remineralization. This property, together with the capacity to interfere with the metabolism of* Streptococcus mutans, *are believed to explain the advantageous clinical findings and therapeutic action.*

## 1 INTRODUCTION

The first clinical caries studies which elucidated the efficacy of xylitol as a sucrose substitute were published in the mid-1970s.[1] The results of these trials have been evaluated and the trials themselves discussed in numerous

# TABLE 1

CLINICAL TRIALS AND FIELD STUDIES TESTING XYLITOL IN THE PREVENTION OF DENTAL CARIES

| Study | Duration of trial | Approx. consumption | Main results |
|---|---|---|---|
| Turku studies, Finland 1972–74 (mostly young adults) I. Feeding study | 2 years | 67 g/day | Xylitol diet reduced caries by at least 85% compared with normal diet. In individual cases a therapeutic (remineralizing) effect was suggested.[1] |
| II. Chewing gum study | 1 year | 6·7 g/day | Xylitol gum reduced caries by at least 82% compared with sucrose gum. A therapeutic effect was suggested.[1] |
| Soviet study Late 1970s (children) | 2 years | 30 g/day | Xylitol candies reduced caries by 73% compared with sucrose candies. The subjects replaced about 50% of normal candy consumption with xylitol products.[11] |
| WHO studies Late 1970s, early 1980s (children) | 2·3–3·0 years (Thailand and Polynesia) | Much less than 20 g/day | The existence of a preventive effect was noted as a result of use of xylitol gum in both countries.[13] |
|  | 3 years (Polynesia) | Up to 20 g/day | Xylitol sweets reduced caries by 37–39% compared with fluoride treatment.[14] |
| First half of 1980s (children) | 2–3 years (Hungary) | 14–20 g/day | Xylitol (mainly in gums) reduced caries by 37–45% compared with control groups.[15,16] |
| Canadian gum study Mid-1980s (children) | 1–2 years | 0·9–3·5 g/day | Xylitol gum reduced caries by at least 52% compared with non-chewing group. During the second year the reduction was about 65%.[19,20] |
| Ylivieska study, Finland 1982–85 (children) Total age groups | 2 years | 7–10 g/day | Xylitol gum reduced caries by 30–57% compared with no gum. A therapeutic effect was suggested.[24] |
| 'High risk' groups | 3 years | 7–10 g/day | Xylitol gum reduced caries by 59–84% compared with no gum (both groups received the same basic fluoride prevention). A therapeutic effect was suggested.[24] |
| Ylivieska follow-up study 1987 (3 years after discontinuation of the use of xylitol gum) | Up to 6 years | No xylitol gum after the trial proper | The teeth had good resistance against caries, continued after discontinuation of the use of xylitol (xylitol had a long-term effect; see test).[26] |

articles.[2-10] These studies gave impetus to several other human and animal trials, and also prompted the World Health Organization to cooperate with various groups of scientists to test the efficacy of xylitol programs under real-life conditions. To date, about ten human clinical caries trials on xylitol have been accomplished; Table 1 shows the impressive results. However, the latest xylitol caries trials have not been reviewed in their entirety. Furthermore, during the past 15 years a large number of microbiological, chemical and other studies on sugar substitutes have been carried out, elucidating the mechanism of action of xylitol in caries prevention. The purpose of this chapter is to assess, in the light of the latest clinical findings, the efficacy of xylitol in the prevention of dental caries, and to outline the mechanism of this action based on known chemical and physicochemical properties of xylitol.

## 2 NEW CLINICAL TRIALS

### 2.1 Soviet Xylitol Study

The effect of partial substitution of sucrose by xylitol on the incidence of dental caries was studied in the state of Kazan in the USSR in the late 1970s.[11] Boarding school children aged 8–14 years were divided into a xylitol group (95 children) and a control group (167 children) for 2 years. The xylitol group subjects were given 30 g of xylitol/day in the form of candies. This xylitol candy intake corresponded to about 50% of the normal daily intake of sweets. The children in both groups otherwise received identical food. They were examined at 6-monthly intervals for caries.

The increase in the number of carious teeth per child was lower in the xylitol group than in the control group ($p < 0.05$) after 6 months. At the end of the 2-year period, 8 teeth out of those 14 attacked in the xylitol group showed initial or 'surface' caries (the terminology used here is based on liberal translation from Russian). In the control group, initial caries (designated as 'spot-like' lesions and 'surface' caries lesions) was observed in 27 teeth of 104 attacked. The difference between the groups was statistically significant ($p < 0.001$). In the xylitol group there was one pulpitic tooth out of the 14 teeth displaying caries. In the control group the corresponding figure was 14 out of 104 carious teeth, this difference being significant ($p < 0.001$). The number of carious teeth per child increased significantly more rapidly in the sucrose group; after 12, 18 and 24 months the differences between the groups were significant ($p < 0.01$ at the two latter examinations). The overall incidence of caries after 2 years was 73% lower in the

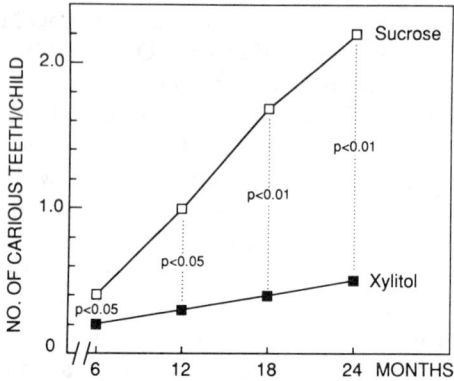

FIG. 1.   Results from the Soviet xylitol study. The increase in the number of carious teeth in children initially aged 8–14 years is shown during consumption of xylitol (30 g daily) or sucrose candies for 2 years. The 30 g daily dose of xylitol in the xylitol group ($n = 95$) substituted about 50% of the daily candy consumption of the children. The control group children ($n = 167$) received sucrose candies only (about 60 g sucrose per day). Based on the data of Galiullin.[11]

xylitol group than in the control group. Figure 1 shows a summary of the caries studies expressed as the number of carious teeth per child over the 2-year period.

Taking account of body weight gains, pulmonary, otolaryngeal, rheumatologic and endocrinologic aspects, and the general physiological development of the children, the groups did not differ significantly in any clinical examinations. Except for dental caries, they did not differ in the incidence of any somatic disease. The original, relatively concise report of Galiullin did not specify all experimental details of the trial. However, personal communication by the Academy of Sciences of the USSR confirmed the general performance and findings of the study. In spite of these developments, most of the xylitol produced in the Soviet Union has to date been used as a sweetener in the diabetic diet; this form of application of xylitol has a relatively long tradition in that country.

## 2.2  WHO Xylitol Studies in Thailand and French Polynesia

The WHO xylitol studies conducted in the Far East comprise one performed in Thailand and two studies carried out in French Polynesia. The caries prevalence differed in Thailand and Polynesia from the 1960s to the 1980s but were rising in both countries.

According to a WHO report, the DMF scores in 12-year-old Thai children were 0·4 and 2·7 in 1960 and 1977, respectively, regarded as low to moderate by WHO standards.[12] In French Polynesia the DMF scores were considerably higher, at 6·5 in 1966 and 10·7 in 1977.[12] The Thai study and the Polynesian study (here designated as the first Polynesian trial; a second was performed later) were planned and realized virtually simultaneously. The subjects consisted of two large groups of 7- and 10-year-old children who were supposed to participate in the studies for up to 3 years. In Thailand, however, the observation time was restricted to 28–32 months in some groups as a result of experimental difficulties. These trials were carried out on chewing gums that contained a mixture of xylitol and sorbitol (51:49) with fluoride (0·25 mg/stick), the daily theoretical polyol dose being 5–7 g per subject. As controls, a fluoride rinsing group (0·2% NaF fortnightly) and a sucrose–fluoride gum group (0·25 mg F/stick) were used. Four pieces of gum were supposed to be consumed on school days. In the first Polynesian study there was an additional fourth group which used a non-fluoridated chewing gum sweetened with the same xylitol–sorbitol mixture.

As is inevitable in most field studies, several problems arose in the Thai trials: (a) effective consumption of the experimental gums was restricted to about 9·5 months in any one year; (b) a delay in the supply of the fluoride–polyol gum; (c) considerable loss of 13-year-old children, and other factors.[13] Regardless of these problems, the existence of a preventive effect in the Thai study was noted in the fluoride and xylitol groups, where the final DMFS counts of the younger subjects were lower than the baseline values of the older children at corresponding ages. The xylitol–sorbitol–fluoride gum was nearly as effective in prevention of caries in the 7- and 10-year-old children as fortnightly rinsing with fluoride; the DMFS increment in the group using the polyol–fluoride gum was 0·94 for the subjects initially 7 years old, and 1·84 in subjects initially 10 years old. The corresponding increments in the fluoride rinsing group were 1·07 and 1·28.

In French Polynesia the 3-year caries (DMFS) increments were high: 2·35 in children initially 7 years old and 4·28 in children initially 10 years old for the xylitol group. The corresponding increments in the fluoride rinsing group were 6·07 and 10·04. Differences between these groups were not given, but the administration of the fluoride-containing xylitol–sorbitol gum was found to be the only method able to prevent caries.[13] The acceptability of chewing gum in caries prevention was promising, although the interest among most participants waned during the course of the study, to the extent that the children did not use all the chewing gum allotted to them.[13]

These two WHO field trials showed that a fluoride-containing xylitol–sorbitol chewing gum also prevented dental caries in circumstances where caries prevalence displayed a rising trend. In French Polynesia, however, the fluoride rinsing program failed to prevent caries. Furthermore, no caries prevention was observed with the xylitol–sorbitol gum that did not contain fluoride.

These two WHO trials contained several features peculiar to most field studies: the randomization of the children into different groups and observance of a blind- or double-blind design were, understandably, not possible.[13] Furthermore, the caries prevalence differed between the groups at the baseline registration, and the inter- and intra-examiner errors and significance of the results were not analyzed in the Thai and the first French Polynesian studies. The studies also differed with regard to planning, duration and the preventive measures used. The distribution of products and their consumption also gave problems. Most field trials have suffered from a large number of dropouts, which ranged from 19·9% in the 2-year Hungarian study (see below) to 82·9% in the first Polynesian study (on 10-year-old subjects). Consequently, although the Thai and the first Polynesian field trials supported the idea of a polyol–fluoride chewing gum as a suitable caries-preventive method, the practical difficulties encountered in the participation of the subjects in the trials do not allow a definite conclusion. Many of the shortcomings of the Thai and the first Polynesian trials were avoided in subsequent field studies (see below).

In a later study carried out in French Polynesia,[14] the experimental arrangements corresponded to a certain extent to those of the study carried out almost simultaneously in Hungary. The children ($n = 746$ at start) were initially 6–12 years old and their caries prevalence was high. The children lived on different islands and were enrolled in a non-randomized, comparative study that lasted for 32 months. The children of one island (Bora Bora; control group; $n = 345$) were supplied with toothbrushes and fluoridated dentifrice, whereas the children of the other two islands (Rangiroa, $n = 240$; Maupiti, $n = 161$) were assigned to the xylitol group. All subjects participated in a similar oral health program, but the xylitol group was provided with sweets sweetened with xylitol, the maximum daily xylitol dose varying between 14 and 20 g per child. Under the conditions on the islands, this can be regarded as a substitution of sugar. The xylitol products included chewing gum, candies, chocolate, ice lollies and gumdrops, which were distributed to the children from schools. Although 468 children completed the study, a total of 278 children were regarded as dropouts. The 195 control group children who completed the study showed an overall mean caries increment of 7·1 DMFS, whereas in the total xylitol group ($n = 273$) the mean caries increment was 4·5 DMFS. The difference between the control and the xylitol group was highly significant ($p < 0.001$). The reduction was calculated to be 37–39% in the xylitol group as compared with the control group. These results agree quite well with the data of the Hungarian WHO study, described below. Overall, the second Polynesian study exploited the experience gathered earlier in that region, and was more successful than the first,[13] showing that partial sucrose substitution by xylitol may be a useful tool in preventing caries in children under conditions where the caries incidence is high.

### 2.3 WHO Xylitol Studies in Hungary

The purpose of the Hungarian WHO trial was to determine the caries-reducing potential of partial substitution of dietary sucrose by xylitol. The study included longitudinal analysis of caries increment, microbiological determinations of salivary aciduric flora, assessment of oral hygiene, and measures related to dietary control. It seems likely that this trial involved addition of xylitol to the diet rather than mere substitution of sucrose by xylitol. Xylitol was used in the form of gums and solid candies, the daily dose being about 14–20 g per child. The xylitol group also used a xylitol-containing fluoride dentifrice without supervision. The subjects were institutionalized 6–11-year-old hearing- and sight-impaired or blind children. There was a 2-year and a 3-year follow-up, in which the xylitol group was compared with a group receiving systemic fluoride (fluoride group) and a group with solely restorative treatment (control group).[15,16] The fluoride group was divided into two subgroups, one receiving fluoridated milk (0·75 mg F per child daily) and using fluoridated dentifrice without supervision, and the other using natural fluoride-containing drinking water (1·2 ppm F). The control children used mainly a fluoride-free dentifrice and a fluoride dentifrice irregularly without supervision. The xylitol group children used three pieces of chewing gum per day—after breakfast, lunch and dinner. Additional xylitol products were available between meals.[17]

The preventive effect of xylitol was approximately the same in all age groups. Overall, xylitol was found to be more effective than systemic fluoride; the 3-year DMFS increment was 4·2 in the xylitol group, 6·5 in the fluoride group and 7·7 in the control group. The corresponding ratios between caries incidence and the surfaces at risk were 4·9, 6·6 and 8·6, respectively. Based on DMFS indices, the caries reduction in the 3-year study amounted to 35% in the xylitol group compared with the fluoride group, and 45% compared with the control group. The differences between groups were highly significant ($p < 0.001$).

The 2-year study was performed on children of whom 67% were participants in the above trial.[16] Initially 1219 children, aged 6–12 years, were included. The xylitol program ($n = 399$) resulted in a 37% lower caries increment and a 40% lower incidence rate than in the control group ($n = 221$). The difference between the xylitol and fluoride group ($n = 356$) was similar but not as distinct.

It must be emphasized that these results were obtained in circumstances where caries prevalence and incidence were high. It was also common for the children to have untreated caries.[15,18] The consumption of xylitol products was not found to reduce the frequency of intake of sucrose products significantly.[17] It is likely, therefore, that the low caries increment in the xylitol group did not result from the absence of sucrose in the diet.

## 2.4 Canadian Xylitol Gum Study

Kandelman et al.[19] carried out a xylitol chewing gum study in Montreal in the mid-1980s; 433 children, initially 8–9 years old with a high caries rate, were divided randomly into two groups receiving two different xylitol levels in chewing gum. One type of gum contained 15% xylitol and the other 65% xylitol, the remaining polyol in the 15% xylitol gum being sorbitol (50%). Because three pieces of gum were supposed to be used per day per child, the total xylitol consumption was calculated to be about 3·3 g in the 65% group and about 0·9 in the 15% group. The gum was distributed 3 times a day by teachers at the schools, who supervised the 5-min chewing period. A third group of children, who did not receive gum from school, acted as a control. All groups were exposed to the same basic prevention program. The entire study lasted for 2 years. After the first 12-month period the adjusted mean net progression of decay was about 3·5 surfaces for the controls compared with about 1·6 for the treated groups (55% decrease). The 65% xylitol gum seemed to give better results than the 15% xylitol gum, although the difference was not significant.

The results from the second 12-month period supported the first year's findings.[20] After 2 years, 274 children who used the xylitol gums showed a significant 65% reduction of the DMF(S) scores compared with the control group which received no gum. The adjusted mean DMF(S) increments were 6·06, 2·39 and 2·09 for control, 15% and 65% xylitol groups, respectively. In line with the first year's findings, the 15% and 65% xylitol groups did not differ significantly. Analyses of tooth surface showed that the net percentage of surfaces with progressive decay differed significantly between the gum groups and the control group.

Some trials have included attempts to assess the degree of plaque growth during the xylitol regimens. The Turku[1] and Hungarian[21] studies showed that xylitol consumption was associated with reduced plaque growth. The Montreal study supported these findings, showing a significant ($p < 0.001$) reduction in plaque index scores in the xylitol groups compared with the control group. In a 3-year sorbitol chewing gum study, no such effect was reported.[22] A review[23] listed several other plaque studies, which showed that in most cases the use of xylitol was associated with lower plaque fresh weights or smaller plaque index values than the use of sorbitol.

Overall, the Montreal study clearly demonstrated a protective effect of the xylitol-containing chewing gums in school preventive programs. Considering the low dosage of xylitol in the 15% xylitol group, the clinical effects caused by xylitol were perhaps the strongest achieved in any xylitol/caries study during the 1970s and 1980s and, if verifiable, suggest that a specific mechanism of action of xylitol in caries prevention must be involved.

### 2.5 Ylivieska Xylitol Gum Studies

The Ylivieska xylitol gum studies[24] were special in several respects. The positive results from the Turku clinical trials had been known since 1974 in Finland and there was a growing consensus among dentists and the general public on the usefulness of xylitol in caries prevention. A field trial on children at caries-active age was still missing. A 2-year field trial was therefore carried out in young children in a locality where caries was relatively well controlled, mostly as a result of fluoride-based prevention—the Ylivieska Health Center in North Finland. The trial[24] was to be carried out in an 'overchallenged' situation while accepted caries-preventive measures were in full use, so the studies were performed on children with 'low or moderate and decreasing caries prevalence', according to WHO criteria.[12]

Another special feature of the Ylivieska studies[24] was the original plan to carry out a re-examination of caries after about 3 years following the termination of supervised xylitol administration in November 1985. Such a retrospective analysis of caries in a nutritional study had not been conducted before. A further special feature was the high degree of cooperation of the children; the total loss of subjects was 10·9% only. A fourth aspect was the continuation of the xylitol gum regimen for a third year on children who were considered to be at 'high risk' with regard to dental caries (all 11-year-olds with DMFT $\geq 5$ and all 12-year-olds with DMFT $\geq 7$ at the baseline examination were allocated to this group).

Because the purpose of the trial was to determine whether xylitol gum would be able to increase the efficacy of existing caries-preventive measures mainly based on fluorides, one group of children received xylitol gum, while no gum was given to the control group children. For ethical reasons it was not possible to give sucrose or sorbitol gum to the control group as part of the study (sorbitol was known to stimulate the growth of *Streptococcus mutans*). However, the control subjects had free access to sucrose- and sorbitol-sweetened products on the market. Some xylitol-containing items were also available at local supermarkets and kiosks, but their use was not consistent. Questionnaires revealed that the results of the study were not caused by significant variation in the intake of sugar products; the only major variable between the two groups was the presence of the xylitol sweetener in the chewing gum of one group. The study had a follow-up of the plaque and salivary levels of *S. mutans*.[25]

The main parts of the Ylivieska studies were carried out in 1982–85. The participants in the first study lasting 2 years were 324 children, initially 11–12 years old. They were divided into a xylitol gum group ($n = 172$) and a control group ($n = 152$). The subjects of this study were born in 1970 and 1971, and since 1972 had organized health programs and dental care on an annual basis. Consequently, all dentinal lesions ($D_2$) present in deciduous and permanent teeth were restored by fillings. The fluoride measures were comprehensive, but it appeared that, at the most, 7·6% of the children used fluoride tablets regularly. Caries-preventive measures also included the use of fissure sealants in the first molars before the trial; no sealants were used during the trial. All the children lived within the same community water supply system, where the fluoride concentration of the water was less than 0·1 ppm.

Evaluation of the results should take into account the fact that the preventive benefit of xylitol was compared with the effect of basic caries prevention procedures used in the control group. It was found that the efficacy of the basic prevention procedures could be statistically improved in children at a caries-active age. Thus the use of 2–3 pieces of xylitol gum daily (7–10 g xylitol/day) improved prevention by 30–60% in the 2-year follow-up and by 50–80% in the 3-year follow-up on children who were considered to be at high risk to dental caries. When the consumption of the gum was equal to or less than 1·5 pieces/day no additional prevention was achieved.

The 3-year follow-up on children with high caries activity also indicated that regular consumption of xylitol gum led to a greater caries prevention in the high risk subjects than in the control group on basic prevention procedures only. The most effective caries-preventive action was achieved during the third study year. The cariostatic effect of xylitol was strongest among the most susceptible teeth or surfaces (i.e. at the age of 11–13 years in fissures of second molars and after the age of 13 on proximal surfaces, especially in the maxillary incisor region considered at high risk for caries).

The experimental groups were homogeneous for many 'background' factors. In 1987, 3 years after the discontinuation of the use of xylitol gum, 90% of the children were re-examined.[26] This showed that the reduction in DMFS increment figures in the xylitol group in 1984–87 was about 51% compared with the control group (figure set at 100%). In teeth erupted during the first xylitol study year, the caries reduction was even better, i.e. about 73%. It has to be emphasized that no differences in the use of xylitol or other sweets during 1984–87 were found between the xylitol and control groups, and that the caries prevention realized in these groups in 1982–84 differed only with regard to the administration of the xylitol chewing gum. These results were totally new and indicated that the teeth of 11–12-year-old children, protected against dental caries using xylitol gum in combination with fluoride prevention, have a good resistance against caries attack, persisting after discontinuation of the use of xylitol. The studies further suggest that such preventive measures can have long-term effects.

The later examination also showed the validity of the relationship between caries incidence and the number of pieces of gum consumed per day; children who used in 1982–84 on the average less than or equal to 1·5 gums per day showed no significant additional benefit after the 3-year pause, in 1987, whereas those who had used on the average 3 pieces of gum per day showed an additional benefit 3 years later. It should be possible to calculate the impact of such additional preventive measures in communities where dental caries remains as a problem.

The saliva or plaque levels of *S. mutans* matched the clinical results. Figure 2 shows an example of the correlation. The more the xylitol gum was used per day, the more effective the caries reduction.[24,25] Once the population of *S. mutans* decreased after the commencement of the xylitol usage, it did not return to its original levels. The results thus indicate that no weakening of the xylitol-associated inhibitory effect was observed in the 2- and 3-year follow-ups. This is in agreement with most previous microbiological and plaque studies, countering the claims of the development of xylitol resistance and decreased cariostatic power in the plaque of xylitol users.

In summary, the Ylivieska studies suggested that xylitol gum, used 2–3 times per day in combination with basic preventive measures with fluorides, constitutes a strong instrument for caries prevention in caries-active age

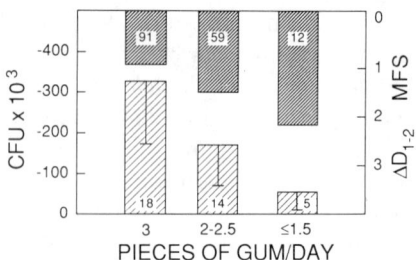

FIG. 2. Decrease (between baseline and 1984) in the plaque CFU values of *S. mutans* in children using on the average 3, 2–2·5 and 1·5 or less pieces of xylitol gum per day, and the corresponding caries incidence in the 2-year Ylivieska xylitol gum study ($n = 162$). A randomized sample ($n = 37$) of the children was analyzed for *S. mutans* (lower columns).

groups and high-risk individuals. The results further indicated the good acceptability of xylitol chewing gum as a caries-preventive agent even in individuals at high risk for caries. The xylitol gum program also had a long-term effect even after discontinuation of the systemic use of the gum.

## 2.6 Planimetric Re-examination of Turku Study Findings

Part of the unpublished data of the Turku xylitol studies[1] was re-evaluated in the mid-1980s using planimetric methods to assess the caries-reducing potential of xylitol.[27–29] It may be recalled that the Turku xylitol studies comprised a 2-year feeding trial and a 1-year chewing gum study. Both suggested that the use of xylitol caused a strong reduction in caries, exceeding 85% compared with the control groups, plus a therapeutic effect.

The purpose of the re-examination was to quantitate changes in the size of proximal caries and buccal white spot lesions to assess the effect of the consumption of sucrose and xylitol on the rate of caries progression. Bite-wing radiographs taken at the beginning (1972) and end (1974) of the 2-year study and the 1-year study (1973–74) were analyzed planimetrically with regard to proximal caries. Similarly standardized color photographs of white spot lesions were taken 7 months after the beginning and at the end of the 2-year study. The method was based on the analysis of the radiographs and photographs projected with 13-fold linear magnification on a planimetry platen with a digitizing surface. The areas of the caries lesions were circumscribed with a cursor connected to a digitizer. The digitizer entered the data on a calculator.

The results were expressed in area (mm²) and showed a highly significant increase in proximal lesion sizes in the sucrose group after 2 years ($p <$

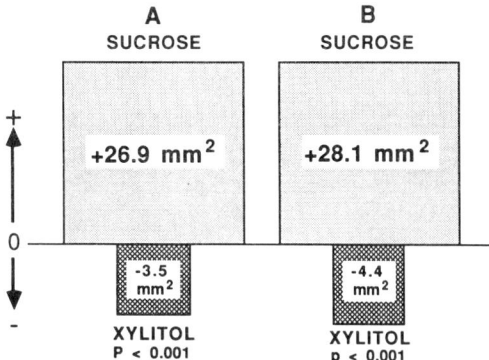

FIG. 3.    Size changes in proximal caries lesions in the 2-year (A) and the 1-year (B) Turku xylitol studies. The values shown (mm$^2$) were calculated from data obtained at 0 and 24 months (A) and 0 and 12 months (B), and give the sum of positive or negative changes detected in all carious tooth surfaces in each experimental group. The number of all carious tooth surfaces was 114 in the sucrose group and 158 in the xylitol group for the 2-year study, and 198 and 183 respectively for the 1-year study.
The size of the squares illustrates the relative total size changes.

0·001), whereas in the xylitol group the proximal lesion sizes remained unchanged. The corresponding difference in lesion size between the xylitol and sucrose groups was highly significant ($p < 0.001$). The progression of proximal caries in the 1-year gum study was practically identical with that in the 2-year feeding study; in the sucrose group the lesions increased in size in absolute values ($p < 0.05$) and in percentages ($p < 0.001$), while in the xylitol group the mean lesion sizes remained unchanged. The area of white spot lesions also decreased in the xylitol group both in absolute values ($p < 0.01$) and in percentages ($p < 0.001$). In the sucrose group these areas increased both in absolute values ($p < 0.05$) and in percentages ($p < 0.01$); these comparisons were made after 17 months of the 2-year trial. The Turku study had also involved testing a stereomicroscopic technique to detect early caries lesions,[28] which gave similar results.

These re-examinations showed collectively that sucrose consumption enlarged the caries lesions, whereas the consumption of xylitol decreased lesion size and arrested caries. The remineralizing effect of xylitol was higher on buccal-free surfaces than on proximal surfaces.

Figure 3 shows an example of these re-examinations. The size of proximal caries lesions in both studies was significantly reduced as a result of the xylitol programs. The reduction was of the same degree in both studies, although the 2-year feeding trial involved about a 10-fold greater daily xylitol consumption than the 1-year chewing gum program.

## 2.7 Comparisons between Long-term Chewing Gum Trials

Only three human caries trials have used chewing gum as the sole vehicle for xylitol. These are the Turku 1-year trial,[1] the Montreal study[19,20] and the Ylivieska trial.[24,25] In all other studies xylitol was used in several different food items. Two gum studies used sorbitol as the sole sweetener.[22,30] In the three xylitol gum studies, a relatively constant frequency of intake of gum was maintained,[1,19,20] or subjects were grouped into different categories depending on the amount of gum used per day.[24] It is premature to draw conclusions from these three trials on the amount of gum needed per day to achieve clinically significant protection against caries. Two of the studies involved children at a caries-active age,[19,20,24] while one study used young adults.[1] They also differed in duration, types of gum used and other details. It is obvious that if a preventive treatment is supposed to be effective it should be effective under different conditions, and the studies showing effects should be performed by independent authors, preferably in different countries. All these requirements have been met.[1,19,20,24] Two of the studies were similar with regard to the type of gum used; the Turku 1-year study[1] and the Ylivieska trial[24] employed rather large ($\sim 5\cdot4$ g) uncoated pieces of gum which, however, differed in composition (the gums in the Turku study contained 50% xylitol, 6% sorbitol and 14·5% other polyols, while the Ylivieska gums contained xylitol as the only sweetener). These differences prevent the drawing of firm conclusions about the correlation between caries incidence and the number of pieces of xylitol-containing gum chewed per day, but the caries reductions achieved in these studies have been summarized in Fig. 4.

None of the caries studies described above was designed to compare xylitol and sorbitol. It is thus still impossible accurately to assess the relative caries-reducing potential of these polyols, although some attempts have been made (Fig. 4). The daily consumption levels of sorbitol and xylitol were 3·6 g and 7–10 g, respectively. Therefore it is difficult to compare the percentage caries reductions achieved, i.e. 10% (sorbitol)[22] and 40–80% (xylitol).[24] Figure 4 also shows the result of another sorbitol gum study which produced no significant caries reduction compared with the control.[30] Based on all human and animal caries studies on sugar substitutes so far carried out, and supported by microbiological and biochemical findings, reasonable evidence exists therefore that xylitol has performed better as a caries-preventive agent. Ignoring the differences between study designs in human trials, sorbitol has resulted in 0–54% caries reduction, whereas the consumption of xylitol has been associated with a 35% to at least an 85% reduction, with observations which indicate

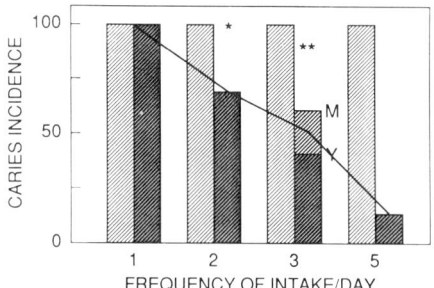

FIG. 4.    Summary of data from studies using chewing gum as a vehicle for xylitol or sorbitol. The caries incidence in the control group in each study is set at 100% (light columns on the left) and the reduction in caries achieved on the gum is shown by the darkened columns. Trials included: frequency of intake = 1 (extrapolated from Ylivieska xylitol data of children using equal or less than 1·5 gums per day; 2-year study[24]); frequency of intake = 2 (extrapolated from Ylivieska xylitol data of children using 2–2·5 gums per day; 2-year study[24]); frequency of intake = 3 (M = Montreal xylitol study on children, 1-year results;[19] Y = Ylivieska trial, 2-year results[24]); frequency of intake = 5 (extrapolated from data of the 1-year Turku xylitol study with young adults;[1] the calculated number of gums used was 4·9). The results from two sorbitol gum studies are shown: *, 2-year Forsyth sorbitol gum study (no reduction in caries[30]); **, 3-year Danish sorbitol gum study (10% reduction compared with control[22]). The positions of the asterisks represent caries reduction reported on sorbitol gum. The solid line shows the average remaining caries incidence in the subjects who participated in the xylitol gum trials.

remineralization of caries lesions in xylitol-consuming subjects. A direct comparison between xylitol and sorbitol is badly needed, however.

It is a difficult task to quantify exactly the caries-preventive effect of xylitol. One objective is to assess the quantity of xylitol needed per day for positive clinical effects, and to study how it could be used successfully along with other nutritive sweeteners, such as Lycasin, sorbitol, maltitol, Palatinit etc.

Some unnecessary controversy arises from the use of different caries criteria in the USA and Europe, and it would be advisable in future trials to consider caries at its various stages and to standardize scoring systems.

# 3 MECHANISM OF ACTION OF XYLITOL IN DENTAL CARIES PREVENTION

## 3.1 Prerequisites

Any attempt to explain the mechanism of action of xylitol in caries prevention must consider the mechanisms of the initiation and spread of dental caries itself. It is beyond the scope of this chapter to discuss the microbiology of dental caries, but several reviews are available.[31-33] It has been concluded that, among the 200–300 species indigenous to human dental plaque, [34,35] only a certain number may be responsible for most human dental decay.[33] It has also been shown that the type of sugar in the diet can influence the composition of the dental microflora[36,37] and its metabolism.[6,38-41] Several $C_6$ carbohydrates, not only sucrose ($C_{12}$), have the potential to play a causative role in dental caries.[42,43]

There is also information about the infectious nature of dental caries.[32,33] To fight this disease, the human body has a versatile system of chemical defense factors, including salivary enzymes attacking microorganisms, specific peptides governing the metabolism of calcium in the mouth or pH changes in plaque, immunoglobulins and other factors. Since dental caries at its early stages is a reversible process, and because the development of advanced lesions may stop, there must be factors in saliva that help to reverse this process (i.e. remineralization). A driving force is the supersaturated level of calcium and phosphate in saliva. The chemical requirements for saliva-induced remineralization include sufficiently high calcium and phosphate levels and pH value, although other factors (fluoride ions, Ca-binding peptides etc.) are also involved. Consequently, attempts to explain the mechanism of action of xylitol should consider the effects of xylitol on cariogenic organisms and on remineralization.

### 3.2 Growth of *S. mutans* and Other Microorganisms in the Presence of Xylitol

The Turku studies showed a significantly lower incidence of *S. mutans* in the plaque of xylitol-consuming young adults relative to subjects consuming sucrose or fructose.[44] Another study showed that a xylitol gum reduced the incidence of *S. mutans* more than a sorbitol, mannitol or fructose gum.[45] In the Hungarian study, the xylitol-consuming children had significantly less *S. mutans* in plaque than the control group or the group receiving fluoride treatment.[46] In the Ylivieska field trial involving the use of only 7–10 g of xylitol per day, the plaque and saliva levels of *S. mutans* were significantly smaller after 2–3-year use of xylitol gum compared with the group receiving no xylitol gum.[25] The use of 3 pieces of xylitol gum per day reduced the occurrence of *S. mutans* more effectively than the use of about 1·5 pieces.[25] The form of xylitol administration may not be of primary importance for these effects to occur, since a xylitol-containing toothpaste also affected the growth of this organism.[47] Other acidogenic or caries-inducive organisms are also affected by xylitol.[6,48] The consumption of a xylitol diet was generally associated with a reduction of acidogenic and aciduric oral flora.[49] Consequently, a prerequisite for positive xylitol effects, i.e. inhibition of *S. mutans* and similar inhibition or inertness of xylitol with regard to several other oral microorganisms, seems to have been demonstrated. These results are in agreement with the general idea that the pentitols constitute a group of unique carbon sources that are metabolized by relatively few microorganisms.[50]

It would be wrong to regard xylitol as a microbiologically fully inert carbohydrate. There are examples of non-oral, especially soil micro-organisms capable of handling xylitol either as an energy source or as an intermediate. Some organisms secrete xylitol as an unwanted by-product.[51] Organisms such as *Candida tropicalis*,[52,53] *C. utilis*,[54] *Rhodotorula toruloides*,[54] *Klebsiella aerogenes*,[55] *Aerobacter aerogenes*,[56] *Saccharomyces cerevisiae*,[54] *Escherichia coli*,[57] *Erwina uredovora*[58] and *Lactobacillus casei*[59-62] serve as examples of microbes that in one way or another are involved in xylitol handling. In general, the importance of xylitol as an energy source for cariogenic organisms may remain small, partly as a result of the principle of evolutionary expediency (see below), and the likelihood that the human basic diet will continue to use mainly hexose-based carbohydrates, masking the lesser metabolism of pentitols in dental plaque.

### 3.3 'Xylitol-sensitive' and 'Xylitol-resistant' Strains of *S. mutans*

Certain members of *L. casei* grow at the expense of ribitol or xylitol, carrying out a heterolactic fermentation producing ethanol, acetic acid and a mixture of D- and L-lactic acid.[59] The growth of a *L. casei* strain at the expense of ribitol was inhibited temporarily if the non-metabolizable xylitol was added to the medium at concentrations of 6 mM or greater.[60] The cessation of growth was caused by the intracellular accumulation of xylitol 5-phosphate (x-ol 5P) because no enzymes existed for its further metabolism.

These concepts were then applied to *S. mutans*, including its adaptation. This was regarded as inevitable, although its clinical impact was considered insignificant.[1,6,39,41,43,44] It was later suggested that culturing of *S. mutans* on glucose in the presence of xylitol resulted in the selection of naturally occurring cells that the authors designated as 'xylitol-resistant' (XR).[63,64] It was claimed that xylitol-sensitive (XS) strains were not all affected to the same degree. It was also suggested that the relative proportions of bacteria from the XR and XS phenotypes were influenced by the presence or absence of refined xylitol in the diet, and that the consumption of xylitol by humans for long periods of time may result in the *in vivo* selection of XR *S. mutans*.[64] Xylitol resistance in laboratory strains was associated with a low constitutive fructose PTS (phosphoenolpyruvate phosphotransferase) activity. Xylitol was phosphorylated and accumulated intracellularly by a constitutive fructose PTS in XS strains. The 'xylitol toxicity' in bacteria was postulated to result from the accumulation of a xylitol phosphate; phosphorylated sugars that are not further hydrolyzed may be toxic.[65,66] The mechanisms by which the constitutive fructose PTS-negative strains arose is not known. Mutation has been considered unlikely because xylitol and its metabolites were shown to be non-mutagenic,[67,68] but there may be an adaptive process.[69] The growth of XR strains was not inhibited by xylitol, whereas the XS strains were, even though both took up xylitol.[70] In the XS cells, xylitol was possibly transported via a PTS leading to an intracellular accumulation of up to 5 mM levels of a $C_5$ intermediate (presumably x-ol 5P or x-ose 5P, i.e. xylitol 5-phosphate and xylulose 5-phosphate).[71,72] These metabolites were not detected in the XR strains.

The works of Reizer,[73,74] Thompson and Chassy,[75,76] Hausman *et al.*[77] and others thus finally led to the concept of the 'futile metabolic cycle', a cycle that promotes the dissipation of phosphoenolpyruvate (PEP), inhibiting the growth of bacteria (Fig. 5). Accordingly, intracellular x-ol 5P first accumulates intracellularly and is gradually hydrolyzed by a phosphatase while an export system expels xylitol. This xylitol cycle thus consumes PEP and serves no energetic purpose.

There is still some uncertainty over these metabolic effects of xylitol, with the suggestion that some isolates of *S. mutans* lose their 'sensitivity' to xylitol in a few successive cultivations in the presence of xylitol. However, the clinical caries data do not support any loss of xylitol sensitivity in *S. mutans*, and some studies have shown strengthened preventive efficacy[1,20,24] towards the end of xylitol administration. Adaptation of *S. mutans* to xylitol-containing media was reported as early as 1975,[78] but subsequently the cells were shown to grow in the presence of xylitol only because they started to use amino acids, peptides and proteins of the medium at an increasing rate, xylitol behaving more or less as an inert carbohydrate.[79] The phenomenon was clearly reversible.[64,69,70,80] The cells should thus retain their xylitol sensitivity in the mouth during habitual use of xylitol. It was further found that the strains studied reacted to xylitol in the medium in distinctly different ways,[78] leading to misunderstandings in reports about the 'loss of xylitol sensitivity'. In fact, once the *S. mutans* levels were brought down by xylitol, they did not rise again even after 3-year use of xylitol.[25]

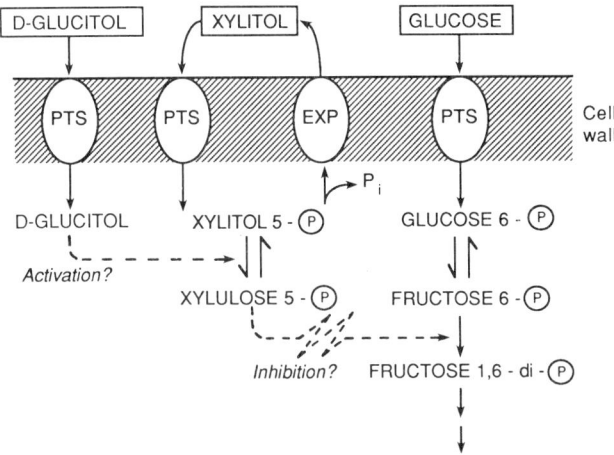

FIG. 5. Possible mechanism of the enhancement of xylitol toxicity in *S. mutans* by D-glucitol. The mechanism involves inhibition by intracellular x-ose 5P of the glycolysis at the phosphofructokinase level. This inhibitory action may stem from D-glucitol-induced concentration of x-ose 5P in the cells. The PEP is here considered to be the primary donor in the phosphorylation of x-ol 5P and glucose 6-phosphate. The 'futile cycle' of xylitol is also shown and comprises the following sequence:
extracellular xylitol → PTS → xylitol 5-P → loss of $P_i$ → expulsion of xylitol.

### 3.4 Enhancement of the Growth Inhibition of Xylitol by D-Glucitol in *S. mutans*

D-Glucitol is normally metabolized by several strains of *S. mutans*.[81] For example, the cells of the *S. mutans* strain OMZ 176 phosphorylate D-glucitol into D-glucitol 6P, which is subsequently transported into the cell. These reactions are catalyzed by a PTS present in the cell wall. A number of studies suggest that the growth of *S. mutans* OMZ 176 on D-glucitol is reduced in the presence of xylitol. Assev and Rölla[81] reported that cells which had been exposed to xylitol displayed a delayed uptake and no metabolism of D-glucitol. The uptake of xylitol was reduced in the presence of D-glucitol with the accumulation of x-ol 5P and/or x-ose 5P in the cells. The x-ose 5P may inhibit glycolysis in cells exposed to xylitol, possibly explaining the D-glucitol-induced increase of the inhibitory action of xylitol on *S. mutans* (Fig. 5). A suitable application would be the use of mixtures of D-glucitol and xylitol in chewing gum,[82] as used in the Turku studies.[1]

### 3.5 Xylitol and the Structure of the Bacterial Cell Wall

Dental plaque developed in the presence of fermentable $C_6$-based sugars (notably sucrose and glucose) contains relatively high amounts of lipoteichoic acids (LTA) which, upon reacting with plaque extracellular glucans, may contribute to the pathogenicity of plaque by enhancing its adhesion to tooth surfaces. The teichoic acids are polymers of glycerol or ribitol in phosphodiester linkage and are responsible for some of the properties of S. mutans. Plaque formed in the presence of sucrose contained more LTA than plaque exposed to xylitol,[83] and more teichoic acids were produced by S. mutans OMZ 176 in the presence of glucose than in the presence of xylitol.[84] A reduction in the amount of LTA should be considered clinically advantageous, since LTA may be involved in cell–surface interactions, and may play a key role in the adhesiveness of sucrose-grown S. mutans.[85] Transferring actively growing cells of S. mutans into xylitol-containing medium caused distinct alterations in bacterial ultrastructure, and they may resemble those observed in lysozyme-treated bacteria.[86]

The effect of D-glucitol and xylitol on the aggregation, adhesion and solubility of polysaccharides produced by S. mutans was studied in the presence of 3% sucrose.[87] D-Glucitol increased the formation of insoluble carbohydrates associated with the cell mass, whereas xylitol decreased the amount and adhesion of these polysaccharides. These findings were explained by alterations in the de novo synthesis of extracellular polysaccharides rather than by direct physicochemical action.[87] The results support the idea that the plaque grown in the presence of xylitol may be looser and more easily removable than plaque grown in the presence of sucrose or D-glucitol.

The use of xylitol gums,[88,89] chewable tablets[90,91] or the consumption of a xylitol diet[41] were associated with reduced quantity or adhesion of plaque, which may be related to the above polysaccharide[87] and LTA findings.[83,84]

### 3.6 Other Microbial Effects of Xylitol

During consumption of xylitol in place of sucrose the plaque micro-organisms may become deprived of a preferred substrate, and may start to synthesize extracellular proteinases and glycosidases for the hydrolysis of proteins, peptides and glycoproteins.[6,41,79] Cells of S. mutans maintained in a medium containing xylitol instead of glucose showed no observable uptake of [$^{14}$C(U)]xylitol, but exhibited a strong increase in the extracellular proteinase activity.[79] Proteinase activity in the whole saliva and dental plaque increased when xylitol was habitually consumed.[41] The activity of plaque aspartate aminotransferase also tended to increase in plaque exposed to xylitol, producing more ammonia than control

TABLE 2

INHIBITION OF MICROBIAL ENZYMES BY POLYOLS[39]

| Enzyme and polyol | $K_i$ (M) |
|---|---|
| D-Xylose isomerase (*Lactobacillus brevis*) | |
| D-Arabitol | 0·13 |
| L-Arabitol | 0·146 |
| Xylitol | 0·0027 |
| D-Glucose isomerase (*Actinoplanes olivocinerus*) | |
| D-Mannitol | 0·2 |
| D-Glucitol | 0·03 |
| D-Arabitol | 0·14 |
| Ribitol | 0·02 |
| Xylitol | 0·024 |
| D-Glucose isomerase (*Actinoplanes* sp.) | |
| D-Mannitol (relative activities) | 100 |
| D-Galactitol | 80 |
| D-Glucitol | 70 |
| Xylitol | 35 |

plaque.[47,90] This ammonia may partly neutralize acids. The concentration of basic amino acids was increased in plaque exposed to xylitol.[92] In general, the few kinetic studies that have been carried out show that pentitols inhibit enzymes involved in hexose or hexitol metabolism, and that xylitol is frequently more inhibitory than other pentitols[39] (Table 2).

## 3.7 Chemical Profile of Xylitol: Reflections in Oral Biology

The clinical and microbiological effects of xylitol (Fig. 6) suggest that it possesses unique chemical properties, for example: (a) Xylitol (and certain other polyols) interact with Ca(II) and other polyvalent cations, forming stable complexes[93-110] which may also form in saliva, dental plaque and caries lesions,[40,111] governing the fate of Ca(II) in those environments. (b) These complexes are formed by displacement of water molecules in the solvation sphere of Ca(II) by OH groups of xylitol,[94,101,112] so xylitol stabilizes Ca(II) in the oral environment. (c) Xylitol also stabilizes native protein structure.[113-121] These may all be dentally desirable properties.

### 3.7.1 Plaque Calcium

Table 3 shows that the consumption of xylitol is associated with increased amounts of soluble calcium in plaque. This may assist in the remineralization of mineral-deficient enamel sites. About half of the total

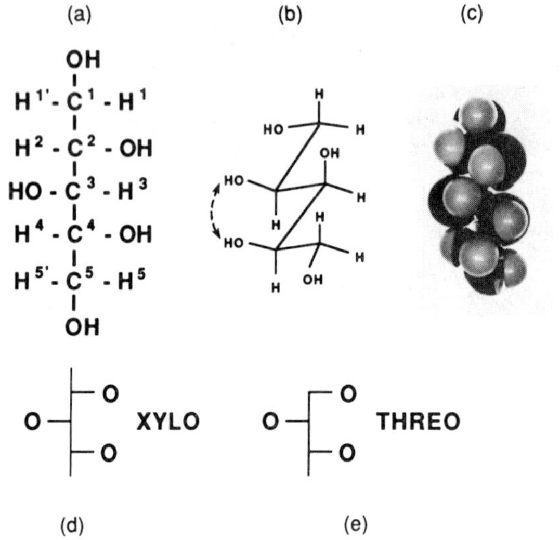

FIG. 6. Structure of xylitol. (a) Numbering of individual carbon and hydrogen atoms. (b) Planar zig-zag formula. (c) 'van der Waals' contour of the molecule (the structures are not to scale and they show approximations only). The double-headed arrow in (b) shows the parallel 1,3-interaction between hydroxyl groups, indicating that rotation(s) around the C–C bonds tend to occur to avoid them, resulting in a bent or sickle conformation. The configurations (e) and (d) show the possible tridentate ligand sets present in xylitol; the *xylo* normally complexes more effectively than *threo* with Ca(II). According to the carbohydrate nomenclature, the *threo* and *xylo* prefixes indicate the presence of two or three asymmetric carbons respectively.

concentration of plaque calcium may be ionized. It is not known how much of the calcium present in 'polyol plaque' is in an ionized form, but polyols may facilitate the solubilization of the insoluble portion of plaque calcium. The increased soluble calcium may inhibit demineralization through a common ion effect, or may enhance remineralization during periods of high pH.

### 3.7.2 Selective Aggregation of Salivary Proteins in the Presence of Carbohydrates

When Millipore-filtered human mixed saliva was incubated at 37°C in the presence of simple carbohydrates, aggregation of salivary Ca–protein complexes took place in the saliva at rates that depended on the structure of the added carbohydrate.[125] D-Xylose, D-ribose and i-erythritol showed no effect whereas xylitol, D-mannitol and D-glucitol strongly inhibited

TABLE 3

CONCENTRATION OF CALCIUM IN DENTAL PLAQUE OF SUBJECTS WHO USED PRODUCTS
CONTAINING XYLITOL

| Study | Calcium | | Remarks |
|-------|---------|---|---------|
| | Xylitol | Control or sucrose | |
| Chewing of xylitol gums (paraffin as control) | $1.22 \pm 0.45$ | $0.78 \pm 0.30$ | In $\mu$g/mg fresh weight ($n = 0$–12; $p < 0.01$). Sorbitol gave similar results.[122] |
| Chewing of xylitol gums | $3.7 \pm 0.5$ | $2.4 \pm 0.2$ | In $\mu$g/mg dry weight ($n = 83$). Gum base gave $3.4 \pm 0.7$. Significance of differences was not given.[123] |
| Rinsing with 0·4M xylitol or sucrose solutions | 0·90 | 0·67 | In $\mu$g/mg protein. Pooled plaque from 11 subjects in each group. 0·01M sodium cyclamate gave 0·60.[124] |

The values shown represent the calcium levels found in the buffer used to extract plaque. Therefore they do not represent true plaque fluid values and they are also considerably lower than the total calcium levels of plaque. To be used to compare the different chewing or rinsing regimens only.

aggregation. Since the precipitable proteins contain calcium, this stabilization may help to keep calcium in a caries lesion or in plaque fluid.

### 3.7.3 Penetration of Polyols in Enamel

Xylitol penetrates fast into the aqueous phase of human enamel,[126] and may inhibit acid dissolution by interfering with the transport of dissolved enamel. More than twice as many xylitol as sucrose molecules diffused through dental enamel.[127] Ca(II) may be solvated by a water–xylitol mixture.[111,126] The difference may be explained in terms of the differences between the radii of the molecules, and by viscosity differences. Xylitol solutions have a lower viscosity than those of sucrose at equimolar concentrations.

Human enamel showed a significant reduction in demineralization after xylitol treatment.[128] Mineral loss after the sucrose and water (control) treatments was about 3 times that after the xylitol treatment.

### 3.7.4 Xylitol in Remineralizing Solutions

The rehardening properties of mucin- or CMC-containing saliva substitutes were enhanced by xylitol, indicating a favorable complex formation with Ca(II).[129] The authors suggested that D-glucitol in rehardening solutions should be replaced with xylitol in all commercial saliva substitutes.

There is no proof that xylitol increases the incorporation of fluoride into enamel but, since xylitol interacts with calcium, saliva and plaque fluid, it may have an indirect effect on fluoride transport in enamel and caries lesions as well. A recent study[130] indicated that the use of xylitol-containing toothpastes was associated with a slightly elevated fluoride content in the enamel compared with glycerol-containing toothpastes. In line with this, glycerol does not form as strong complexes with Ca(II) as xylitol does. Related to this is an earlier finding that the saliva of xylitol-consuming subjects contained slightly less fluoride than the saliva of sucrose-consuming subjects.[41] This was explained in terms of decreased solubility of enamel in the xylitol group; fluoride was retained in the solid phase, 'where it was supposed to be'. These findings need verification, however.

Xylitol has been tested in a self-administered mouth-rinse designed to induce remineralization,[131] and in a mucin-containing artificial saliva preparation for xerostomic patients.[132] In Finland some dental schools and pharmacies have prepared xylitol-containing remineralizing solutions to alleviate problems resulting from hyposalivation.


## 4 CONCLUSION

The first claim for a remineralizing (therapeutic effect) of xylitol, based on human studies, was made in 1975,[1] and believed to be confirmed later.[27–29] Following a round of verifying trials in different countries,[11,14–16,19] the original findings[1] received convincing support in latest studies.[19,20,24,26]

Based on chemical, biochemical, microbiological and physicochemical studies, it seems that the clinical effects of xylitol on dental caries can be explained by (1) the inertness of xylitol as a substrate for most oral microorganisms and its inhibitory effect on *S. mutans*, and (2) its physicochemical action in the mouth. Xylitol also stimulates the secretion of saliva, promoting various salivary defensive factors.

'Xylitol plaque' has improved characteristics over sucrose plaque or sorbitol plaque. Cariogenic bacteria have developed specific enzyme and transport systems for the utilization of hexose-type carbohydrates. The $C_5$-based carbohydrates, especially polyols, are relatively unsuitable energy sources for cariogenic microorganisms. No true adaptation to use xylitol has been observed in mixed plaque flora, and it is unlikely that strong xylitol-utilizing cariogenic strains will develop under the competitive conditions that prevail in the human mixed oral flora.

One likely factor in the efficacy of xylitol in caries limitation is its 'antimetabolic' effect on strains of *S. mutans* which take up this polyol but cannot utilize it. The inhibition of glycolytic enzymes by xylitol may thus be the basis for most of the dental effects of xylitol, but this does not explain its remineralizing potential. In conclusion, the various properties outlined here help to clarify the mechanism of action of xylitol in caries control.

## REFERENCES

1. Scheinin, A. & Mäkinen, K. K., Turku sugar studies I–XXI. *Acta Odontol. Scand.* (Suppl. 70), **33** (1975) 3–348.
2. Anon., Xylitol as a sucrose substitute: relation to dental caries. *Nutr. Rev.*, **39** (1981) 368–71.
3. Wang, Y.-M. & van Eys, J., Nutritional significance of fructose and sugar alcohols. *Ann. Rev. Nutr.*, **1** (1981) 437–75.
4. Silverstone, L. M., Johnson, N. W., Hardie, J. M. & Williams, R. A. D., *Dental Caries: Aetiology, Pathology and Prevention.* Macmillan Press, London, 1981, pp. 119–221.
5. Loesche, W., *Dental Caries: A Treatable Infection.* University of Michigan Dental Publications, Ann Arbor, MI, 1981, pp. 227–81.
6. Mäkinen, K. K. & Scheinen, A., Xylitol and dental caries. *Ann. Rev. Nutr.*, **2** (1982) 133–50.
7. Bär, A., Xylitol. In *Alternative Sweeteners*, ed. L. O'B. Nabors & R. C. Gelardi. Marcel Dekker, New York, 1986, pp. 185–216.
8. Linke, H. A. B., Sugar alcohols and dental health. *World Rev. Nutr. Diet.*, **47** (1986) 134–62.
9. Bär, A., Caries prevention with xylitol. *World Rev. Nutr. Diet.*, **55** (1988) 1–27.
10. Mäkinen, K. K., Sweeteners and prevention of dental caries with special reference to xylitol. *Oral Health*, **88** (1988) 57–66.
11. Galiullin, A. N., Evaluation of the caries-preventive action of xylitol. *Kazan Med. J.*, **67** (1981) 16–18 (in Russian).
12. World Health Organization, *Prevention Methods and Programmes for Oral Diseases*, Technical Report Series 713. WHO, Geneva, 1984.
13. Barmes, D., Barnaud, J., Khambonanda, S. & Sardo Infirri, J., Field trials of preventive regimes in Thailand and French Polynesia. *Int. Dent. J.*, **35** (1985) 66–72.
14. Kandelman, D., Bär, A. & Hefti, A., Collaborative WHO xylitol field study in French Polynesia. I. Baseline prevalence and 32-month caries increment. *Caries Res.*, **22** (1988) 55–62.
15. Scheinin, A., Bánóczy, J., Szöke, J., Esztári, I., Pienihäkkinen, K., Scheinin, U., Tiekso, J., Zimmermann, P. & Hadas, E., Collaborative WHO xylitol field

studies in Hungary. I. Three-year caries activity in institutionalized children. *Acta Odontol. Scand.*, **43** (1985) 327–47.

16. Scheinin, A., Pienihäkkinen, K., Tiekso, J., Bánóczy, J., Szöke, J., Esztári, I., Zimmermann, P. & Hadas, E., Collaborative WHO xylitol field studies in Hungary. VII. Two-year caries incidence in 976 institutionalized children. *Acta Odontol. Scand.*, **43** (1985) 381–7.

17. Bánóczy, J., Scheinin, A., Pados, R., Ember, G., Kertész, P. & Pienihäkkinen, K., Collaborative WHO xylitol field studies in Hungary. II. General background and control of the dietary regimen. *Acta Odontol. Scand.*, **43** (1985) 349–57.

18. Pienihäkkinen, K., Gábris, K., Nyárasdy, I., Rigó, O., Scheinin, A. & Bánóczy, J., Collaborative WHO xylitol field studies in Hungary. III. Longitudinal counts of lactobacilli and yeasts in saliva. *Acta Odontol. Scand.*, **43** (1985) 359–65.

19. Kandelman, D. & Gagnon, G., Clinical results after 12 months from a study of the incidence and progression of dental caries in relation to consumption of chewing gum containing xylitol in school preventive programs. *J. Dent. Res.*, **66** (1987) 1407–11.

20. Kandelman, D. & Gagnon, G., Effect on dental caries of xylitol chewing gum; two-year results. *J. Dent. Res.* (Special Issue), **67** (1988) 172, Abstr. 472.

21. Szöke, J., Pienihäkkinen, K., Esztári, I., Bánóczy, J. & Scheinin, A., Collaborative WHO xylitol field studies in Hungary. V. Three-year development of oral hygiene. *Acta Odontol. Scand.*, **43** (1985) 371–6.

22. Möller, I. J. & Poulsen, S., The effect of sorbitol-containing chewing gum on the incidence of dental caries, plaque and gingivitis. *Comm. Dent. Oral Epidemiol.*, **1** (1973) 58–67.

23. Mäkinen, K. K. & Isokangas, P., Relationship between carbohydrate sweeteners and oral diseases. *Progr. Food Nutr. Sci.*, **12** (1988) 73–109.

24. Isokangas, P., Alanen, P., Tiekso, J. & Mäkinen, K. K., Xylitol chewing gum in caries prevention: a field study in children. *J. Am. Dent. Assoc.*, **117** (1988) 315–20.

25. Mäkinen, K. K., Söderling, E., Isokangas, P., Tenovuo, J. & Tiekso, J., Oral biochemical status and depression of *Streptococcus mutans* in children during 24- to 36-month use of xylitol chewing gum. *Caries Res.*, **23** (1989) 261–7.

26. Isokangas, P., Tiekso, J., Alanen, P. & Mäkinen, K. K., Long-term effect of xylitol chewing gum on dental caries. *Comm. Dent. Oral Epidemiol.*, **17** (1989) 200–3.

27. Rekola, M., A planimetric evaluation of enamel caries progression with special reference to the consumption of xylitol. PhD thesis, University of Turku, Turku, Finland, 1985.

28. Rekola, M., Changes in buccal white spots during 2-year consumption of dietary sucrose or xylitol. *Acta Odontol. Scand.*, **44** (1986) 285–90.

29. Rekola, M., Approximal caries development during 2-year total substitution of dietary sucrose with xylitol. *Caries Res.*, **21** (1987) 87–94.

30. Glass, R. L., A two-year clinical trial of sorbitol gum. *Caries Res.*, **17** (1983) 365–8.

31. Navia, J. M., Experimental dental caries. In *Animal Models in Dental Research*. University of Alabama Press, Birmingham, 1977, pp. 257–97.

32. Van Houte, J., Bacterial specificity in the etiology of dental caries. *Int. Dent. J.,* **30** (1980) 305–26.
33. Loesche, W., Role of *Streptococcus mutans* in human dental decay. *Microbiol. Rev.,* **50** (1986) 353–80.
34. Moore, W. E. C., Holdeman, L. V., Smibert, R. M., Good, I. J., Burmeister, J. A., Palcanis, K. G. & Ranney, R. R., Bacteriology of experimental gingivitis in young adult humans. *Infect. Immun.,* **38** (1982) 651–7.
35. Socransky, S. S., Tanner, A. C. R., Haffajee, A., Hillman, J. D. & Goodson, J. M., Present status of studies on the microbial etiology of periodontal diseases. In *Host–Parasite Interactions in Periodontal Disease,* ed. R. J. Genco & S. E. Mergenhagen. American Society of Microbiology, Washington DC, 1982, pp. 1–12.
36. Skinner, A. & Woods, A., An investigation of the effects of maltose and sucrose in the diet on the microbiology of dental plaque in man. *Arch. Oral Biol.,* **29** (1984) 323–6.
37. Van Houte, J., Carbohydrates, sugar substitutes and oral bacterial colonization. In *Health and Sugar Substitutes,* ed. B. Guggenheim. Karger, Basel, 1979, pp. 199–204.
38. Mäkinen, K. K., The role of sucrose and other sugars in the development of dental caries: a review. *Int. Dent. J.,* **22** (1972) 363–86.
39. Mäkinen, K. K., Biochemical principles of the use of xylitol in medicine and nutrition with special consideration of dental aspects. *Experientia,* Suppl. 30 (1978) 1–160.
40. Mäkinen, K. K., New biochemical aspects of sweeteners. *Int. Dent. J.,* **35** (1985) 23–35.
41. Mäkinen, K. K. & Scheinin, A., Turku sugar studies. VII. Principal biochemical findings on whole saliva and plaque. *Acta Odontol. Scand.,* Suppl. 70, **33** (1975) 129–71.
42. Brown, A. T., The role of dietary carbohydrates in plaque formation and oral disease. *Nutr. Rev.,* **33** (1975) 353–61.
43. Mäkinen, K. K., Xylitol and oral health. *Adv. Food Res.,* **25** (1979) 137–58.
44. Gehring, F., Mäkinen, K. K., Larmas, M. & Scheinin, A., Turku sugar studies. X. Occurrence of polysaccharide-forming streptococci and ability of the mixed plaque microbiota to ferment various carbohydrates. *Acta Odontol. Scand.,* Suppl. 70, **33** (1975) 223–37.
45. Loesche, W. J., Grossman, N. S., Earnest, R. & Corpron, R., The effect of chewing xylitol gum on the plaque and saliva levels of *Streptococcus mutans. J. Am. Dent. Assoc.,* **108** (1984) 587–92.
46. Bánóczy, J., Orsós, M., Pienihäkkinen, K. & Scheinin, A., Collaborative WHO xylitol field studies in Hungary. IV. Saliva levels of *Streptococcus mutans. Acta Odontol. Scand.,* **43** (1985) 367–70.
47. Mäkinen, K. K., Söderling, E., Hurttia, H., Lehtonen, O.-P. & Luukkala, E., Biochemical, microbiologic, and clinical comparison between two dentifrices that contain different mixtures of sugar alcohols. *J. Am. Dent. Assoc.,* **111** (1985) 745–51.
48. Mäkinen, K. K., Xylitol. In *Foods, Nutrition and Dental Health,* ed. J. J. Hefferren & H. M. Kohler. Pathotox Publishers, park Forrest South, IL, 1981, pp. 83–96.

49. Larmas, M., Mäkinnen, K. K. & Scheinin, A., Turku sugar studies. VIII. Principal microbiological findings. *Acta Odontol. Scand.*, Suppl. 70, 33 (1975) 123–216.

50. Mortlock, R. P., Catabolism of unnatural carbohydrates by microorganisms. *Adv. Microb. Physiol.*, 13 (1976) 1–58.

51. Maleszka, R. & Schneider, H., Fermentation of D-xylose, xylitol and D-xylulose by yeasts. *Can. J. Microbiol.*, 28 (1982) 360–3.

52. Hahn-Hägerdahl, B., Jonsson, B. & Lohmeier-Vogel, E., Shifting product formation from xylitol to ethanol in pentose fermentations using *Candida tropicalis* by adding polyethylene glycol. *Appl. Microbiol. Biotechnol.*, 21 (1985) 173–5.

53. Lohmeier-Vogel, E. & Hahn-Hägerdahl, B., The utilization of metabolic inhibitors for shifting product formation from xylitol to ethanol in pentose fermentations using *Candida tropicalis*. *Appl. Microbiol. Biotechnol.*, 21 (1985) 167–72.

54. Hsiao, H.-Y., Chiang, L.-C., Ueng, P. P. & Tsao, G. T., Sequential utilization of mixed monosaccharides by yeasts. *Appl. Environ. Microbiol.*, 43 (1982) 840–5.

55. Neuberger, M. S. & Hartley, B. S., Minicircular ColWl-related DNA in strains of *Klebsiella aerogenes* selected for fast growth on xylitol. *J. Gen. Microbiol.*, 118 (1980) 171–7.

56. Wu, T. T., Lin, E. C. C. & Tanaka, S., Mutants of *Aerobacter aerogenes* capable of utilizing xylitol as a novel carbon source. *J. Bacteriol.*, 96 (1968) 447–56.

57. Wu, T. T., Growth of a mutant of *Escherichia coli* K-12 on xylitol by recruiting enzymes for D-xylose and L-1,2-propanediol metabolism. *Biochim. Biophys. Acta*, 428 (1976) 656–63.

58. Doten, R. C. & Mortlock, R. P., Characterization of xylitol-utilizing mutants of *Erwina uredovora*. *J. Bacteriol.*, 161 (1985) 529–33.

59. London, J. & Chace, N. M., Pentitol metabolism in *Lactobacillus casei*. *J. Bacteriol.*, 140 (1979) 949–54.

60. London, J. & Hausman, S., Xylitol-mediated transient inhibition of ribitol utilization by *Lactobacillus casei*. *J. Bacteriol.*, 150 (1982) 657–61.

61. London, J. & Hausman, S., Purification and characterization of the III$^{Xtl}$ phosphocarrier protein of the phosphoenolpyruvate-dependent xylitol: phosphotransferase found in *Lactobacillus casei* C183. *J. Bacteriol.*, 156 (1983) 611–19.

62. Trahan, L., Neron, S. & Bareil, S., Preparation and purification of xylitol-5-phosphate from a cell extract of *Lactobacillus casei* Cl-16. *Appl. Envir. Microbiol.*, 54 (1988) 570–3.

63. Vadeboncoeur, C., Trahan, L., Mouton, C. & Mayrand, D., Effect of xylitol on the growth and glycolysis of acidogenic oral bacteria. *J. Dent. Res.*, 62 (1983) 882–4.

64. Trahan, L. & Mouton, C., Selection for *Streptococcus mutans* with an altered xylitol transport capacity in chronic xylitol consumers. *J. Dent. Res.*, 66 (1987) 982–8.

65. Engelsberg, E., Anderson, R. L., Weinberg, R., Lee, N., Hoffee, G., Huttenhauer, G. & Boyer, H., L-Arabinose-sensitive, L-ribulose 5-phosphate 4-epimerase-deficient mutants of *Escherichia coli*. *J. Bacteriol.*, 84 (1962) 137–46.

66. Lengeler, J., Mutations affecting transport of the hexitols D-mannitol, D-glucitol and galactitol in *Escherichia coli* K12 isolation and mapping. *J. Bacteriol.*, **124** (1975) 26–38.

67. Batzinger, R. P., Suh-Yun, L. O. & Beuding, E., Saccharin and other sweeteners: mutagenic properties. *Science*, **198** (1977) 944–6.

68. Krishnan, R., Wilkinson, I., Joyce, L., Rofe, A. M., Bais, R., Conyers, R. A. J. & Edwards, J. B., The effect of dietary xylitol on the ability of rat caecal flora to metabolize xylitol. *Aust. J. Exp. Biol. Med. Sci.*, **58** (1980) 639–52.

69. Gauthier, L., Vadeboncoeur, C. & Mayrand, D., Loss of sensitivity to xylitol by *Streptococcus mutans* LG-1. *Caries Res.*, **18** (1984) 289–95.

70. Assev, S. & Scheie, A. A., Xylitol metabolism in xylitol-sensitive and xylitol-resistant strains of streptococci. *Acta Path. Microbiol. Immunol. Scand.*, Sect. B, **94** (1986) 239–43.

71. Assev, S. & Rölla, G., Evidence for presence of a xylitol phosphotransferase system in *Streptococcus mutans* OMZ 176. *Acta Path. Microbiol. Immunol. Scand.*, Sect. B, **92** (1984) 89–92.

72. Assev, S. & Rölla, G., Further studies on the growth inhibition of *Streptococcus mutans* OMZ 176 by xylitol. *Acta Path. Microbiol. Immunol. Scand.*, Sect. B, **94** (1986) 97–102.

73. Reizer, J., Novotny, M. J., Panos, C. & Saier, M. H., Jr, Mechanism of inducer expulsion in *Streptococcus pyogenes*: a two-step process activated by ATP. *J. Bacteriol.*, **156** (1983) 354–61.

74. Reizer, J., Deutscher, J., Sutrina, S., Thompson, J. & Saier, M. H., Sugar accumulation in Gram-positive bacteria: exclusion and expulsion mechanisms. *TIBS*, January (1985) 32–5.

75. Thompson, J. & Chassy, B. M., Novel phosphoenolpyruvate-dependent futile cycle in *Streptococcus lactis*; 2-deoxy-D-glucose uncouples energy production from growth. *J. Bacteriol.*, **151** (1982) 1454–85.

76. Thompson, J. & Chassy, B. M., Intracellular hexose-6-phosphate:phosphohydrolase from *Streptococcus lactis*; purification, properties and function. *J. Bacteriol.*, **156** (1983) 70–80.

77. Hausman, S. Z., Thompson, J. & London, J., Futile xylitol cycle in *Lactobacillus casei*. *J. Bacteriol.*, **160** (1984) 211–15.

78. Knuuttila, M. L. E. & Mäkinen, K. K., Effect of xylitol on the growth and metabolism of *Streptococcus mutans*. *Caries Res.*, **9** (1975) 177–89.

79. Knuuttila, M. L. E. & Mäkinen, K. K., Extracellular hydrolase activity of the cells of the bacterium *Streptococcus mutans* isolated from man and grown on glucose or xylitol. *Arch. Oral Biol.*, **26** (1981) 899–904.

80. Söderling, E. & Pihlanto-Leppälä, A., Uptake and expulsion of $^{14}$C-xylitol by xylitol-cultured *Streptococcus mutans* 25175 *in vitro*. *Scand. J. Dent. Res.* (1989) in press.

81. Assev, S. & Rölla, G., Sorbitol increases the growth inhibition of xylitol on *Streptococcus mutans* OMZ 176. *Acta Path. Microbiol. Immunol. Scand.*, Sect. B, **94** (1986) 231–7.

82. Söderling, E., Mäkinen, K. K., Chen, C.-Y., Pape, H. R., Jr, Loesche, W. & Mäkinen, P.-L., Effect of sorbitol, xylitol or xylitol/sorbitol chewing gums on dental plaque. *Caries Res.* (1989) in press.

83. Rölla, G., Oppermann, R. V., Bowen, W. H., Ciardi, J. E. & Knox, K. W., High

amounts of lipoteichoic acid in sucrose-induced plaque *in vivo. Caries Res.*, **14** (1980) 235–8.

84. Assev, S., Vegarud, G. & Rölla, G., Addition of xylitol to the growth medium of *Streptococcus mutans* OMZ 176; effect on the synthesis of extractable glycerol-phosphate polymers. *Acta Path. Microbiol. Immunol. Scand.*, Sect. B, **93** (1985) 145–9.

85. Rölla, G., Iversen, O.-J. & Bonesvoll, P., Lipoteichoic acid—the key to the adhesiveness of sucrose-grown *Streptococcus mutans*. In *Secretory Immunity and Infection*, ed. J. R. McGhee, J. Mestecky & J. L. Babb. Plenum Press, New York, 1978, pp. 607–17.

86. Tuompo, H., Meurman, J. H., Lounatmaa, K. & Linkola, J., Effect of xylitol and other carbon sources on the cell wall of *Streptococcus mutans. Scand. J. Dent. Res.*, **91** (1983) 17–25.

87. Söderling, E., Alaräisänen, L., Scheinin, A. & Mäkinen, K. K., Effect of xylitol and sorbitol on polysaccharide production by and adhesive properties of *Streptococcus mutans. Caries Res.*, **21** (1987) 109–16.

88. Larmas, M., Scheinin, A., Gehring, F. & Mäkinen, K. K., Turku sugar studies. XX. Microbiological findings and plaque index values in relation to 1-year use of xylitol chewing gums. *Acta Odontol. Scand.*, Suppl. 70, **33** (1975) 321–36.

89. Rekola, M., A comparison of the effects of xylitol and sorbitol sweetened chewing gums on dental plaque. *Proc. Finn. Dent. Soc.*, **78** (1982) 128–33.

90. Pakkala, U., Liesmaa, H. & Mäkinen, K. K., The use of xylitol in the control of oral hygiene in mentally retarded children. *Proc. Finn. Dent. Soc.*, **77** (1981) 271–7.

91. Rekola, M., Comparative effects of xylitol- and sucrose-sweetened chew tablets and chewing gums on plaque quantity. *Scand. J. Dent. Res.*, **89** (1981) 393–9.

92. Mäkinen, K. K., Lönnberg, P. & Scheinin, A., Turku sugar studies. XIV. Amino acid analysis of saliva. *Acta Odontol. Scand.*, Suppl. 70, **33** (1975) 277–86.

93. Durette, P. L. & Horton, D., Conformational analysis of sugars and their derivatives. *Adv. Carbohydr. Chem. Biochem.*, **26** (1971) 49–125.

94. Angyal, S. J., Complex formation between sugars and metal ions. *Pure Appl. Chem.*, **35** (1973) 131–46.

95. Angyal, S. J., Complexing of polyols with cations. *Tetrahedron*, **30** (1974) 1695–702.

96. Angyal, S. J., Greeves, D. & Mills, J. A., Complexes of carbohydrates with metal cations. III. Conformations of alditols in aqueous solution. *Aust. J. Chem.*, **27** (1974) 1447–56.

97. Angyal, S. J. & Mills, J. A., Complexes of carbohydrates with metal cations. XIV. Separation of sugars and alditols by means of their lanthanum complexes. *Aust. J. Chem.*, **38** (1985) 1279–85.

98. Angyal, S. J. & Davies, K. P., Complexing sugars with metal ions. *Chem. Commun.* (1971) 500–1.

99. Kieboom, A. P. G., Spoormaker, T., Sinnema, A., van der Toorn, J. M. & van Bekkum, H., $^1$H-NMR study of the complex formation of alditols with multivalent cations in aqueous solutions using praseodymium(III) nitrate as shift reagent. *Rec. J. Royal Neth. Chem. Soc.*, **94** (1975) 53–9.

100. Kieboom, A. P. G., Buurmans, H. M. A., van Leeuwen, L. K. & van Benschop, H. J., Stability constants of (hydroxy) carboxylate– and alditol–calcium(II) complexes in aqueous medium as determined by a solubility method. *Rec. J. Royal Neth. Chem. Soc.*, **98** (1979) 393–4.

101. Briggs, J., Finch, P., Matulewicz, M. C. & Weigel, H., Complexes of copper(II), calcium and other metal ions with carbohydrates; thin-layer ligand exchange chromatography and determination of relative stabilities of complexes. *Carbohydr. Res.*, **97** (1981) 181–8.

102. Beattie, J. K. & Kelso, M. T., Equilibrium and dynamics of the binding of calcium ions to sorbitol (D-glucitol). *Aust. J. Chem.*, **34** (1981) 2563–8.

103. Tamaki, Y., Metal compounds of sucrose and polyhydric alcohols. V. The complex compounds between sugars or polyhydric alcohols and metals. Reaction between the hydroxides of alkaline earth and polyhydric alcohols. *Kogyo Kagaku Zasshi*, **70** (1967) 949–52 (in Japanese).

104. Hauser, H., Levine, B. A. & Williams, R. J. P., Interactions of ions with membranes. *Trends Biochem. Sci.* (1976) 278–81.

105. Chalmers, R. A. & Sinclair, A. G., Organic molybdate complexes. *J. Inorg. Nucl. Chem.*, **29** (1967) 2065–80.

106. Huttunen, E., A comparative study on the chelation of pentitols with boric and germanic acids. *Finn. Chem. Lett.* (1979) 48–52.

107. Roy, G. L., Laferriere, A. L. & Edwards, J. O., A comparative study of polyol complexes of arsenite, borate and tellurate ions. *J. Inorg. Nucl. Chem.*, **4** (1957) 106–14.

108. Brown, D. H. & MacPherson, J., Molybdenum-(V) and -(VI) complexes with some naturally occurring ligands. *J. Inorg. Nucl. Chem.*, **34** (1972) 1705–10.

109. Mikešová, M. & Bartušek, M., Reaction of molybdate and tungsten with oxalate, mannitol and sorbitol. *Coll. Czech. Chem. Commun.*, **43** (1978) 1867–77.

110. Searle, F. & Weigel, H., Interaction between polyhydroxy compounds and vanadate ions; electrophoresis and composition of complexes. *Carbohydr. Res.*, **85** (1980) 51–9.

111. Mäkinen, K. K. & Söderling, E., Solubility of calcium salts, enamel and hydroxyapatite in aqueous solutions of simple carbohydrates. *Calcif. Tissue Int.*, **36** (1984) 64–71.

112. Lewin, S., *Displacement of Water and Its Control of Biochemical Reactions*. Academic Press, London, 1974, Chapters 1–4.

113. Gekko, K. & Satake, I., Differential scanning calorimetry of unfreezable water in water–protein–polyol systems. *Agric. Biol. Chem.*, **45** (1981) 2209–17.

114. Gekko, K. & Morikawa, T., Preferential hydration of bovine serum albumin in polyhydric alcohol–water mixtures. *J. Biochem.*, **90** (1981) 39–50.

115. Gekko, K. & Morikawa, T., Thermodynamics of polyol-induced thermal stabilization of chymotrypsinogen. *J. Biochem.*, **90** (1981) 51–60.

116. Gekko, K., Calorimetric study on thermal denaturation of lysozyme in polyol–water mixtures. *J. Biochem.*, **91** (1982) 1197–204.

117. Gekko, K., Mechanism of polyol-induced protein stabilization; solubility of amino acids and diglycine in aqueous polyol solutions. *J. Biochem.*, **90** (1981) 1633–41.

118. Back, J. F., Oakenfull, D. & Smith, M. B., Increased thermal stability of proteins in the presence of sugars and polyols. *Biochemistry*, **18** (1979) 5191–6.

119. Gerlsma, S. Y., The effects of polyhydric and monohydric alcohols on the heat-induced reversible denaturation of chymotrypsinogen. *Eur. J. Biochem.*, **14** (1970) 150–3.

120. Gerlsma, S. Y. & Stuur, E. R., The effect of polyhydric and monohydric alcohols on the heat-induced reversible denaturation of lysozyme and ribonuclease. *Int. J. Pept. Protein Res.*, **4** (1972) 377–83.

121. Gerlsma, S. Y. & Stuur, E. R., The effects of combining two different alcohols on the heat-induced reversible denaturation of ribonuclease. *Int. J. Pept. Protein Res.*, **6** (1974) 65–74.

122. Mäkinen, K. K., Läikkö, I., Rekola, M. & Scheinin, A., Die Wirkung von Xylit und Sorbit auf die Biochemie der Plaque. II. Biochemische Veränderungen in Plaque und Gesamtspeichel in Relation zu intensivem, einmonatigem Konsum von Xylit- und Sorbit-Kaugummis. *Kariesprophylaxe*, **3** (1980) 103–13.

123. Grenby, T. H., Bashaarat, A. H. & Gey, K. H., A clinical trial to compare the effects of xylitol and sucrose chewing gums on dental plaque growth. *Br. Dent. J.*, **152** (1982) 339–43.

124. Hurttia, H., Multanen, V.-M., Mäkinen, K. K., Tenovuo, J. & Paunio, K., Effects on oral health of mouthrinses containing xylitol, sodium cyclamate and sucrose sweeteners in the absence of oral hygiene. III. Composition and bone resorbing potential of dental plaque. *Proc. Finn. Dent. Soc.*, **80** (1984) 20–7.

125. Söderling, E. & Mäkinen, K. K., Aggregation of human salivary Ca-proteinates in the presence of simple carbohydrates *in vitro*. *Scand. J. Dent. Res.*, **94** (1986) 125–31.

126. Arends, J., Christoffersen, J., Schuthof, J. & Smits, M. T., Influence of xylitol on demineralization of enamel. *Caries Res.*, **18** (1984) 296–301.

127. Tarján, I. & Lindén, L. A., Isotope studies on the permeability of the dental enamel to sucrose and xylitol. *J. Int. Ass. Dent. Child.*, **13** (1982) 53–6.

128. Smits, M. T. & Arends, J., Influence of extraoral xylitol and sucrose dippings on enamel demineralization *in vivo*. *Caries Res.*, **22** (1988) 160–5.

129. Vissink, A., 's-Gravenmade, E. J., Gelhard, T. B. F. M., Panders, A. K. & Franken, M. H., Rehardening properties of mucin- or CMC-containing saliva substitutes on softened human enamel. *Caries Res.*, **19** (1985) 212–18.

130. Smits, M. T., Xylitol and dental caries. PhD thesis, University of Groningen, The Netherlands, 1987.

131. Featherstone, J. D. B., Cutress, T. W., Rodgers, B. E. & Dennison, P. J., Remineralization of artificial caries-like lesions *in vivo* by a self-administered mouth-rinse or paste. *Caries Res.*, **16** (1982) 235–43.

132. Duxbury, A. J., Hayes, N. F., Thakkar, N. S., Wastell, D. G., Kelly, A. & Leach, F. N., Clinical trial of a mucin-containing artificial saliva. *IRCS Med. Sci.*, **13** (1985) 1197–8.

*Chapter* 14

# RESEARCH ON LACTITOL AND DENTAL HEALTH: A REVIEW*

T. H. Grenby

*Department of Oral Medicine and Pathology,*
*United Medical and Dental Schools,*
*Guy's Hospital, London, UK*

## SUMMARY

*Research on the dental properties of lactitol as a bulk sweetener to replace dietary sugar is reviewed under three headings:*

*(A) Microbiological experiments in vitro.*
*(B) Investigations in laboratory animals.*
*(C) Studies in man.*

*It was found that lactitol was not easily metabolised by acidogenic and polysaccharide-forming oral micro-organisms, that its enamel-demineralising potential in vitro and its cariogenicity in laboratory rats were low and that intra-oral acid development and dental-plaque formation from lactitol in man were substantially lower than from sucrose. The indications for dental health are promising.*

## INTRODUCTION

The three main health benefits put forward for new sweeteners to replace sugars in foods and drinks are (a) the assistance they can give in weight control, (b) their usefulness in diabetic low-sugar regimes and (c) their improved dental properties compared with sugars.

* Based on material prepared for the *International Dental Journal*, **39** (1989) 25–32. Reproduced with permission © 1989 Fédération Dentaire Internationale.

Depending on which sweetener is under consideration, these three purposes assume different degrees of importance. In case (c)—the promotion of dental health—the *intense* or non-nutritive sweeteners have excellent properties because the only feature they have in common with sugars is their sweet taste and, unlike the sugars, they are not fermentable to cariogenic acids by oral micro-organisms. Although they are suitable for replacing sugars in drinks, they lack the 'bulking' powers that are a basic function of sugars in many types of solid foods. *Bulk* sweeteners show an advantage in this respect, but they are all carbohydrate-type compounds or carbohydrate derivatives, so that their dental properties are more open to question than those of the intense sweeteners, and research is needed to assess their cariogenicity and associated features.

Among the strongest contenders as bulk sweeteners, for use in confectionery, chewing-gum, desserts, bakery products and many other items of food, are the polyols—hydrogenated derivatives of some of the common sugars, to which they are closely related. A very large amount of dental research has been done on xylitol as a low-cariogenic sugar-substitute, with somewhat less on sorbitol and mannitol, but these are all monosaccharide polyols derived from C5 or C6 sugars. Sucrose, by far the most common sugar in the western diet and widely believed to be the most cariogenic, is a disaccharide (C12). Increasing attention is now being paid to the production and properties of a range of disaccharide polyols, one of which is lactitol. A limited amount of research has been completed on its dental qualities and forms the subject of this review.

## BACKGROUND

Lactitol (4-*O*-(*β*-galactosyl)-D-glucitol) is produced by the hydrogenation of lactose, which is available in ample quantities from the dairy industry, making the formulation of foods with lactitol a viable economic proposition. The store of published information on the preparation and properties of lactitol is growing rapidly. A detailed description was given by Van Velthuijsen[1] and in a European Patent application,[2] followed by a summary by Linko[3] and a more recent review by Den Uyl,[4] presenting data on its purity, sensory evaluation, physical, chemical, microbiological and physiological behaviour, together with an outline of its applications in a variety of foods.

The metabolic and physiological effects of lactitol are reviewed comprehensively by Van Velthuijsen,[5] who concluded that it is toxicologically safe and, as it is not digested in the small intestine but is fully fermented by the microflora of the large intestine, it has useful nutritional and therapeutic properties. These aspects are summarised by Ziesenitz and Siebert,[6] with reference to its low rate of hydrolysis in the small intestine, its contribution to colonic-hydrogen production, the provision of less utilisable energy than from sucrose and its cathartic action, which has been put to medical use in the treatment of acute hepatic or portal system encephalopathy, where its advantages over lactulose have been shown in Refs 7–10 and in a complete review by Morgan.[11]

These findings on the medical advantages of lactitol are complemented by other up-to-date work on its digestion and metabolism. Van Es *et al.*[12] observed a smaller contribution to body energy from lactitol than from sucrose. Human intestinal-biopsy specimens hydrolysed lactitol with only 1·3% of the activity they showed towards lactose and isomaltose.[13] From these and other studies, including a jejunal-perfusion technique,[14] it was concluded that lactitol is not absorbed by the human small intestine and, although gastrointestinal side-effects can be observed as the dose is increased, 40 g per day can easily be tolerated.

Other biological information on lactitol has been submitted to the WHO from a variety of sources, covering its absorption, distribution, excretion, hydrolysis, degradability, toxicology, mutagenicity and teratology, etc., but this is almost all in the form of reports which have not been published. Some of the main conclusions were:

(1) Lactitol was not metabolised as a carbohydrate, as it was not absorbed in the small intestine and was split only very slowly by enzymes.

(2) Lactitol was extensively degraded in rats, presumably mainly by the intestinal microflora.

(3) Toxicity following single or repeated large doses was very low. The most prominent effect was some increase in caecum weight.

(4) 24 g of lactitol per day was tolerated satisfactorily in man.

(5) Lactitol was not mutagenic in microbial systems.

The FAO/WHO has assigned an acceptable daily intake (ADI) of 'not specified' to lactitol, meaning that, on the basis of the data available, the levels necessary to achieve its desired biological effect are considered not to represent a hazard to health and for this reason the establishment of an ADI is not necessary.

## DENTAL STUDIES

Three main types of experiment are used to evaluate the cariogenicity of foods and food ingredients:

(A) Studies *in vitro*.

(B) Experiments on laboratory animals.

(C) Clinical trials and investigations on human subjects.

In ideal circumstances, a food would be tested by all three techniques and a verdict on its cariogenic potential would then be arrived at, but, in practice, owing to limitations imposed by the diverse properties of foods and by economic, ethical and many other considerations, the amount of testing that can be done is restricted and less data are usually available from clinical studies than from experimentation *in vitro* and on laboratory animals. One particular difficulty in reviewing the dental research on lactitol is that some of the early work is in the form of unpublished reports, but classifying what is available into divisions (A)–(C):

### (A) Studies *In Vitro*

Among the earliest reports are those of Havenaar[15,16] on the utilisation of lactitol by oral micro-organisms. The first point investigated was whether lactitol can be fermented into acids by any of the organisms, as acid production from sugars and its attack on dental enamel are key features of the caries process. Using a peptone/yeast extract medium under anaerobic incubation conditions with multiple subculturing and detecting fermentation by a pH drop to 5·6 or lower, Havenaar[15] found a number of plaque bacteria that could metabolise lactitol as a substrate, including *Streptococcus mutans*, which is known to possess cariogenic activity, and certain strains of *Streptococcus sanguis*, *Bifidobacterium*, *Lactobacillus*, *Actinomyces israelii*, *Actinomyces viscosus* and *Actinomyces naeslundii* (50–90% of the strains tested). Also, 90–100% of cultures of *Streptococcus milleri* were observed to ferment lactitol.[16]

However, these first experiments did not establish the speed of fermentation, which, because of the limited time that sugars and sweeteners remain in the mouth, is a determinant of their cariogenicity. It was later shown[15] with unadapted serotypes of *S. mutans* that the pH did not fall to below 5·0 until incubation had been continued for 72 h, leading to the conclusion that lactitol could be fermented only slowly by these organisms. Next, adaptation of strains of *S. mutans* to metabolise lactitol was investigated by repeated culturing (10 times) into lactitol-containing media.

This increased the speed of acid production somewhat, although it was still slower than that from glucose, but it did not alter the final pH. Moreover, subsequent culturing in a glucose medium showed instability in the adaptation.[16]

In addition to acid production, another important property of cariogenic oral bacteria is the capacity to synthesise extra-cellular polysaccharide from carbohydrate substrates, because of the contribution that the polysaccharide can make to the structure and biochemical behaviour of the dental plaque. Some elementary tests on this were carried out by examining viscosity and the adherence of growing cells of various strains of *S. mutans* and *S. sanguis* to glass surfaces, but no evidence for polysaccharide synthesis was found.[15] However, after adaptation to lactitol media, *S. mutans* showed some temporary capacity to form *intra*-cellular polysaccharide from lactitol, up to about 25% of that from glucose.[16]

Havenaar concluded that, from the microbiological point of view, lactitol could be a useful sugar substitute, though it had some limitations. No decision could be given as to whether it was preferable to sorbitol.

Although mainly concerned with animal experiments, some of the data of Gehring[17] related to microbiological studies *in vitro*. Measuring the pH over periods up to 48 h in growth media inoculated with *S. mutans*, Gehring observed much slower acid development on lactitol than on a range of carbohydrates, including lactose, fructose and sucrose. There was even less fermentation of lactitol by *Streptococcus salivarius*, *S. milleri*, *Streptococcus mitis* and *Lactobacillus casei*. Mixed cultures of human salivary micro-organisms also produced little or no acid in media containing lactitol, in contrast to substantial pH falls in the presence of sucrose, fructose and lactose. These were short-term experiments, up to 5 h, but differences between lactitol and the other materials had begun to appear after as little as 10 min.

A more recent study on lactitol *in vitro*[18] compared it with five other bulk sweeteners, incubated in otherwise carbohydrate-free media with mixed cultures of human dental-plaque micro-organisms, measuring acid development by both pH and titration against dilute sodium hydroxide, insoluble polysaccharide synthesis and the attack of the acid on enamel mineral, as determined by calcium and phosphorus analysis. The six different sweeteners fell into three groups, with the highest acid generation, polysaccharide production and enamel demineralisation from glucose and sucrose, less from sorbitol and mannitol and least of all from lactitol and xylitol (Fig. 1). Whereas media containing glucose, sucrose, sorbitol and mannitol all had final pH values in the range 4·0–5·2 after 24 h incubation,

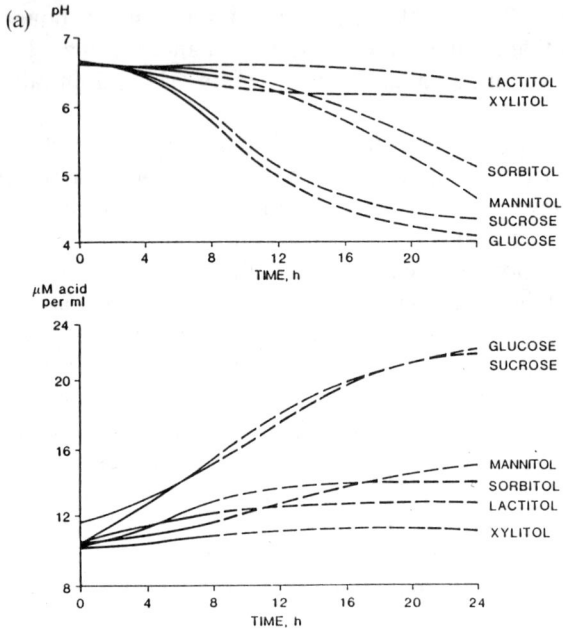

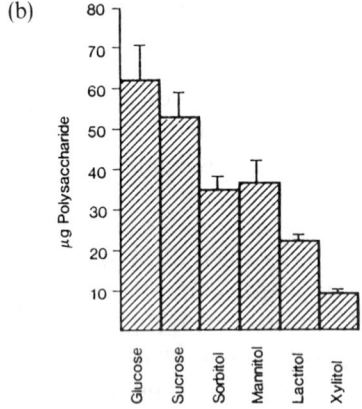

FIG. 1.   Comparison of lactitol with five other bulk sweeteners in incubations of standardised cultures of human dental-plaque micro-organisms in peptone liquid media.[18] (a) Acid developing during incubations. pH (above) and titratable acid (below). Mean values of three sets of determinations, (b) insoluble polysaccharide synthesised after 24-hour incubation. Mean ± SE of six sets of determinations.

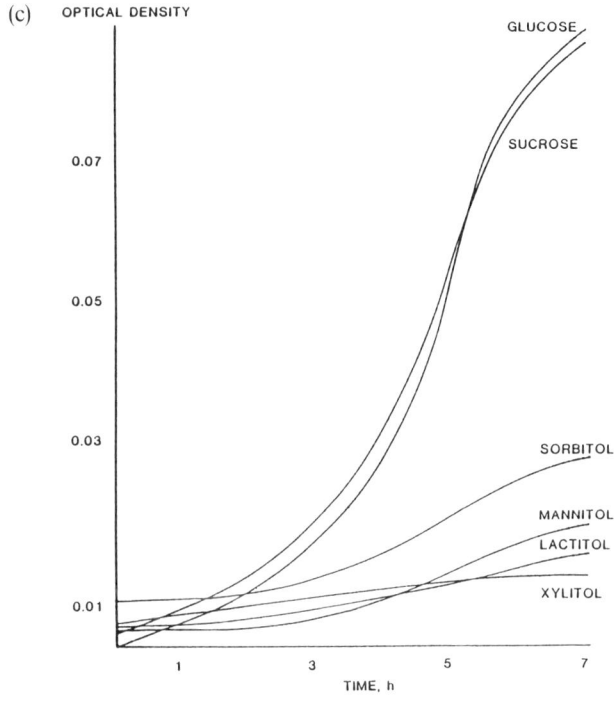

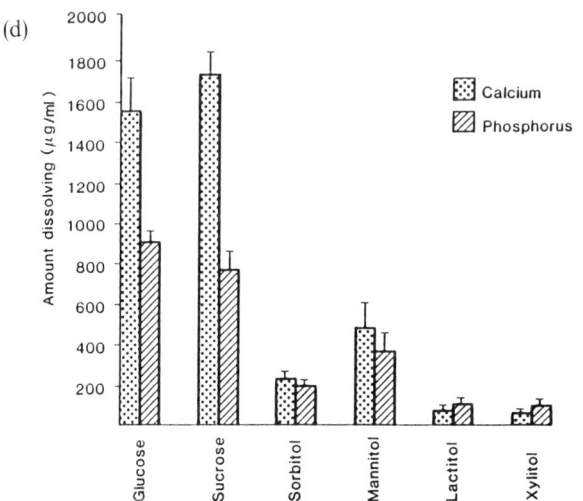

FIG. 1.—*contd.* (c) Microbial growth measured turbidimetrically at 540 nm at hourly intervals during the first 7 h of incubation. Mean values of three sets of determinations, (d) calcium and phosphorus dissolving from powdered enamel in incubation experiments. Mean ± SE of six sets of data.

the pH value in the lactitol media remained high, at approximately 6·7, with consequent very low enamel demineralisation figures, significantly below those of all the other bulk sweeteners, except xylitol.

There is good agreement between these three separate sets of studies on the dental properties of lactitol *in vitro*, illustrating its low fermentability by a variety of oral micro-organisms. The explanation of this may lie in the lack of induced anabolic enzymes in oral micro-organisms or in competitive inhibition for active sites on enzymes, although specific information on lactitol, as opposed to other polyols, is lacking.[19]

## (B) Experiments on Laboratory Animals

Three sets of studies are available on laboratory animal research. Lactitol was one of a number of sugar substitutes tested by Gehring.[17] A group of 12 Sprague–Dawley rats was given 9 or 18 small 'meals' per day of a diet containing 30% lactitol in the first week, 45% in the second week and 56% for the remainder of the 6-week experimental period. At the end of that time the fissure-caries score (mean 21·5) was significantly below those of rats on a variety of other diets containing sucrose, fructose or lactose (means all in

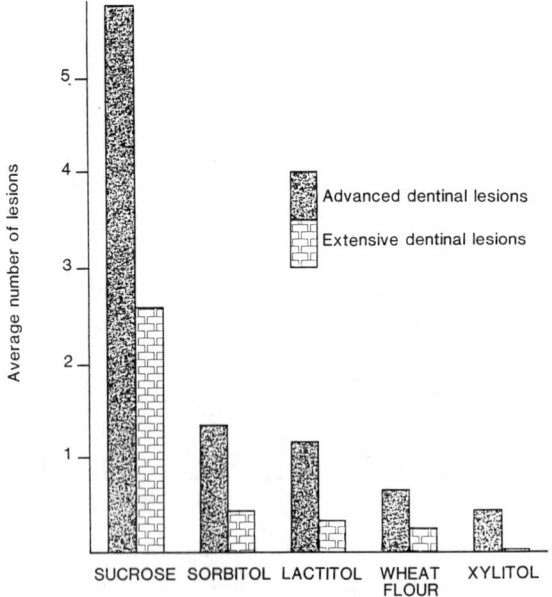

FIG. 2.   Carious fissure lesions in groups of 12 rats fed for three weeks on diets containing polyols at 25%, compared with sucrose and wheat flour controls.[20]

the range of 47·5 to 61·1), although it remained higher than that of a control group receiving a basal diet containing 40% of raw starch and no sweeteners (mean 4·5).

The next report of a cariogenicity test of lactitol in rats was by Van der Hoeven.[20] In a single programmed-feeding trial, 12 animals were given a diet containing 25% lactitol at the rate of 18 portions of 0·5 g per day for three weeks. Other regimes tested at the same time contained 25% of sucrose, sorbitol, xylitol or wheat flour, so that statistical comparisons of the numbers of carious lesions in the fissures of the molar teeth could be made. The results were presented as a table, but are shown here in graphical form in Fig. 2. It can be clearly seen that substitution of lactitol for sucrose significantly reduced the caries increment. Lactitol was slightly, but not significantly, more cariogenic than wheat flour. It appeared that the caries increments on lactitol and the other polyols reflected their fermentation rates by oral bacteria, such as S. mutans. From an examination of weight gains, etc., Van der Hoeven concluded that the general health of the rats in the sorbitol and xylitol groups was not optimal, whereas the weight gains on the lactitol regime were not significantly below those of the sucrose and wheat flour controls.

The most recent studies of the dental properties of lactitol in rats tested it at lower levels in the diet, closer to the average dietary content of sucrose in developed countries, and also in the form of a finished human food product, rather than as a raw ingredient in a blended animal diet.[21]

In the first trial, lactitol, xylitol and sucrose were compared at a level of 16% in the diets fed to matched groups of 22 caries-active Osborne–Mendel rats for eight weeks from weaning (Fig. 3). The caries scores averaged 1·9 and 1·7 on the lactitol and xylitol regimes, respectively, compared with 7·5 on the sucrose control—a highly-significant difference. Counts of the total number of carious lesions showed a similar difference, and the average score per lesion was also higher on the sucrose regime than on the two polyols ($p < 0.0002$), with hardly any of the lesions on the polyol diets having progressed beyond the very early 'white-spot' stage. The only distinction detected between the two polyols in their dental effects was in the amount of plaque on the teeth, which was significantly lower on the xylitol than on the lactitol regime.

During the first two to three weeks, water intake increased and food intake decreased on the polyol regimes, compared with the sucrose control. The animals adapted more quickly to the lactitol than to the xylitol regime and tolerated it better, but total weight gains on both polyols were significantly below those on the sucrose regime. Measurements of food and energy

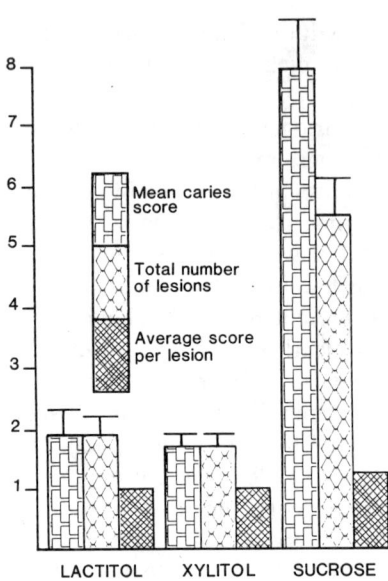

FIG. 3.  Dental caries in rats fed on diets containing 16% of lactitol, xylitol or
sucrose for eight weeks: means and standard errors.[21] Differences between polyol
and sucrose regimes: all $p < 0.0002$.

consumption, carcase analysis and tissue energy retention showed greater
fat deposition, energy retention in the body and conversion of food energy
into fat on the sucrose than on the two polyol regimes.

The second trial compared two types of shortbread biscuits—
conventional biscuits containing 16·6% of sucrose and an experimental
batch made with 16·6% of lactitol. These were incorporated at 66% in
pulverised, blended diets fed to groups of 21 rats, so that the final level of
lactitol in the test diet was 11%. Again, after a period of 8 weeks, caries
attack, as recorded by four separate sets of measurements, showed highly
significant reductions on replacing the sucrose in the biscuits by lactitol
(Fig. 4). The mean caries score fell by 75% and the average score per lesion
by 40%, while the average number of cavities fell from 4·7 to below one per
animal. This time plaque scores were also significantly lower on the lactitol
than on the sucrose biscuits regime.

Differences in food and water intake between the lactitol and sucrose
groups were smaller than in the earlier experiment, but the average weight
gains on the lactitol regime were again consistently lower than those on the
sucrose control.

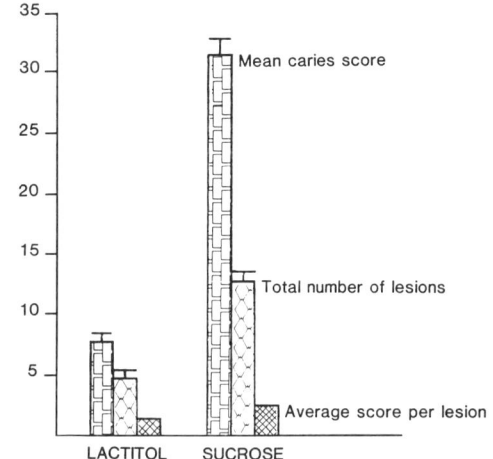

FIG. 4. Dental caries in rats fed for eight weeks on diets containing biscuits made with 16·6% lactitol or sucrose: means and standard errors.[21] Differences between lactitol and sucrose regimes: all $p < 0.001$.

To summarise this animal work, the three sets of studies, done in different ways in three separate laboratories, all show the low cariogenicity of lactitol compared with sucrose. The main points are encapsulated in Table 1, showing the low caries figures on the lactitol regimes and the significance of the differences between the lactitol and sucrose figures. A reduction in caries attack was seen at dietary levels as low as 11% and also after processing (baking) the lactitol into a snack food (biscuits), but, in addition, metabolic differences between lactitol and sucrose were detected and the implications of these have to be evaluated.

## (C) Investigations in Man

A technique that has become the standard guide to acceptance for the 'zahnschonend' classification of certain foods in Switzerland—i.e. intra-oral pH telemetry, in which the pH drop at the surface of a tooth is measured after its exposure to the food in the mouth—is said to have proved that lactitol is hypoacidogenic.[22] However, very few details are given, apart from a single curve of the pH of five-day-old dental plaque in a lone subject after rinsing of the mouth with 15 ml of 10% lactitol solution. The pH fell from a peak of about 6·8 to a minimum of approximately 6·0, whereas the minima for lactose and sucrose were approximately 5·0 and 4·2, respectively.

Further studies using this technique have been included in unpublished

TABLE 1

SUMMARY OF FINDINGS ON THE CARIOGENICITY OF LACTITOL IN LABORATORY RATS.
REDUCTION IN CARIES ON SUBSTITUTING LACTITOL FOR SUCROSE IN THE DIET,
EXPRESSED AS A PERCENTAGE OF THE CARIES ATTACK ON SUCROSE CONTROL DIET, AND
STATISTICAL SIGNIFICANCE OF THE DIFFERENCE

| Authors | No. of rats per group | Level of lactitol in diet (%) | Method of recording caries | Percentage reduction: lactitol versus sucrose | Significance of difference |
|---|---|---|---|---|---|
| Gehring (1978)[17] | 12 | 30/45/56 | Carious fissures/ severity | 58 | $p < 0.001$ |
| Van der Hoeven (1986)[20] | 12 | 25 | Advanced dentinal | 79 | $p < 0.01$ |
| | | | Extensive dentinal | 87 | $p < 0.01$ |
| Grenby & Phillips (1989)[21] | 22 | 16 | Caries scores | 75 | $p < 0.000\,2$ |
| | | | Number of lesions | 65 | $p < 0.000\,1$ |
| Grenby & Phillips (1989)[21] | 21 | 11 | Caries scores | 75 | $p < 0.001$ |
| | | | Number of lesions | 62 | $p < 0.001$ |
| | | | Gross cavities | 81 | $p < 0.001$ |

reports by Mühlemann[23] and Graf.[24] Mühlemann tested chocolate
containing lactitol in six subjects under varying conditions and observed
little or no intra-oral pH drop, while Graf tested marzipan made with
23·9% lactitol and 21% isomalt instead of sucrose, by methods established
earlier.[25,26] Some of the results are shown in Fig. 5, from which the
consistently high inter-dental plaque pH on the experimental marzipan,
compared with the fall in pH from a 10% sucrose control, can easily be seen.
The inference is that lactitol would be classified as 'safe for teeth', because
the level of acidity developing in the plaque from it is well above the critical
pH at which enamel begins to demineralise in the dental caries process.

Another study reported was a trial of lactitol in sweets.[27] Thirty 18–20-
year-old subjects were divided into two groups and given either
conventional sucrose sweets, or sweets made with lactitol as the only bulk
sweetener, to eat in addition to their normal diet over a three-day period.

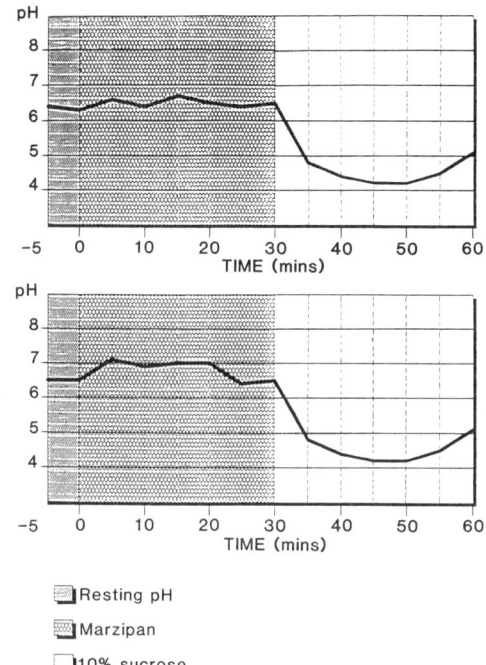

FIG. 5.   Telemetric recording of the pH of human interdental four-day-old plaque in two subjects after chewing marzipan made with lactitol and isomalt (first 30 min) and rinsing with a 10% sucrose solution (second 30 min).[24]

Their teeth were cleaned of all deposits at the start and they refrained from any oral hygiene procedures during the trial. They ate on average 9–10 sweets per day. At the end of the three days plaque accumulating on the teeth was measured by instantaneous examination after applying disclosing solution, a photographic method and finally a gravimetric technique. The plaque scores were always lower on the lactitol than on the sucrose sweets (Fig. 6), with significant difference in both the photographic and the gravimetric results. The area of plaque covering the teeth was cut down by over 50% and the weight of the plaque was more than 33% lower on the lactitol than on the sucrose sweets. The composition of the plaque varied too, with less soluble carbohydrate, glucose and sucrose, but relatively more protein, calcium and phosphorus, in the samples collected from the lactitol sweets group than from the sucrose sweets group.

   The reduction in the amount of plaque on the lactitol sweets may be taken as indirect evidence of a dentally beneficial property, particularly in

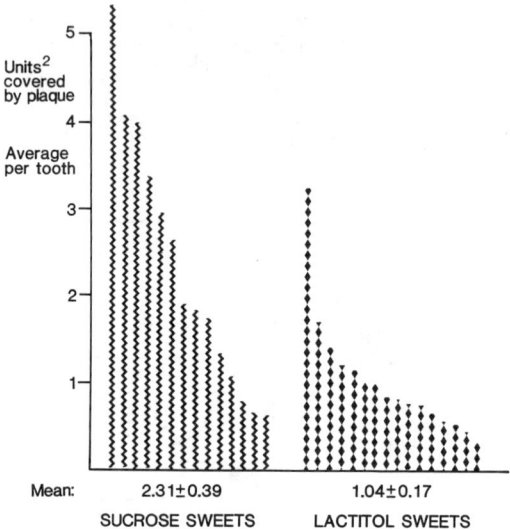

FIG. 6. Dental plaque in subjects eating either sucrose or lactitol sweets for three days. Measurement of the area of plaque covering the labial and buccal surfaces of the teeth by a photographic method. Subjects ranked in order of decreasing plaque scores.[27]

view of the correlation that has been observed in earlier studies of a similar type between individuals' caries experience (decayed, missing and filled tooth surfaces—DMFS) and their plaque scores.

A questionnaire included in the trial showed that other factors to be taken into account in the use of these re-formulated products are their organoleptic and metabolic properties. In particular, 15 out of the 16 subjects on the lactitol sweets experienced some degree of flatulence or gastro-intestinal disturbance, although this was seldom severe.

## COMMENTS

All three types of experiment should be taken into consideration in arriving at a verdict on the dental properties of lactitol. The animal trials (B above) all indicated its low cariogenicity and the main human trial in (C) showed its lower plaque-forming capacity compared with sucrose, without providing any explanation for these findings. The reasons must be sought in the data from the microbiological experiments *in vitro*, summarised in section (A). It is clear that lactitol cannot be metabolised by acidogenic and plaque-forming oral micro-organisms in the same way as sugars. It is not easily fermented to the acids that can attack the teeth in the caries process. Such adaptation of oral bacteria to metabolise lactitol as has been observed does not take place readily and does not seem to be permanent.

The results of all three types of research support the classification of lactitol as a non-dentally harmful bulk sweetener. In a review of the dental advantages of a range of bulk sweeteners,[28] xylitol is cited as having the best dental properties, so it may be used as a 'yardstick' for evaluating the cariogenicity of lactitol. This was done in several of the trials reviewed above, using xylitol as a 'negative' control alongside sucrose as positive control. In all cases the behaviour of lactitol was closer to that of xylitol than to sucrose. In the animal trials, Van der Hoeven[20] found fewer fissure lesions on xylitol than on lactitol at 25% in the diet, but on cutting the polyol content to 16%, the caries scores, lesion counts and severity of the lesions were so close on the two polyol regimes as to be indistinguishable.[21]

Additionally, in the studies *in vitro*, lactitol and xylitol proved to have superior dental properties to two other polyols, sorbitol and mannitol, with lower acid and polysaccharide formation by oral micro-organisms and less attack on enamel mineral.[18] Lactitol is the only one of these polyols that is derived from a disaccharide (C12) rather than a monosaccharide sugar. It might therefore be expected that oral micro-organisms would metabolise it less readily than C5 or C6 substrates, but Linke[19] points out that the majority of polyols can be taken up by the bacterial cell, are phosphory- lated, enter the glycolytic pathway and are finally metabolised into acidic end-products, such as lactic acid. Furthermore, there is the possibility that, with wider use of polyols in foods, the oral microflora may adapt to metabolise them.[29] On the other hand, Havenaar *et al.*[16] found that the strains of oral bacteria they tested did not adapt easily to lactitol and it has been suggested that, as long as there is an ample supply of fermentable sugars in the diet, significant facultative adaptation of the oral microflora to utilise polyols as a primary energy source is unlikely.

The dental prospects for the use of lactitol, therefore, seem favourable. Since the trial of lactitol sweets in man,[27] advances have been made in improving the organoleptic properties of lactitol confectionery, with the formulation of a range of new products including boiled sweets, gums/pastilles, chocolate and chewing-gum. The roughness of texture of the lactitol sweets experienced in the human dental plaque trial can be avoided by blending the lactitol with other bulk sweeteners and the level of sweetness can be adjusted by the use of intense sweeteners. However, its gastrointestinal effects may be a more intractable matter, as lactitol passes

378    T. H. GRENBY

through the small intestine without hydrolysis or absorption. As with the
other polyols, toleration gradually improves with time and should not be a
problem on a diet in which only a limited number of items are re-
formulated with lactitol.

Ideally more information is needed on adaptation of the oral microflora
to lactitol and also on its metabolism in the mammalian body, but the
investigations reviewed here indicate that lactitol has dental properties
equal to, or superior to, those of most of the other polyols put forward as
bulk sweeteners to replace sucrose.

## REFERENCES

1. Van Velthuijsen, J. A., Food additives derived from lactose: lactitol and lactitol palmitate. *J. Agric. Food Chem.*, **27** (1979) 680–6.
2. Chemie Combinatie Amsterdam. European Patent Application No. 0039981 C.V. CCA, 1981.
3. Linko, P., Lactose and lactitol. In *Nutritive Sweeteners*, ed. G. G. Birch & K. J. Parker. Applied Science, London, 1982, pp. 109–31.
4. Den Uyl, C. H., Technical and commercial aspects of the use of lactitol in foods as a reduced-calorie bulk sweetener. In *Developments in Sweeteners—3*, ed. T. H. Grenby. Elsevier Applied Science, London, 1987, pp. 65–81.
5. Van Velthuijsen, J. A., Lactitol: chemical and biological properties. In *Hepatic Encephalopathy: Management with Lactulose and Related Carbohydrates*, ed. H. O. Conn & J. Bircher. Medi-Ed Press, East Lansing, MI, USA, 1988, pp. 213–36.
6. Ziesenitz, S. & Siebert, G., The metabolism and utilization of polyols and other bulk sweeteners compared with sugar. In *Developments in Sweeteners—3*, ed. T. H. Grenby. Elsevier Applied Science, London, 1987, pp. 109–49.
7. Morgan, M. Y. & Hawley, K. E., Lactitol versus lactulose in the treatment of acute hepatic encephalopathy in cirrhotic patients: A double-blind, random-ised trial. *Hepatology*, **7** (1987) 1278–84.
8. Heredia, D., Caballeria, J., Arroyo, V., Ravelli, G. & Rodes, J., Lactitol versus lactulose in the treatment of acute portal systemic encephalopathy (PSE). *J. Hepatology*, **4** (1987) 293–8.
9. Anon., Lactitol. *The Lancet*, (11 July 1987) 81–2.
10. Patil, D. H., Westaby, D., Mahida, Y. R., Palmer, K. R., Rees, R., Clark, M. L., Dawson, A. M. & Silk, D. B. A., Comparative modes of action of lactitol and lactulose in the treatment of hepatic encephalopathy. *Gut*, **28** (1987) 255–9.
11. Morgan, M., Lactitol for the treatment of hepatic encephalopathy. In *Hepatic Encephalopathy: Management with Lactulose and Related Carbohydrates*, ed. H. O. Conn & J. Bircher. Medi-Ed Press, East Lansing, MI, USA, 1988, p. 237.
12. Van Es, A. J. H., De Groot, L. & Vogt, J. E., Energy balances of eight volunteers fed on diets supplemented with either lactitol or saccharose. *Brit. J. Nutr.*, **56** (1986) 545–54.

13. Nilsson, U. & Jagerstad, M., Hydrolysis of lactitol, maltitol and Palatinit® by human intestinal biopsies. *Brit. J. Nutr.*, **58** (1987) 199–206.
14. Patil, D., Grimble, G. K. & Silk, D. B. A., Lactitol, a new hydrogenated lactose derivative: intestinal absorption and laxative threshold in normal human subjects. *Brit. J. Nutr.*, **57** (1987) 195–9.
15. Havenaar, R., Microbiological investigations on the cariogenicity of the sugar substitute lactitol. Report to CCA Biochem b.v. Gorinchem, The Netherlands, 1976, pp. 1–23.
16. Havenaar, R., Huis in't Veld, J. H. J., Backer Dirks, O. & de Stoppelaar, J. D., Some bacteriological aspects of sugar substitutes. In *Health and Sugar Substitutes—Proc. ERGOB Conf., Geneva, 1978*, ed. B. Guggenheim. Karger, Basel, Switzerland, 1978, pp. 192–8.
17. Gehring, F., Prufung der Kariogenität von Lactose. Report VO(EG) 723/78, 1978, pp. 1–20.
18. Grenby, T. H., Phillips, A. & Mistry, M., Studies of the dental properties of lactitol compared with five other bulk sweeteners *in vitro*. *Caries Res.*, **23** (1989) 315–19.
19. Linke, H. A. B., Sweeteners and dental health: The influence of sugar substitutes on oral microorganisms. In *Developments in sweeteners—3*, ed. T. H. Grenby. Elsevier Applied Science, London, 1987, pp. 151–88.
20. Van der Hoeven, J. S., Cariogenicity of lactitol in program-fed rats. *Caries Res.*, **20** (1986) 441–3.
21. Grenby, T. H. & Phillips, A., Dental and metabolic effects of lactitol in the diet of laboratory rats. *Brit. J. Nutr.*, **61** (1989) 17–24.
22. Imfeld, T. N., *Identification of Low Caries Risk Dietary Components*. Karger, Basel, Switzerland, 1983, pp. 122–3 & 127.
23. Mühlemann, H. R., Gutachten uber die zahnschonenden Eigenschaften von Lacty Schokolade und Lacty Milchschokolade. Report by Dental Institute, University of Zurich, Switzerland, to CCA Biochem b.v., Gorinchem, The Netherlands, 1977, 10-M.
24. Graf, H., Intraorale pH-telemetrie. Produkt: Marzipan. Zur Erlangen der Pradikate 'Zahnschonend' und 'zahnfreundlich' durch das Bundesamt für Gesundheitswesen. Report from Bern University, Switzerland, 22 April 1988.
25. Graf, H. & Mühlemann, H. R., Telemetry of plaque pH from interdental area. *Helv. Odont. Acta*, **10** (1966) 94–101.
26. Graf, R. & Graf, H., Simplified in-vivo pH measurements of oral microbial deposits. *Helv. Odont. Acta*, **15** (1971) 42.
27. Grenby, T. H. & Desai, T., A trial of lactitol in sweets and its effects on human dental plaque. *Brit. Dent. J.*, **164** (1988) 381–5.
28. Havenaar, R., Dental advantages of some bulk sweeteners in laboratory animal trials. In *Developments in Sweeteners—3*, ed. T. H. Grenby. Elsevier Applied Science, London, 1987, pp. 189–211.
29. Linke, H. A. B., Sugar alcohols and dental health. *World Rev. Nutr. Diet.*, **47** (1986) 134–62.

# INDEX